Laboratory Evaluations in Molecular Medicine

Nutrients, Toxicants, and Cell Regulators

J. Alexander Bralley, Ph.D., C.C.N.
Richard S. Lord, Ph.D.

IAMM
Institute for Advances in Molecular Medicine

The Institute for Advances in Molecular Medicine
Norcross, GA

Library of Congress Cataloging-in-Publication Data

Laboratory Evaluations in Molecular Medicine/J. Alexander Bralley and Richard S. Lord
Includes bibliographic references and index
ISBN 0-9673949-1-0
I. J. Alexander Bralley, 1952--. II. Richard S Lord, 1942--

International Standard Book Number 0-9673949-1-0
Printed in the United States of America 1 2 3 4 5 6 7 8 9 0
Printed on acid-free paper

Preface

The title of this book has been vigorously debated by those involved with its development. The original title, Laboratory Evaluations in Nutritional Medicine, seemed appropriate since many of the tests and recommended therapies involve nutrients. This, however, did not fully portray the scope of the book. Molecular medicine, on the other hand, focuses on gene expression and cell regulation processes. Consequently, its literature has most often centered on medical genetic issues and a book title incorporating this term might lead the reader to expect a different sort of content than is actually presented.

Yet, this book rightly falls in the domain of molecular medicine, since it describes laboratory analyses that look at the molecular level of activity underlying disease symptoms. In fact, Linus Pauling in the 1950s originally developed the concept of molecular medicine and consistently discussed the important role nutrition plays in maintenance of health and disease treatment. We focus on nutrients responsible for tissue growth, maintenance, repair, and protection; hormones that regulate molecular activity; and toxicants, both endogenous and exogenous, that disrupt the molecular basis of tissue health.

While we do not discuss genetic testing per se, the impact of the cellular environment on genetic expression and the impact of the genes on cellular requirements (biochemical individuality) are described in certain tests where the connection is clearly established. For example, laboratory testing can detect the genetic "flaw" that creates elevated plasma homocysteine levels in 20% of the population. The physician can use this information to prescribe greater amounts of vitamins B_6, folate and B_{12} in the patient's diet to reduce homocysteine levels and thereby neutralize the effects of this genetic predisposition to heart disease. This approach to genetic predisposition can have broad preventive application to other gene expressions, such as breast cancer and Down's syndrome, which are influenced by nutrient molecules. Laboratory evaluations will play an increasingly important role in this exciting new medicine and we expect future editions of this book will include newer testing for gene/nutrient/toxicity interactions as the science unfolds.

Since many of the principles underlying the nutritional aspects of molecular medicine have not been included to any significant extent in medical school curricula in the past, this book has been written for healthcare practitioners who want to delve more deeply into this area and use some of the wealth of its resources for patient care. It is also intended as an introductory textbook for students of medicine and related disciplines. It has, in fact, evolved from lectures we have given to students and practitioners in various settings. It is also the result of the many technical monographs we have written over the years to describe the individual tests offered at MetaMetrix Clinical Laboratory.

This book, for the first time, presents a comprehensive overview of laboratory tests currently available to assist healthcare practitioners in designing nutritional therapies to treat health problems. To a great extent, we have attempted not to dwell on the function of nutrients in the body. There are many excellent books on biochemistry and nutritional medicine and we assume the reader already has a good working knowledge of these topics. Since there may be several types of tests for any given nutrient, we have attempted to clarify the factors to consider in test selection. Chapter 1 introduces the concepts of the nutritional component of molecular medicine. Chapters 2 through 5 explain the assessment of vitamins, minerals, amino acids, and fatty acids. Chapter 6 deals with the interpretation of urinary organic acids as metabolic markers that relate to nutrients, toxicants, and intestinal microflora. The relationships between nutrients and toxicants are considered within these chapters. Since nutrient status is intimately dependent on digestion and assimilation, Chapter 7 is devoted to the evaluation of nutritional support for gastrointestinal function. Chapter 8 focuses on the subject of evaluating individual capacities for dealing with toxicants that impair health and can be modulated by nutritional intervention. Chapter 9 covers testing for oxidant stress and the benefits of antioxidant nutrients in avoiding accelerated aging. Chapter 10 takes a brief look at hormonal interactions with nutritional factors. Chapter 11 provides an overview of how three compounds can be used to inexpensively access significant functional factors that determine nutrient needs. Chapter 12 attempts to integrate many of the concepts of molecular medicine evaluation by citing examples and recommending options for intervention based on multiple testing profiles. The appendices are both quick reference guides as well as budding chapters for future editions of this book.

This book could not have been created without the assistance of many skilled and dedicated individuals at MetaMetrix. Editing of something this technical is always a chore, and Molly Rodriguez has done an excellent job in this regard. Layout and redrawing of figures is also tedious and time consuming. Andrew Stine and Kristen McCullough have produced an exceptional document. We would also like to thank the other staff at MetaMetrix for putting up with the disruptions in their daily routines as a result of birthing this baby. And finally, Carolyn Bralley, who has consistently supported and helped in this endeavor, we are deeply grateful.

J. Alexander Bralley, Ph.D., C.C.N.
and Richard S. Lord, Ph.D.
April, 2001

About The Authors

J. Alexander Bralley, Ph.D., CCN, C.E.O. of Metametrix Clinical Laboratory, received his B.S. in biology from the University of Illinois in 1975 and his Ph.D. in medical sciences, specializing in neuroscience, from the University of Florida College of Medicine in 1980. He has worked and published in the area of nutritional biochemistry since graduation. For the past 17 years he has been C.E.O. and Laboratory Director of MetaMetrix Clinical Laboratory in Atlanta, an internationally recognized clinical laboratory specializing in nutritional and metabolic profiling. He has been instrumental in developing tests and interpretation guidelines for innovative nutritional and metabolic profiling analyses. Dr. Bralley has gained extensive experience in the use of nutritional therapies to treat chronic illnesses through consultating with physicians and clients, reviewing the medical literature and attending and speaking at conferences. He lectures widely on the use of laboratory testing to define nutrient needs for the individual. He is a member of several professional organizations for nutrition and clinical laboratory science; he sits on several boards including the editorial review boards for Alternative Medicine Review and the Journal of Applied Nutrition, the executive board for the Clinical Nutrition Certification Board, and the certification exam review board for International and American Association for Clinical Nutrition. His ongoing research interests lie in examining the relationship between nutritional, metabolic, and toxic factors and how they affect chronic disease processes and influence health. In 1997, Dr. Bralley founded NuBasX Nutritional Medical Clinic, aimed to address the nutritional, metabolic, and toxicologic aspects of chronic illness. The NuBasX (pronounced "new basics") mission is to develop nutritional and metabolic treatment modalities with the approach of treating the cause not the symptoms of disease and providing quality medical service alternatives for the patient. In 2001, Dr. Bralley formed a joint venture with two internationally recognized physicians in metabolic and molecular genetic diagnostic testing to create Horizon Molecular Medicine Laboratory. This venture brings a wealth of experience and expertise in the rapidly expanding area of molecular medicine and genomics to the companies started by Dr. Bralley.

Richard S. Lord, Ph.D., Director of the Department of Science and Education at MetaMetrix Clinical Laboratory, received his Ph.D. in Biochemistry from the University of Texas at Austin in 1970 with studies of methods for protein tertiary structure measurements. From 1970 to 1973 he was an NIH postdoctoral fellow at the University of Arizona conducting studies of the mechanism of insulin self association and from there he moved to a staff fellowship at NIH in Bethesda, MD to work in the Institute of Arthritis and Metabolic Disease. Dr. Lord served as Professor and Chairman of Biochemistry at Life College for 10 years where he concurrently taught Advanced Clinical Nutrition and worked as part of the team that put in place a new undergraduate degree program in nutrition. His publications include articles in professional journals as well as books and laboratory manuals for college curricula. He has served as Clinical Laboratory Director at various facilities and has developed methods for analysis of trace elements, fatty acids, and organic acids.

Contents in Brief

Table of Contents

Chapter 1

Basic Concepts of Laboratory Evaluations in Molecular Medicine

Introduction

One lesson in medicine we have learned over the last century is that our successes often give rise to new challenges. Tremendous strides in reducing mortality from acute illness and extending the life expectancy have yielded an increase in chronic, degenerative diseases [1]. The power of antibiotics in saving lives has encouraged the hope of a "magic bullet" for treating symptoms, a model which has, to a large extent, failed to produce the same success with diseases that are multifactor in origin and chronic in nature. Additionally, the very use of these drugs has created "super bugs" resistant to antibiotics [2]. Advances in food production and distribution have resulted in an abundance of processed foods altered from their original form and that have high calorie-to-micronutrient ratios. Technological progress has created literally thousands of new chemicals that challenge the environment and our own metabolic processes [3].

As more of the world's population experiences the benefits of our successes, overcoming the new challenges we have created requires a broader, more holistic approach that takes into account the web-like interconnectedness of human metabolism and its relation to the environment. The field of molecular medicine cannot be advanced without attention being paid to the capacity of the human body to adapt to change, heal, and express its genetic potential if it is provided the raw materials -- the basic macronutrients, micronutrients, and conditionally essential nutrients -- that it requires for healthy metabolic function.

Nutrition may be the single most important component of health maintenance, since diet is associated with many diseases including cancer, diabetes, hypertension, heart disease, stroke, cirrhosis of the liver, childhood developmental and behavioral disorders, and coeliac disease [4, 5]. Yet optimal nutrition, the level of nutrient intake that maintains the best possible health, is highly variable from person to person. The concepts of biochemical individuality and the genetotropic theory of disease, as first described by Roger Williams, a discoverer of pantothenic acid (vitamin B_5), continue to exert a major influence on this emergent model of human health and the development of molecular medicine.

Biochemical individuality influences not only the individual need for nutrients but also the expression of nutrient insufficiency [6]. Although this has been recognized for centuries, the concept has been virtually ignored by the current medical system. Diseases such as beriberi, pellagra, and scurvy, all single nutrient deficiency diseases, are often thought of as specific sets of symptoms and disease progressions. This, however, is not the case. Biochemical individuality plays a key role in the expression of functional deficits as they develop even in a single nutrient deficiency disease. An excerpt from an officer's account of the British sailing ship Anson's four year voyage around the world (1740-1744) where 626 of 961 men died of scurvy, illustrates that even centuries ago widely varying responses to vitamin C deficiency were apparent:

> *"This disease, so frequently attending all long voyages, and so particularly destructive to us, is surely the most singular and unaccountable of any that affects the human body. For its symptoms are inconstant and innumerable, and its progress and effects extremely irregular: for scarcely two persons have the same complaints, and where there is found some conformity in symptoms, the order of their appearance has been totally different. However, though it frequently puts on the form of many other diseases, and is therefore not to be described by any exclusive and infallible criterions..."* -- Richard Walter

Because most essential nutrients play such fundamental roles in cellular metabolism, as insufficiencies develop the effects can be system wide. Variation in disease expression is exponentially compounded when multiple nutrient deficiencies are involved, which is more often the case in chronic illnesses. Any and every aspect of body function can potentially become compromised depending on the patient's own unique biochemical expression.

The corollary of this situation is that nutritional and metabolic evaluations of patients with chronic illnesses also exhibit this variation, making it difficult to define a disease by any set of laboratory markers. Conventional wisdom in diagnostic medicine requires that each diagnosed disease have its own unique set of signs, symptoms, and laboratory indicators. Diabetics exhibit elevated blood glucose, prostate cancer patients have elevated serum PSA, and elevated white blood cells can indicate an infection. This same thinking has been applied to the diagnosis of various chronic disorders. Is there a metabolic or nutritional deficit pattern to fibromyalgia that would distinguish it from chronic fatigue or attention deficit disorder?

Since nutrient insufficiencies affect multiple organ systems in ways dependent on the history of each patient, one would expect varying outcomes for similar patterns of nutrient depletions, rather than strong definitive patterns of nutrient status with each chronic illness. Decades of experience with nutritional and metabolic profiling calls into question the validity of categorizing and labeling chronic illness with a diagnosis, since there does not appear to be a unique set of diagnostic laboratory criteria matching a specific disease diagnosis. A specific chronic illness is not diagnostically defined by a unique set of laboratory values. Consequently, the clinician is not really treating fibromyalgia, chronic fatigue, inflammatory bowel, etc., but rather the individual nutritional and metabolic imbalances and toxicities that prevent normal function. These are expressed in each person as a different set of symptoms, perhaps the weakest links in a person's genetic makeup. Toxic lead exposure, for example, may be expressed in some individuals as inflammatory disorders, in others as immune disorders, and still others as neurological disorders [7, 8]. Disease labels are conveniences that allow reductionism so our minds can categorize treatments; yet this labeling has inhibited our ability to effectively treat nutrient-related disorders. Patients are better served by approaching chronic disease as a manifestation of underlying nutritional and metabolic imbalances and/or toxicities. Correction of the imbalances yields elimination of the disease, just as the various expressions of vitamin C deficiency in scurvy are reversed once the patient is replete in the vitamin.

In this book the terms "deficiency" and "insufficiency" are often used interchangeably. In general, however, "deficiency" relates to frank deficiency disease situations (i.e., pellagra, scurvy, beriberi), whereas "insufficiency" is used to express non-optimal nutriture that can adversely affect function. Therefore, in most cases, the clinicians in developed countries are dealing with nutrient insufficiency and not deficiency disease. The classic nutrient deficiency diseases are not usually seen in developed countries; however, the need for additional nutrient supplementation for optimal function can be clearly demonstrated for many conditions. Homocysteinemia, for example, is treated with high doses of vitamin B_6, folate, and vitamin B_{12}. In this case, there is an insufficiency of these nutrients to provide optimal homocysteine metabolism. This does not represent a classical deficiency state, rather the need for a nutrient in sufficient amounts, usually higher than RDA levels, to maintain optimal function. Using a broader definition of "deficiency" as any state in which dietary intake has failed to meet demand, both patients with vitamin B_{12}-responsive homocysteinemia and those with vitamin B_{12} deficiency anemia are deficient in vitamin B_{12}.

Once the concept of nutrient insufficiencies manifesting as the etiology of chronic illness is understood, the relevant clinical question is how best to detect early states of insufficiency to restore and maintain wellness. We have attempted to compare laboratory evaluations with regard to the sensitivity for detecting early states of nutrient depletion. Biochemical pathways involving large flux of metabolites are generally the first stage at which nutrient effects are

FIGURE 1-1. Stages of Development of Nutrient Insufficiency Diseases

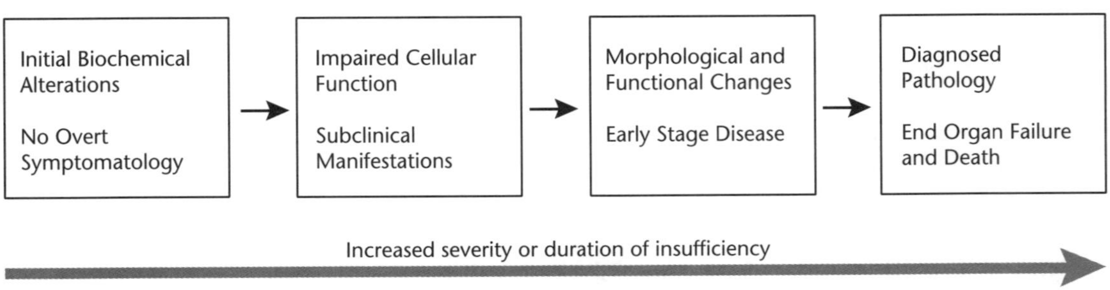

seen. There is a progression from this stage to the end organ failure that is usually observed when a classic diagnosis is made (Figure 1-1).

Initial conditions of insufficiency are seen at a biochemical level often without any overt symptom expression. This is why asymptomatic and symptomatic individuals may have similar laboratory markers for nutrient insufficiencies and metabolic imbalances. As the duration of the insufficiency increases, additional changes appear in cellular function that can be seen as subclinical manifestations, such as behavioral and mental/emotional instabilities. As the insufficiency progresses into later stages, morphological and functional changes occur that can be defined as early stage disease. Finally, diagnosed pathology is expressed with end organ failure and ultimately death.

Many factors influence nutrient status in the individual (Figure 1-2), such as diet, digestion, absorption, disease states, age, medications, stress, toxic exposure, activity levels, genetics, and biological differences. Laboratory tests can provide valuable indicators of many of these factors.

FIGURE 1-2. Components Affecting Nutritional Status

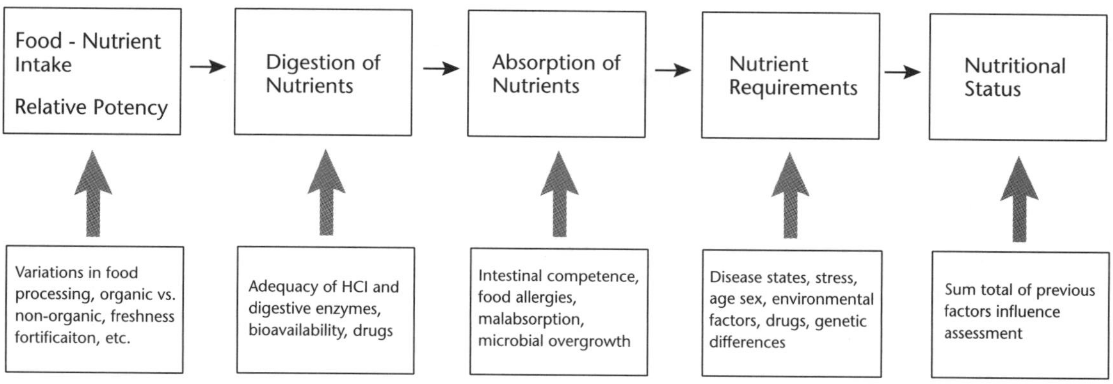

Issues in Assessment

Not all tests for any particular analyte are equivalent, so it is important for the clinician to be aware of four variables involved in testing: the analyte, the specimen, patient preparation, and instrumentation. The analyte can be a nutrient, toxicant, or metabolic control molecule. The choice of specimen, such as blood, urine, stool, hair, or other tissue, is critical to the interpretation and clinical utility. Preparation of the patient is critical. Is the patient fasting? Was a challenge compound used? Has the patient been taking a nutrient that is being measured? And, finally, the analytical method itself can affect the sensitivity and specificity of the measurement. For example, homocystine measured using a post-column ninhydrin derivatization liquid chromatographic method is much less sensitive than other pre-column derivatization methods [9]. Both methods produce a result for the analyte, but clinically relevant values are less than the limits of detection for the former method. Further, improvements in analytical instrumentation, such as tandem mass spectrometry, will allow measurement of analytes at very low concentrations that were not possible previously [10]. This is opening up a whole new area of environmental chemical toxicology and body burden assessment.

Methods of Assessment

The following discussion of possible strengths and limitations of various standard laboratory approaches to assessment reflects these issues and provides a framework for more detailed chapters that follow.

1. Static measurement of a nutrient, nutrient metabolite, or toxicant.

As is the case with every laboratory measure, the physiology of the body needs to be taken into account when interpreting a laboratory value. A blood glucose level, for example, is interpreted differently depending upon when the patient ingests his last meal. In addition, because nutrient insufficiency is initially expressed at the metabolic level, blood levels of the nutrient are often normal in early insufficiency stages, so measurements of nutrient concentrations can be misleading. Magnesium, for example, is found primarily intracellular with only 1% of body magnesium found in the serum, where its level is tightly regulated. Consequently serum levels of magnesium, which change only after there is a significant depletion in the tissues, is not the best measurement when the purpose is to evaluate magnesium insufficiency in a patient [11]. This is often the case when the nutrient is primarily intracellular. Serum vitamin E, on the other hand, provides a useful measure of nutrient status since a considerable portion of total body vitamin E resides in serum as it protects blood lipids from oxidation. Other macronutrient sources, such as amino acids and fatty acids, are dynamically metabolized in the body. Fasting plasma measurements of these nutrients reach a level of homeostatic stability that allows for consistent interpretation of data in repeat testing. Toxic metals and other chemicals in serum can be useful as indicators of an ongoing exposure but are not necessarily reflective of the body burden of the toxicant [12-14].

Hormonal evaluation is particularly problematic. While this book does not exhaustively address this area, many clinicians using nutritional therapies regularly assess various hormonal issues that affect function. Hormone levels that vary over time need to be assessed appropriately. Serum levels taken at different points in time may be a better indicator of proper function than a one-time measure. The use of 24-hour urine obviates the need for multiple blood draws, since it is used to measure total daily output. Salivary hormone levels may be valid for certain hormones and not others. Again, timing is an issue. Consideration of the physiological functioning of the hormone being assessed should always be a factor in determining useful and reliable laboratory testing procedures.

2. Enzyme stimulation assays.

Intracellular enzyme stimulation assays were developed to provide a more reliable, functional indication of nutrient needs. However, usually only one enzyme is assayed for a particular vitamin cofactor, such as erythrocyte transketolase for thiamine insufficiency or glutathione reductase for riboflavin insufficiency [15]. One drawback to this type of assessment is that not all enzymes that require the cofactor are measured. Other enzymes that use the same cofactor may need more or less of the essential nutrient. If another enzyme requires more of a particular nutrient for optimal function, this may not be detected by measuring only one enzyme activity.

3. Abnormal levels of metabolites.

Measurements of intermediary metabolite elevations are functional assays since they detect the failure of a nutrient-dependent metabolic pathway. As with the stimulation assays, these functional markers are only assessing one nutrient-dependent metabolic pathway and may not give a complete picture of nutrient need. A stronger indicator of nutrient need is obtained when multiple metabolites are assessed that use the same cofactors for their metabolism, as is the case in B vitamin insufficiency states and multiple ketoacid elevations. Different dehydrogenase enzymes metabolize various ketoacids. Elevation of several ketoacids is a strong indicator of a global B vitamin insufficiency state decreasing the functioning of multiple enzymes.

4. Loading tests or saturation measures.

Tests that measure the retention of a nutrient load utilize the dry sponge concept, that is, if a nutrient is deficient or needed by the tissues, it will be more retained by the body and less will be excreted upon loading than in the nutrient-replete condition. Magnesium loading then is more of a functional need assessment tool, giving the clinician a better picture of sufficiency status of the patient [16].

5. Challenge tests.

Chelation mobilization tests for toxic metals are very useful to assess the body burden for a toxic element. This is a way to evaluate the need for detoxification treatments for patients with chronic illnesses. When a water-soluble challenge compound is administered orally it has primary impact on the liver and kidney. Thus, these tests reveal primarily hepatic and renal toxic metal status. The use of challenge compounds to assess a metabolic pathway is particularly valuable as it allows the clinician to directly assess how well a specific critical system function is operating. Caffeine clearance, acetaminophen, and salicylate challenges have proven valuable in assessing important detoxification pathways. Loading with methionine before homocysteine testing may enhance the sensitivity of detecting homocysteinemia [17].

6. Other procedures.

The assessment of nutritional and toxicological factors in healthcare would not be complete without an evaluation of the digestive tract. Stool analysis does not fit handily into any of the above categories, yet information gained from the stool can be very useful clinically. Microscopic and chemical analysis of stool reflects the efficiency of the digestive tract, nutritional status, or inflammatory condition. Microbiological assessment examines the balance of gut ecology that can have an important impact on health. Toxic metabolites from certain microflora overgrowing in the gut can be measured in the urine [18]. Culture and sensitivity measures provide the clinician with the information to restore normal microflora balance.

Food itself can have a significant negative impact on health if a patient is allergic or intolerant to the food. Analysis of the immunologic responses to foods has been very useful in determining clinically significant allergies and intolerances. Serum antibody measures for both immediate and delayed sensitivity reactions provide the clinician with useful data to guide diet and nutritional recommendations that affect patient response.

Since biochemical imbalances created by nutritional insufficiencies precede clinical signs and symptoms, laboratory testing can provide early warnings of nutrient-related clinical problems. Nutrition affects multiple metabolic processes, and laboratory evaluation of nutritional status can take many forms. The chapters that follow provide the clinician with insight regarding the scientific basis of tools to evaluate chronically ill patients and to assist in developing safe, cost-effective, and reliable treatments.

References

[1] Stamler, J., *et al.*, Low risk-factor profile and long-term cardiovascular and noncardiovascular mortality and life expectancy: findings for 5 large cohorts of young adult and middle-aged men and women. *Jama*, 282(21). 2012-8., 1999.

[2] Kennedy, M., Overuse of antibiotics leads to "super bugs". *Wmj*, 99(5). 22-3, 28., 2000.

[3] Miller, C.S. and T.J. Prihoda, The Environmental Exposure and Sensitivity Inventory (EESI): a standardized approach for measuring chemical intolerances for research and clinical applications. *Toxicol Ind Health*, 15(3-4). 370-85., 1999.

[4] Matek, Z., M. Jungvirth-Hegedus, and S. Kolacek, Epidemiology of coeliac disease in children in one Croatian county: possible factors that could affect the incidence of coeliac disease and adherence to a gluten-free diet (Part II). *Coll Antropol*, 24(2). 397-404., 2000.

[5] Kris-Etherton, P.M., A new role for diet in reducing the incidence of cardiovascular disease: evidence from recent studies. *Curr Atheroscler Rep*, 1(3). 185-7., 1999.

[6] Velazquez, A., Biotin deficiency in protein-energy malnutrition: implications for nutritional homeostasis and individuality. *Nutrition*, 13(11-12). 991-2., 1997.

[7] Milanov, I. and P. Kolev, Clinical and electromyographic examinations of patients with tremor after chronic occupational lead exposure. *Occup Med (Lond)*, 51(3). 157-162., 2001.

[8] Hwang, Y.F., *et al.*, Chronic industrial exposure to lead in 63 subjects. I. Clinical and erythrokinetic findings. *Southeast Asian J Trop Med Public Health*, 7(4). 559-68., 1976.

[9] Mizobuchi, N., *et al.*, Isotachophoretic analyses of cystine, homocystine and cystathionine in urines from patients with inborn errors of metabolism. *J Chromatogr*, 382. 321-5., 1986.

[10] Vishwanathan, K., *et al.*, Determination of arginine and methylated arginines in human plasma by liquid chromatography-tandem mass spectrometry. *J Chromatogr B Biomed Sci Appl*, 748(1). 157-66., 2000.

[11] Seelig, M.S., *Magnesium Deficiency in the Pathogenesis of Disease*. 1980, New York: Plenum Medical Books.

[12] deFur, P.L. and L. Foersom, Toxic chemicals: can what we don't know harm us? *Environ Res*, 82(2). 113-33., 2000.

[13] Stephenson, J., Health agencies update: toxic chemicals and defects. *JAMA*, 284(3). 296., 2000.

[14] Zabrodskii, P.F. and V.G. Germanchuk, Role of corticosterone in realization of immunosuppressive effects in acute poisoning with toxic chemicals. *Bull Exp Biol Med*, 129(5). 552-5., 2000.

[15] Adelekan, D.A. and D.I. Thurnham, Glutathione peroxidase (EC 1.11.1.9) and superoxide dismutase (EC 1.15.1.1) activities in riboflavin-deficient rats infected with Plasmodium berghei malaria. *Br J Nutr*, 79(3). 305-9., 1998.

[16] Rob, P.M., *et al.*, Can one really measure magnesium deficiency using the short-term magnesium loading test? *J Intern Med*, 246(4). 373-8., 1999.

[17] Tsai, M.Y., *et al.*, Short-term variability in the measurement of plasma homocysteine, fasting and post-methionine loading. *Clin Biochem*, 34(1). 49-52., 2001.

[18 Chalmers, R.A., H.B. Valman, and M.M. Liberman, Measurement of 4-hydroxyphenylacetic aciduria as a screening test for small-bowel disease. *Clin Chem*, 25(10). 1791-4, 1979.

Chapter 2

Vitamins

TABLE 2-1. Summary of Vitamin Status Evaluations

Vitamin	Common Deficiency Tests	Positive Result	Adult Repletion*
A (Retinol)	Vitamin A – serum	Low	5000 – 10000 IU
B_1 (Thiamin)	α-Keto acids – urine	High	50 – 200 mg
	Erythrocyte Trasketolase Index	High	
B_2 (Riboflavin)	α-Keto acids – urine	High	100 – 1000 mg
	Ethylmalonate – urine	High	
	EGR Acitivity Coefficient		
B_3 (Niacin)	Lactate & pyruvate – urine	High	100 – 1000 mg
B_5 (Pantothenate)	α-Keto acids – urine	High	100 – 1000 mg
B_6 (Pyridoxine)	Homocysteine – plasma	High	50 – 200 mg
	Kynurenate – urine	High	
	EGOT Index	High	
B_{12} (Cyanocobalamin)	Methylamalonate – urine	High	100 – 1000 mcg
	Vitamin B_{12} – serum	Low	
C (Ascorbate)	Vitamin C – serum	– –	1000 – 5000 mg
K	Prothrombin – plasma	High	500 – 1000 mcg
	Vitamin K – serum		
D	25-Hydroxyvitamin D – serum	Low	400 – 800 IU
	Bone-specific peptide – urine	High	
E (Tocopherol)	Serum vitamin E	Low	200 – 1600 IU
Biotin	α-Hydroxisovalerate – urine	High	500 – 5000 mcg
Folate	Homocysteine – plasma	High	200 – 800 mcg
	Folate – serum		
Carnitine	Adipate, Butyrate, Ethylmalonate – urine	High	200 – 800 mcg
Coenzyme Q_{10}	Coenzyme Q_{10} – serum	Low	10 – 300 mg
	Succinate – urine	High	
	Hydroxymethylglutarate – urine		

*Daily oral dose for overcoming deficiency.

Introduction

There are various ways of acquiring data about vitamin status. Concentrations of vitamins can be measured in serum or blood cells. The excretory products formed from vitamins may be measured in urine. Changes in response to added vitamins may be measured as specific enzyme activities in blood or growth of leukocytes in cell culture. Finally, functional adequacy of a particular vitamin can be revealed by the urinary levels of specific metabolic intermediates controlled by the action of the vitamin. For routine clinical purposes, the most useful assay gives a clear answer to the question of whether body pools are adequate to meet current demands. Table 2-1 shows a summary of the laboratory evaluations of vitamin status that are discussed in this chapter.

Methods

Vitamins encompass several classes of compounds. The challenge of determining adequacy of body stores is complicated by the diversity of structures, the presence of multiple forms, and the availability of specimens reflecting the whole body vitamin stores. No single assay can give adequate information about all vitamins. The availability of functional tests has increased dramatically as the technology for detection of metabolic markers of vitamin deficiency has become more widely available. Also, the knowledge of metabolic intermediates that can serve as markers of functional deficiencies has grown [1].

Serum Concentrations

The first widely available tests of vitamin concentrations in serum were microbiological growth assays. Protozoan organisms whose growth is dependent on the presence of vitamins are grown in media that supply all but the vitamin being tested. The assays are very labor intensive and, because of the large number of variables affecting growth of whole cells, the reliability of results is less than desired. These methods have been replaced by immunoassays with greater sensitivity and reproducibility and lower cost.

Stimulation Assays

Adding saturating amounts of a vitamin to an enzyme whose activity is dependent on it causes an increase in activity, which corresponds to how poorly the enzyme was initially supplied with the vitamin. This type of measurement is called an enzyme stimulation assay. Erythrocyte transketolase activity, for example, is dependent on thiamin pyrophosphate (TPP). The activity is measured without any added TPP and again after adding excess TPP. The increase in activity may be expressed as a ratio or percentage. Abnormally high increases in activity indicate degrees of deficiency.

An inherent limitation of the assay is that only one enzyme in erythrocytes is examined so total body-functional status of the vitamin may not be revealed.

Another type of stimulation assay involves whole cell responses. Leukocytes are removed from the patient sample and grown in media with increasing concentrations of the vitamin to be assayed. Vitamins must be added in their active cofactor forms. Changes in growth rate‡ at increasing levels of added nutrients provide a measure of the initial state of nutrient adequacy. Cells from a vitamin-deficient patient will require more added vitamin to achieve maximum growth than those from one who is well nourished. This approach has great potential because it measures the total cellular response to the nutrient [2]. The limitation is that, as with protozoan growth assays, there are many variables controlling the growth of whole leukocytes, and current techniques present extreme challenges to control all possible variables in the growth media. Because the technique depends on the initial viability of the cells, specimen transport also adds a significant variable. William Shive and colleagues largely contributed to the development of this test. The test is available for most vitamins and other classes of essential nutrients.

Liquid Chromatography

For certain vitamins, the highly reliable and accurate methods of liquid chromatography are applicable. Profiles of vitamins A, E, Carotene, and Coenzyme Q_{10} may be obtained by this method. Simple extractions of serum efficiently isolate the fat-soluble components that need no further preparation prior to injection. The methods are "rugged," meaning they can work well even when performed by laboratory personnel with limited training.

Gas Chromatography

The coupling of mass spectrometric detection with a gas chromatograph is the method used for measuring most of the metabolic markers described in this chapter. The technology offers universal detection and high sensitivity, but it has a large labor component for sample preparation and requires expensive equipment that must be maintained by highly trained personnel.

Vitamin A (Retinol) and β-Carotene

Active forms: All-*trans* retinoic acid, 11-*cis*-retinal
Biochemical role: Cell differentiation, Vision
Example: 11-*cis*-retinal + light → all-*trans*-retinoic acid

Vitamin A is one of the essential nutrients most likely to be supplied inadequately in the diets of the North American population [3].

‡ Since DNA formation is a requisite for growth, measuring the incorporation of radioactively labeled nucleotides into DNA is a convenient way to monitor growth. All other factors must be strictly controlled, however, to reliably ascribe changes in nucleotide incorporation to addition of a single nutrient.

Deficiencies result from poor food choices and depletion of vitamin A from processed foods. Hypovitaminosis A is a major nutritional problem throughout the world, especially in developing countries. In these populations, severe vitamin A deficiency leads to xerophthalmia, the most common cause of blindness in young children [4].

Specific maintenance roles for Vitamin A have been reported for vision, bone growth, cancer, cataracts, skin integrity, atherosclerosis, macular degeneration, and spermatogenesis. In addition to the mechanism of retinoic acid in vision, the vitamin is known to be involved in sub-cellular membrane function and mucopolysaccharide synthesis [5]. Vitamin A status has no known impact on circulating enzyme activities or excretion of metabolic intermediates, so there is no biochemical measurement of functional vitamin A deficiency. Measurement of vitamin A concentration in serum is still the best method to assess individual nutritional status of this vitamin. Serum concentrations of vitamin A fall during low dietary intake intervals [6].

A patient with both a long-standing low intake of vitamin A and a low tissue reserve will have a serum concentration below 20 µg/dl. A dose response relationship exists between oral supplementation and serum vitamin A in humans. Once normal tissue levels are obtained, daily intake of vitamin A has been maintained at levels of 10,000 IU for five years without significant long-term rise in serum retinol [7]. Serum retinol levels below 30 µg/dl indicate depleted liver stores of vitamin A, while levels between 30 and 60 reflect increasing liver stores. Maintenance of optimal tissue functions will depend on the absence of chronic inflammatory and immune challenges. Adult men achieve a level of approximately 60 µg/dl at an intake of 2400 µg retinol per day [8]. Serum levels above 150 µg/dl indicate excessive vitamin A intake with potential toxic responses. Acute, chronic, and teratogenic syndromes of vitamin A toxicity have been associated with very high doses. Danger of such toxic responses is especially high during early pregnancy [9]. Excessive intake of vitamin A will be revealed as an elevated serum concentration.

Vitamin A is a member of a family of closely related compounds known as carotenoids. β-Carotene supplementation resulted in significant dose-related decreases in serum cholesterol and triglyceride fractions, indicating antihyperlipidemic effects in spontaneously hypertensive rats [10]. Methods have been developed for the measurement of up to 50 different carotenoids in plasma [11, 12]. Although little useful clinical information is gained from such exhaustive analysis, the measurement of β-carotene along with serum vitamin A is useful because β-carotene (also known as provitamin A) is converted into vitamin A in the liver. β-Carotene also may have antioxidant or other roles independent of vitamin A, but it mainly serves as a pool that may be converted into vitamin A. Body pools of the two compounds are maintained independently so that, unless vitamin A levels are initially low, increased intake of β-carotene may not affect higher levels of vitamin A [13].

Vitamin A deficiency inhibits thyroid function [14]. Conversely, thyroid hormone is involved in carotenoid metabolism and transport. Elevated β-carotene and retinol concentrations in plasma are often associated with hypothyroidism [15, 16]. In the United States, almost 75% of average dietary retinol equivalents is supplied as preformed vitamin A and the other 25% is in the form of β-carotene[17].

Vitamin B1 (Thiamin)

Active form: Thiamin pyrophosphate (TPP)
Biochemical role: Decarboxylation
Example: Pyruvate → Acetyl-CoA + CO_2

Many reviews of thiamin start with a statement about thiamin deficiency occurring only among populations where rice is the staple food. The ease with which thiamin is depleted by processing rice creates an increased risk of deficiency of this vitamin among such populations. The potential need for thiamin to achieve optimal health for individuals in any society is demonstrated in the growing list of known thiamin-responsive disorders of metabolism, which includes hyperkinetic cardiomyopathy, megaloblastic anemia, and maple syrup urine disease [18]. Genetically induced alterations of thiamin-dependent enzymes can cause weakened binding of the highly water-soluble vitamin B_1, resulting in abnormally rapid urinary losses. Such alterations result from subtle genetic variability within the normal population. Thiamin deficiency will occur when thiamin intake is insufficient to maintain tissue levels high enough to achieve enzyme activities necessary for all biochemical demands. Tissue and serum levels that are in the "normal range" may be insufficient in such individuals. Biochemical markers of functional sufficiency are the most useful measure of adequacy. For example, three family members with dysautonomic symptoms comparable to those seen in classical beriberi required high doses of thiamin tetrahydrofurfuryl disulfide to restore biochemical function as measured by transketolase activity [19] (see below).

The most sensitive and definitive measure of thiamin adequacy is obtained from detection of increased levels of amino acids and their keto-acid analogs that are excreted in the absence of adequate tissue levels of thiamin. Because the catabolism of amino acids is a heavily utilized pathway in the liver, the breakdown of branched chain amino acids (BCAAs) affords an opportunity for detecting interruptions in the pathway caused by inadequacy of thiamin. Leucine, isoleucine, and valine are deaminated to the corresponding keto acids prior to the first thiamin-dependent step. Increased plasma or urine isoleucine or the appearances of significant levels of the branched chain keto acids (BCKAs) in urine are markers of thiamin deficiency [20]. Thiamin deficiency also results in increased blood pyruvate, but this effect is not specific for thiamin [21]. The combination most useful in assessing an individual need for thiamin is isoleucine, pyruvate, α-ketoisovalerate, α-ketoisocaproate, and α-keto-β-methylvalerate (Table 2-2). The latter three are the keto forms of valine, leucine, and isoleucine, respectively. Further information may be found in the chapter on Organic Acids.

TABLE 2-2. Markers for Vitamin B_1 Deficiency

Test	Unit	Subjects	Risk of Deficiency		
			High	Medium	Low
Isoleucine, plasma	µM	Adult	> 180	160 - 180	< 160
Isoleucine, urine	µM	2-18 yr	> 175	125 - 175	< 125
Isoleucine, urine	µmol/day	All	> 90	70 - 90	< 70
α-Ketoisovalerate, urine	µg/mg creat.	All	> 1.8	1.5 - 1.8	< 1.5
α-Ketoisocaproate, urine	µg/mg creat.	All	> 2.4	2.0 - 2.4	< 2.0
α-Keto-β-methylvalerate	µg/mg creat.	All	> 1.6	1.2 - 1.6	< 1.2
Erythrocyte transketolase	% Increase	All	>25%	15 - 24%	<14%

Erythrocytes heavily utilize the pentose phosphate pathway to produce reducing equivalents for maintaining reduced glutathione. This pathway involves two steps catalyzed by a transketolase enzyme that requires thiamin pyrophosphate (TPP). This enzyme may be used to perform a stimulation assay from a whole blood specimen. A "TPP stimulation" result greater than 14% demonstrates a significant risk of thiamin deficiency [22]. Limits have been proposed for higher risk of deficiency (see Table 2-2). This approach is useful, but it is focused on a single enzyme in erythrocytes. In contrast, a better test for revealing functional thiamin deficiency may be the detection of elevated branched chain keto-acids in urine. They are formed from BCAAs that are generated in the catabolic pathway of amino acids, which is active in many tissues.

A portion of a load of thiamin administered orally or intramuscularly will appear in urine within a few hours if the tissues are replete with thiamin. Administration of 5 mg of thiamin parenterally should result in the appearance of at least 20 µg in urine over the next four hours [23]. The parenteral thiamin load test is primarily useful as an indicator of intake, rather than functional sufficiency of thiamin.

The test used most frequently for population surveys is measurement of urinary thiamin concentration. Urinary excretion of thiamin decreases proportionately with thiamin intake. This fact has led to measurements of thiamin levels in urine as indicators of adequacy in nutrition surveys. Both 24 hour and random urine tests have been used to assess thiamin intake. Random urine readings, expressed as mg per gm of creatinine to correct for dilution, are directly related to intake in the range of 0.2 to 0.8 mg thiamin/100 Cal/day [24]. When results are expressed per gm creatinine, children have much higher excretion rates than adults, necessitating the use of a sliding scale for reference limits (Table 2-3).

TABLE 2-3. Urinary Excretion of Thiamin

Test	Unit	Subjects	Risk of Deficiency		
			High	Medium	Low
Loading urine	µg/4 hr	Adults	< 20	20 - 79	> 80
Timed excretion	µg/24 hr	Adults	< 40	40 - 99	> 100
	µg/6 hr	Adults	< 10	10 - 24	> 25
Random urine	µg/gm creat.	Adults	< 25	25 - 55	> 55
		13 - 15 yr	< 27	27 - 65	> 65
		10 - 12 yr	< 60	60 - 180	> 180
		7 - 9 yr	< 70	70 - 180	> 180
		4 - 6 yr	< 85	85 - 120	> 120
		1 - 3 yr	< 120	120 - 175	> 175

Vitamin B2 (Riboflavin)

Active forms: Flavin mononucleotide (FMN)
Flavin adenine dinucleotide (FAD)
Biochemical role: Redox – Central energy pathways
Example: Succinate + FAD → Fumarate + FADH2

Vitamin B_2 may be evaluated by assays of urinary markers, erythrocyte enzyme activation, or urinary vitamin output (Table 2-4). Excretion of ethylmalonate, β-hydroxyisovalerate, and several acyl glycine compounds is evidence of inadequate activity of the flavin-binding enzyme system, multiple acyl dehydrogenase (MAD) [25]. Individuals who excrete high levels of these compounds and show symptoms of easy fatigability and exercise intolerance have been successfully treated with riboflavin [26]. Similarly, elevated urinary methylsuccinate and suberylglycine are associated with riboflavin-responsive mitochondrial oxidation impairment [27]. Treatments usually involve 100 mg riboflavin with other related nutrients such as biotin, niacin, and carnitine.

In erythrocytes, oxidized glutathione is recycled to the reduced form by the action of the enzyme, erythrocyte glutathione reductase (EGR). The stimulation of activity of this enzyme by the addition of the flavin adenine dinucleotide (FAD) coenzyme is a good functional measure of riboflavin adequacy. The value is expressed as an activity coefficient that is the ratio of activity with added FAD divided by the activity without added FAD.

TABLE 2-4. Vitamin B_2 Deficiency Tests

Marker Compound	Deficiency Indication
Methylsuccinate, Ethylmalonate, and Suberylglycine	Levels above normal limits
EGR Activity Coefficient	> 1.2
Urinary Riboflavin	< 80 (Adult)

Controlled human studies have established a correlation between dietary intake of riboflavin and urinary excretion of the vitamin. Individuals consuming less than the recommended amount of riboflavin excrete less than 100 µg of riboflavin per day. Urinary levels tend to reflect recent dietary intake, not whole body status, however, and thus are prone to large variation. Children excrete more riboflavin per gram of creatinine than adults do, so age-adjusted reference ranges are necessary.

Vitamin B3 (Niacin)

Active forms: Nicotinamide adenine dinucleotide (NAD)
Nicotinamide adenine dinucleotide phosphate (NADP)
Biochemical role: NAD^+: Redox – Central energy pathways
$NADP^+$: Biosynthesis
Example: Lactate + NAD^+ → Pyruvate + NADH + H^+

Biochemical markers are especially helpful in assessment of niacin adequacy due to the difficulty in establishing dietary intake and the very wide range of symptoms related to niacin deficiency. The active forms of niacin, nicotinamide adenine dinucleotide (NAD) and nicotinamide adenine dinucleotide phosphate (NADP) function as coenzymes in oxidation/reduction reactions. The biological responses to niacin, nicotinic acid, and its amine form, niacinamide, are virtually equivalent (except at supra physiological doses) because of the ease with which the liver produces the amine from the acid. Dietary analysis is complicated by the fact that nicotinamide is frequently bound in unavailable forms in foods, especially corn. The amino acid, tryptophan, can serve as a precursor to niacin in humans, but the conversion is dependent on adequate status of both vitamin B_6 and iron [28]. The optimum conversion of tryptophan to niacin yields 1 mg from 60 mg of amino acid.

Urinary nicotinic acid is relatively uninfluenced by dietary intake of niacin or tryptophan. Testing based on excretion of catabolic products of niacin (*i.e.,* N-methylnicotinamide) has been developed, but the test is not sensitive enough for use as a functional marker of mild deficiency states. Simultaneous elevation of lactate and pyruvate in blood or urine is a biochemical marker for niacin deficiency [29]. The pathways involved here are the oxidation of lactate to pyruvate and the conversion of pyruvate to acetyl-CoA, both of which are dependent on niacin in the form of NAD. These are active pathways and are good reflections of critical metabolic functions of niacin. Further discussion of these markers is presented in Chapter 6 – Organic Acids.

Niacin deficiency is one of the potential causes of increased fecal fat [30], possibly due to impairment of intestinal protein absorption. The maintenance of intestinal epithelial cells is highly dependent on niacin availability [31].

Vitamin B5 (Pantothenic Acid)

Active form: Coenzyme A (CoA)
Biochemical role: Carrier for acyl groups
Examples: Acetyl-CoA, Fatty acyl-CoA

Pantothenic acid or pantothenate derives its name from the observation that it is found in all life forms. It is found, therefore, in all whole foods. The essential nutrient, pantothenic acid, is converted into its active form, coenzyme A, in the liver. Because of the centrality of coenzyme A in metabolic function, pantothenic acid deficiency is one of the most difficult nutrient conditions to establish conclusively. Animals placed on diets without pantothenic acid display loss of appetite, slow growth, skin lesions, weakness, and death [32, 33]. When body stores of coenzyme A are exhausted, all tissues experience massive decline in activity of energetic and biosynthetic pathways [34]. The pervasive nature of the effects on critical organs results in rapid, simultaneous decline of many functions, including the central nervous system.

Pantothenic acid is converted into coenzyme A, the acyl carrier in keto acid and fatty acid oxidation. All oxidation pathways for carbohydrate, fat, and protein require coenzyme-A to carry acetate, keto-acid, or fatty acid intermediates during the key oxidative steps. The finding of multiple intermediates from pathways that require this carrier function is indication that inadequate pantothenic acid supply is limiting optimum metabolism. Intermediates commonly reported in urinary organic acid analysis that provide such clues are pyruvate, α-ketoisocaproate, α-ketoisovalerate, α-keto-β-methylvalerate, and α-ketoglutarate. Urinary excretion of pantothenic acid less than 1.0 mg/day is also indicative of inadequate intake [35].

Vitamin B6 (Pyridoxine)

Active form: Pyridoxal-5-phosphate (P-5-P)
Biochemical role: Amino acid metabolism
Example: Homocysteine + Serine → Cystathionine

There are approximately 60 enzymes involved in amino acid metabolism that require the active form of vitamin B_6 (pyridoxal-5-phosphate, P-5-P) for their function. Other P-5-P-requiring enzymes are needed for conversions of carbohydrates and fats. Sensitive tests for the detection of vitamin B_6 are necessary because of the various factors affecting status (Table 2-5). Smoking further increases the incidence of vitamin B_6 deficiency [36].

TABLE 2-5. Factors Affecting Vitamin B_6 Status

Type	Example
Factors that decrease supply	Marked loss in food proccessing Ease of depletion from body stores Increased needs with pregnancy
Antagonistic compounds that increase need	Antimetabolite drugs • Birth control pills • Isoniazid and cycloserine (tuberculosis) • Hydralazine (hypertension) • Phelelzine (depression) • Penicillamine (arthritis & lead poisoning) Antimetabolite environmental chemicals • Hydrazines from smoking • Rocket fuels • Maleic hydrazide (herbicide) • Succinic acid-2, 2-dimethylhydrazide (fruit ripening agent) • Tartrazine (FD&C yellow No. 5)

Vitamin B_6 deficiency is a part of the nutritional triad that leads most directly to failure in the metabolism of homocysteine. The other factors, vitamin B_{12} and folate, are discussed and illustrated in Figure 2-1. The B_{12}-folate system allows the recovery of homocysteine as methionine, while the P-5-P pathway is necessary for the conversion of homocysteine to cysteine [37]. Significant portions of the population in developed countries carry a genetic determinant that causes weaker cofactor binding to the enzymes, cystathionine β-synthase and 5,10-methylene-H4folate reductase. Individuals with defective enzymes require greater concentrations of the coenzyme to achieve necessary enzyme activity levels to protect from homocysteine accumulation [38]. A limitation of homocysteine as a marker for B_6 deficiency is the lack of specificity due to concurrent involvement of the B_{12} and folate enzymes in the control of homocysteine.

FIGURE 2-1. Nutrient Requirements for Homocysteine Metabolism

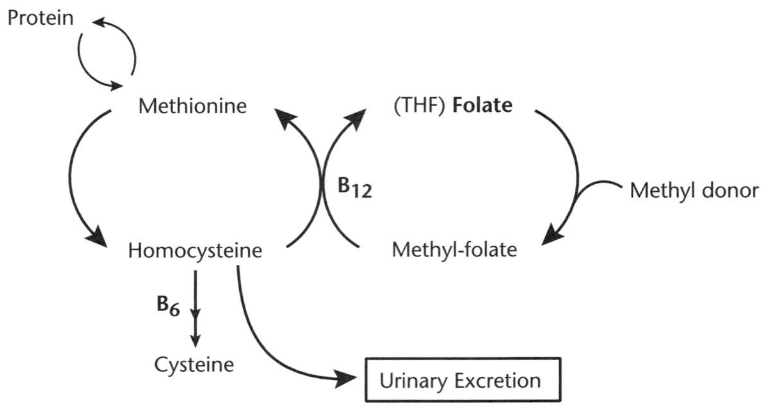

Specific biochemical markers of vitamin B_6 status are xanthurenate, kynurenate, and quinolinate, which are metabolic acids produced from tryptophan. The breakdown of these three compounds depends on enzymes that function only when P-5-P is bound. Inadequate P-5-P results in accumulation of these markers that spill into urine. Their presence above normal limits in urine is evidence of vitamin B_6 deficiency (Table 2-6).

Erythrocyte transaminase enzymes provide another method for assessing vitamin B_6 status. This sensitive and specific assay has been frequently used in investigations of vitamin B_6 status [39]. Erythrocyte glutamate-oxaloacetate transaminase (EGOT) and glutamate-pyruvate transminase (EGPT) activities decline with inadequate intake of vitamin B_6. Their activities can be restored in the laboratory by addition of P-5-P, and the index obtained by expressing the increase as a ratio of the stimulated to unstimulated reaction rates provides a reliable measure of vitamin B_6 depletion

TABLE 2-6. Vitamin B_6 Deficiency Tests

Marker Compound	Deficiency Indication
Plasma Homocysteine (HCys)	> 10 nmol/mL
Urinary Homocysteine	> 25µg/mg creatinine
Urinary Xanthurenate, Kynurenate, or Quinolinate	Above reference limit
EGOT index	> 1.5
EGPT index	> 1.25

(Table 2-6). One reason this assay has not found wide clinical application is its relative expense as a single nutrient marker.

Vitamin B_6 deficiency affects the metabolism of polyunsaturated fatty acids. Patients whose plasma fatty acids profile shows an elevated ratio of alpha linolenic acid to docosahexaenoic acid may be deficient in vitamin B_6 [40].

Vitamin B_{12} (Cobalamin)

Active forms: Methylcobalamin, and 5′-Deoxyadenosylcobalamin (Coenzyme B_{12})

Biochemical role: Methyl group transfer

Examples: Methylmalonyl-CoA → Succinyl-CoA

Homocysteine → Methionine

Deficiencies of vitamin B_{12} usually result from inadequate dietary intake or impaired absorption from lack of intrinsic factor in gastric secretions. Because of the close relationship between the biochemical role of vitamin B_{12} and folic acid, the clinical and hematological results of deficiencies of the two nutrients are indistinguishable. The relative time scale for various manifestations of folate deficiency (Figure 2-2) may be applied similarly to vitamin B_{12}. Tests specific for vitamin B_{12} are serum concentration measurement and elevated urine methylmalonate (MMA) excretion. Urine MMA is a measure of functional adequacy, while serum concentration measures the amount available for use. The distinction is important because of the possibility that a "normal" serum concentration may be inadequate to saturate vitamin B_{12}-requiring enzymes to carry out essential metabolic functions in some individuals. Because of its specificity and sensitivity, the test for methylmalonate elevation has been called the gold standard for assessing tissue vitamin B_{12} status (Table 2-7) [41].

TABLE 2-7. Markers of Vitamin B_{12} Deficiency

Marker Compound	Deficiency Indication
Methylmalonate, urine	> 3 μg/mg creatinine
Homocysteine, urine	> 25 μg/mg creatinine
Homocysteine, plasma	> 10 nmol/mL
Serum vitamin B_{12} (RIA)	< 100 pg/mL

When used for neonatal screening, elevated urinary MMA levels reveal infant vitamin B_{12} deficiency. Two cases of deficiency in otherwise non-symptomatic newborns have been reported, the mother being vegetarian in one case and having pernicious anemia in the other [42]. This is an example of the use of metabolic testing to detect preventable vitamin deficiency cases before the onset of symptoms that might otherwise persist undetected for many years.

Although the detection of elevated homocysteine in plasma or urine is not as specific as methylmalonic aciduria for vitamin B_{12}, the increasing use of this test as a screen for nutrient-related cardiovascular risk will allow the discovery of many cases needing the vitamin. Vitamin B_{12} and folate supplementation in these individuals will not only reduce cardiovascular risk but, equally importantly, will allow the early correction of other vitamin B_{12} deficiency effects.

The detection of vitamin B_{12} deficiency raises other questions regarding metabolic impact and origin of the deficiency. The accumulation of MMA caused by vitamin B_{12}-deficiency can cause a disruption of normal glucose and glutamic acid metabolism [43]. The Schilling test, used to determine if vitamin B_{12} deficiency is due to malabsorption, is described in the chapter on GI function.

Folate

Active form: Tetrahydrofolate (THF)
Biochemical role: Methyl group transfer
Example: Homocysteine → Methionine

Megaloblastic anemia resulting from folate deficiency is found in up to 5% of pregnant women in developed countries [1]. Since megaloblastic anemia is a sign that develops long after biochemical lesions (Figure 2-2), functional folic acid deficiency is likely to be present in a large percentage of the population. This rationale is supported by the variety of causes of folic acid deficiency, including inadequate dietary intake, impaired absorption, excessive tissue demands, and metabolic derangement [44].

FIGURE 2-2. **Folate Deficiency – From Biochemical to Clinical Expression**

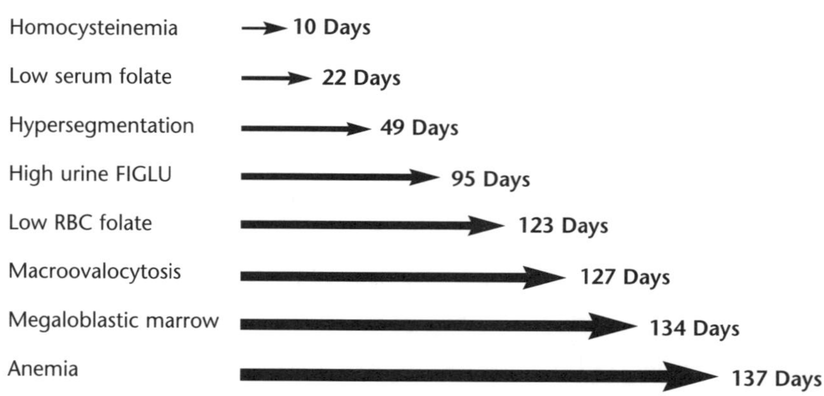

Homocysteine accumulation is a sensitive marker of folate inadequacy at the biochemical level because of the great flux of metabolites from methionine through homocysteine and cysteine to supply the constant demand for glutathione. Since elevated homocysteine is not necessarily due to folate deficiency, measurement of serum folate may be used as a confirmatory test (Table 2-8). Other signs appear at later stages in the succession of steps whereby a biochemical pathway inefficiency progresses to cellular malfunction and apoptosis, and finally to tissue failure. The most susceptible tissue is bone marrow, again owing to the constant high demand for replenishment of mature erythrocytes. Anemia is the end-stage disease of the biochemical lesion resulting from folate deficiency.

A serum folate concentration of less than 0.3 ng/ml is evidence of deficiency, but false normal values are frequently due to the response of serum levels to transient dietary intake changes that do not correspond to tissue repletion of the nutrient. There is a marked increase in demands for the vitamin during the third trimester of pregnancy. Short-term deficiencies may not produce hematological changes in the mother but are likely to contribute to birth defects [45].

Erythrocyte folate may be used to evaluate deficiency, but the period of deficiency that must pass before abnormal values are observed is longer than that causing elevated homocysteine. Leukocytes contain much higher levels of folate, but the changes parallel those for erythrocytes and, thus, are not valuable early markers of deficiency (Table 2-8). Formiminoglutamic acid (FIGLU) is an intermediate in the degradation of histidine accumulates in the later stages of folate deficiency, especially when histidine is administered as an oral load [46].

TABLE 2-8. Folic Acid Deficiency Tests

Marker Compound	Deficiency Indication
Homocysteine, urine	> 25 µg/mg creatinine
Homocysteine, plasma	> 10 nmol/mL
Serum folate	< 3 ng/ml
FIGLU	> 50 mg/day following 5-25 gm His
Erythrocyte folate	< 160 ng/ml

Biotin

Active form: Biocytin
Biochemical role: Carboxylation
Example: Pyruvate → Oxaloacetate

Biotin is the B-complex vitamin most dramatically affected by oral antibiotics. The reason is two-fold. First, most foods are poor sources of biotin. (Exceptions of rich biotin supply are egg yolk, kidney, liver, and some cheeses.) Second, the bacteria of the healthy gut can produce biotin, adding significantly to dietary supply [47].

The active form of biotin, biocytin, is linked to the amino group of enzymes that operate in biochemical pathways of gluconeogenesis, fatty acid synthesis, and amino acid catabolism [48]. Genetic variations in the enzymes involved in the release and recovery of biotin contribute to a functional biotin deficiency state. Other factors operate to modify the flow of biotin from the gut, including bacterial populations that can synthesize the nutrient and the capacity for intestinal absorption. Because of the many biochemical roles of biopterin enzymes, the clinical manifestations are highly varied. Abnormalities of hair growth and skin integrity and impaired immune function are some early signs of biotin deficiency [49].

Evidence in support of a functional marker of biotin adequacy has appeared only relatively recently [50]. Urinary β-hydroxyisovalerate elevation results from biotin deficiency due to the required cofactor function of biotin failing to assist in the degradation of this catabolic product of valine. Values of β-hydroxyisovalerate above 20 µg/mg creatinine in overnight urine indicate potential biotin deficiency. Further discussion of the biotin and β-hydroxyisovalerate relationship may be found in Chapter 6, Organic Acids.

Vitamin C (Ascorbic Acid)

Active form: Ascorbic acid
Biochemical role: Antioxidant, hydroxylation
Examples: Lysine-Collagen → Hydroxylysine-Collagen

Because of its short physiological half-life and chemical instability, vitamin C is one of the most problematic nutrients to assess in terms of optimal intake for an individual. Intake must be relatively constant to maintain tissue levels in a range where short-term effects on amino acid metabolism and long-term effects on collagen synthesis are optimal. There are no functional biochemical markers specific for vitamin C status. Most studies of vitamin C have relied on serum or plasma levels that show a linear relationship with intake over a limited range. Plasma ascorbic acid reaches a maximum concentration at about 1.4 mg/dl. Further intake results in rapid renal clearance. With deprivation of vitamin C, plasma ascorbic acid decreases rapidly, and clinical signs of scurvy will appear at continued levels of less than 0.2 mg/dl [51]. Under controlled conditions, measurements of plasma, tissue, and urinary vitamin C in healthy patients on various doses of vitamin C reveal bioavailability and tissue saturation. After keeping subjects on low vitamin C intake for several weeks, such studies have demonstrated complete tissue uptake of doses up to 200 mg with declining bioavailability at single doses above 500 mg [52].

In controlled human trials, early signs of scurvy appeared when plasma ascorbic acid levels fell below 0.3 mg/dl. Most frequently associated signs were petechiae, gum changes, hyperkeratosis, and arthralgia. Ascorbic acid disappears from urine early in the process of depletion. Leukocyte levels fall more slowly than plasma ascorbate; they drop most dramatically only when signs of scurvy appear. Determination of leukocyte ascorbate has other drawbacks -- the assay is technically difficult and requires large volumes of blood. Improper handling of the leukocytes will cause a loss of ascorbate resulting in falsely low values. Also, ascorbic acid is an extremely labile vitamin and extensive efforts must be made to preserve the vitamin, especially in urine where exposure to air increases the rate of oxidation.

Claims of correlation of ascorbic acid tissue saturation and the rate of disappearance of the intradermally injected dye, 2,6-dichloroindophenol, or the disappearance of the dye after topical application to the tongue have been shown to be invalid [53], [56].

Vitamin D (Cholecalciferol)

Active forms: 1,25-Dihydroxycholecalciferol,
1,25-Dihydroxyvitamin D3, Calcitriol
Biochemical role: Regulation of calcium
Example: Calcium-binding protein (CBP) synthesis

Vitamin D is unlike the B-vitamins in that it does not function to activate enzymes, but rather to regulate DNA expression and protein synthesis. The proteins synthesized under the action of vitamin D, in turn, regulate the absorption and distribution of calcium, including the rate of deposition in bone remodeling. The process of bone formation is under complex control by several factors, only one of which is vitamin D. Assays of collagen peptide markers for bone loss [57] are described under the calcium section of the chapter on minerals. Urinary or plasma proline and hydroxyproline elevations are also found in individuals with increased bone turnover resulting from inadequate calcium supply (see Chapter 4).

The most frequently used assay for this vitamin is serum 25-hydroxyvitamin D (25-OH-D), the most abundant metabolite of the vitamin and the direct precursor of the active form. Because of variations in methodologies, normal ranges are somewhat site-dependent, but most laboratories report a low cutoff around 25 nmol/L. Cases requiring massive doses of vitamin D to achieve normal 25-OH-D levels have been reported [58]. Such cases illustrate the range of individual requirements for vitamin D. Iron deficiency also lowers 25-OH-D, possibly by impairing small intestinal absorption [59]. Measurements of 25-OH-D may be performed by enzyme-linked immunoassay [60] or radioimmunoassay [61].

Severe deficiency of vitamin D indirectly causes increased serum activity of alkaline phosphatase because of the role of this enzyme in bone mineralization. Alkaline phosphatase is also under the influence of a variety of factors, including several disease processes [1] and status of zinc [62] and vitamin C [63]. The calcium-binding protein, osteocalcin, is secreted by osteoblasts, and its serum concentration is elevated in a number of conditions with high bone turnover. Osteocalcin, however, is not as sensitive a marker of bone turnover as serum alkaline phosphatase in Paget's disease [64]. A part of this lack of sensitivity may be the co-dependence on vitamin K status (see below) for osteocalcin formation. In elderly females with vitamin-D deficiency, secondary hyperparathyroidism is associated with increased serum osteocalcin levels indicating an increased bone formation; these conditions might contribute to the bone disease of geriatric patients [65]. Other measurements that reflect vitamin D deficiency are low urinary calcium excretion and low serum calcium and phosphorus.

Supplementation of vitamin D in milk products helps to prevent high deficiency rates. There is reason for concern over excessive intake of vitamin D due to effects on vascular calcification [66], especially for individuals who frequently consume milk products.

Vitamin E (Tocopherol)

Active forms: α-, β-, γ-, and δ-Tocopherol, α-, β-, γ-, and δ-Tocotrienol
Biochemical role: Antioxidant, especially membrane and plasma lipoprotein PUFA protection
Example: Protection of low-density lipoprotein from oxidation

Vitamin E is transported in plasma in the lipoproteins, and it serves as the most important membrane protective antioxidant and free radical scavenger in the body [67]. Although α-tocopherol is the major component in foods and human tissues, the β, γ, and δ isomers are sometimes included in profiles to show detail of tissue composition. Experimental vitamin E deficiency is difficult to produce in humans because of the intricate system of checks and balances in the antioxidant cascade. Although symptoms of vitamin E deficiency are subtle, many clinical effects are well documented. Cancer, heart disease, hemolytic anemia, neurologic disease (including ataxia), and other neuropathies are some of the more frequently associated human disorders related to vitamin E deficiency [68].

Plasma or serum α-tocopherol should be above 12 mg/L. Plasma tocopherol levels tend to be higher or lower in proportion to cholesterol and lipoprotein levels. This behavior is to be expected since all of the tocopherol in serum is associated with the lipoprotein fractions. For lipid extracts of serum, vitamin E deficiency is indicated when the ratio of serum tocopherol to total lipids is less than 0.8 mg/g. Because of the ease of simultaneous analysis, vitamins A, E, β-carotene, and coenzyme Q_{10} are frequently reported as a profile of fat-soluble vitamins. Concentrations of these vitamins in serum are measured by high performance liquid chromatography [11, 69, 70].

Vitamin K (Phylloquinone)

Active form: Protein-bound vitamin K
Biochemical role: Gamma glutamyl carboxylation
Example: Formation of active osteocalcin

Vitamin K deficiency results in an increase in the time required for blood to clot. The usual clinical manifestation is a tendency to hemorrhage. The measurement of prothrombin time in plasma reveals the vitamin K-dependent activation of prothrombin and other clotting factors. Prothrombin times longer than 12 seconds can indicate vitamin K deficiency (Table 2-9). The biochemical action dependent on vitamin K involves unique gamma-carboxylation of glutamyl residues of specific proteins. The result of the carboxylation is increased binding of calcium that is known to be required in blood clot formation and bone remodeling [71]. Vitamin K status is especially important in the elderly because of dietary and absorptive difficulties, frequently complicated by drug therapies. Monitoring vitamin K can help reduce osteoporotic bone fractures by identifying individuals whose bone loss is due to vitamin K deficiency [72, 73]. Calcium loss in such cases can be reduced by up to 50% with vitamin K supplementation [74].

Functional measures of vitamin K status are becoming available. Radioimmunoassay for undercarboxylated osteocalcin and measurement of γ-carboxyglutamate (Gla) in urine are two ways to see whether the biochemical functions of the vitamin are adequate (Table 2-9). These markers are sensitive indicators of vitamin K status [75].

TABLE 2-9. Vitamin K Tests

Test	Deficiency Indication
Prothrombin time	> 12 sec
Phylloquinone, plasma	< 1 nmol/L
Gla, urine	< 3 μmol/mmol creatinine
Under-carboxylated osteocalcin, plasma	> 20% Total Osteocalcin

Conditionally Essential Vitamins

Carnitine

The action of carnitine may be described as a key that opens the gate of energy flow in most cells. With the notable exception of brain tissue, the dominant central energy pathway in human cells oxidizes fatty acids to generate ATP. The enzymes of this pathway reside inside the mitochondria. Fatty acid molecules that are substrates for the enzymes, however, must gain entry to the mitochondria. The entry is accomplished by attachment of fatty acids to carnitine in the mitochondrial trans-membrane space.

Carnitine is a water-soluble compound that is abundant in animal muscle tissue, including red meats (Latin: carne). Carnitine is synthesized from lysine and, therefore, it is not strictly classified as an essential nutrient. Evidence of its conditional essentiality comes from the finding of low plasma carnitine in patients on long-term total parenteral nutrition [76]. Skeletal muscle myopathy has been successfully treated with carnitine [77]. Measurement of acyl- and free carnitine is used to detect secondary carnitine deficiencies in certain inborn errors of organic acid metabolism [78]. Carnitine has been advocated as a "mitochondrial metavitamin" therapy for adult onset diabetes [79].

Functional carnitine deficiency is revealed by elevated urinary excretion of adipic, suberic, and ethylmalonic acids [80]. Impairment of fatty acid β-oxidation is accompanied with high levels, as reported in Reye-like syndrome [81] and other mitochondrial impairment disorders [82].

Coenzyme Q_{10}

The primary function of coenzyme Q_{10}, also called ubiquinone, is shuttling electrons in the mitochondrial inner membrane. This pathway is frequently referred to as the electron transport or oxidative phosphorylation part of the central energy pathway. The electrons are received directly from succinate and choline or indirectly from several other substrates (Figure 2-3). Coenzyme Q_{10} (CoQ_{10}) then moves from one electron carrier complex to the other to ultimately deliver the electrons to oxygen one at a time in a never-ending cycle of oxidation and reduction. This function is critical for the fundamental energy generation that powers all cell function, and the tissue that has the largest, most critical energy demand -- the

FIGURE 2-3. Relationship of Coenzyme Q_{10} to Oxidizing Pathways

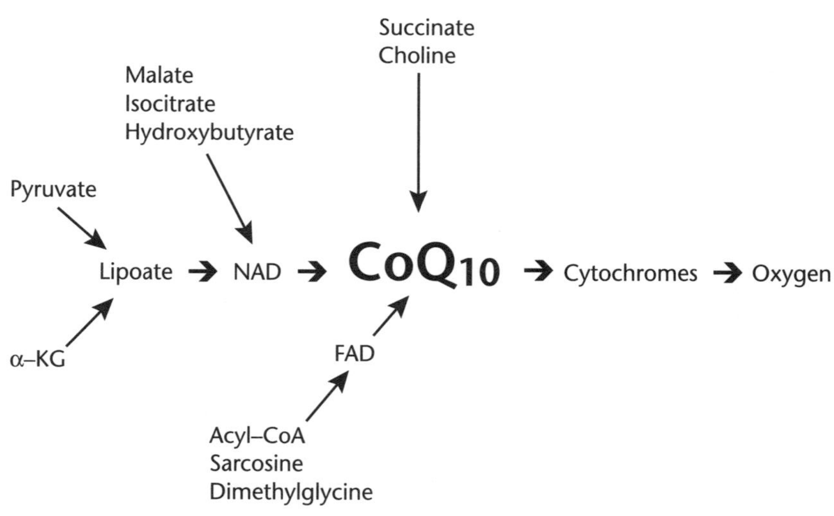

heart -- will first show effects of conditional deficiency of the coenzyme. The exercise-stimulated increase of CoQ_{10} in heart and skeletal muscle is out of proportion with the changes in other electron transport system components, indicating a direct role of the coenzyme in regulation of mitochondrial capacity for forming ATP [83]. There also may be a role of CoQ_{10} as an antioxidant, independent of its function in mitochondrial electron transport [84].

Various evidence points to the status of CoQ_{10} as a conditionally essential nutrient. It is found in significant amounts in all nutrient-dense foods. Plants contain a slightly different compound that may serve the electron transport function in animals, though with lower efficiency. Patients with mitochondrial encephalomyopathy due to electron transport complex I and IV deficiencies show improved mitochondrial function and an increased capacity for fat metabolism when supplemented with 150 mg/d of coenzyme CoQ_{10} [85]. Lactate and pyruvate elevations are reduced in patients with mitochondrial myopathy when they are given daily oral doses of 120 mg of CoQ_{10} [86].

The clinical interpretation of abnormal serum CoQ_{10} levels may be enhanced by simultaneous measurement of markers that monitor biosynthetic and functional roles of the cofactor. CoQ_{10} synthesis is dependent on the availability of hydroxymethylglutarate (HMG). Low levels of HMG slow the rate of synthesis (Table 2-10). Statin drugs block the conversion of HMG to cholesterol and to coenzyme Q_{10}. HMG accumulates and spills in urine at abnormally high levels that signals reduced CoQ_{10} biosynthesis (Table 2-10). Impairment of the synthetic pathway does not necessarily mean functional impairment.

TABLE 2-10. **Reinforcing Tests for Coenzyme Q_{10} Deficiency**

Test	Abnormality	Indication
HMG, urine	Low	Metabolic block before HMG
	High	Metabolic block after HMG
CoQ_{10}, serum	Low	Sign of depletion of tissue CoQ10
Pyruvate, succinate, fumarate, malate, urine	High	Functional insufficiency to meet energy pathway demands

A functional impairment at the level of mitochondrial coenzyme Q_{10} electron transfer will affect succinate, malate, fumarate, and pyruvate, which are the energy pathway intermediates. Coenzyme Q_{10} deficiency causes elevation of these intermediates (Table 2-10) [87]. The direct transfer of electrons from succinate to the flavin monoucleotide (FMN) reductase enzyme in the electron transport system is slowed when the electron shuttle action of coenzyme Q_{10} is inadequate to meet demands.

Chapter 12 presents further detail about the interpretation of multiple profile data for vitamins and the other nutrient classes covered here.

References

[1] Sauberlich, H.E., *Assessment of nutritional status.* 1999, Boca Raton: CRC. 486.

[2] Shive, W. and K.S. Matthews, Nutritional requirements for growth of human lymphocytes. *Annu Rev Nutr,* 8. 81-97, 1988.

[3] Raica, N., *et al.*, Vitamin A concentration in human tissues collected from five areas in the United States. *Am J Clin Nutr,* 25. 291, 1972.

[4] McLaren, D., H. Oomen, and H. Escapini, Ocular manifestations of vitamin A deficiency in man. *Bull W H O,* 34. 357, 1966.

[5] Ganguly, J., *Biochemistry of vitamin A.* 1989, Boca Raton: CRC. 221.

[6] Patwardhan, V.N., Hypovitaminosis A and epidemiology of xerophthalmia. *Am J Clin Nutr,* 22(8). 1106-18, 1969.

[7] Stauber, P.M., *et al.*, A longitudinal study of the relationship between vitamin A supplementation and plasma retinol, retinyl esters, and liver enzyme activities in a healthy elderly population. *Am J Clin Nutr,* 54(5). 878-83, 1991.

[8] Hodges, R. and H. Kolder, Experimental vitamin A *deficiency in human volunteers, in Summary of proceedings of workshop on biochemical and clinical criteria for determining human vitamin A nutriture,* J. Beri, Editor. 1971, Food and Nutrition Board, National Academy of Sciences-National Research Council: Washington, DC. p. 10.

[9] Shils, M., J. Olson, and M. Shike, eds. *Modern nutrition in health and disease.* 8 ed. 1994, Lea & Febiger: Philadelphia. 1657.

[10] Tsai, A.C., H.A. Mazeedi, and M.S. Mameesh, Dietary beta-carotene reduces serum lipid concentrations in spontaneously hypertensive rats fed a vitamin A-fortified and cholesterol-enriched diet. *J Nutr,* 122(9). 1768-71, 1992.

[11] Sowell, A.L., *et al.*, Retinol, alpha-tocopherol, lutein/zeaxanthin, beta-cryptoxanthin, lycopene, alpha-carotene, trans-beta-carotene, and four retinyl esters in serum determined simultaneously by reversed-phase HPLC with multiwavelength detection. *Clin Chem,* 40(3). 411-6, 1994.

[12] Su, Q., K.G. Rowley, and K. O'Dea, Stability of individual carotenoids, retinol and tocopherols in human plasma during exposure to light and after extraction. *J Chromatogr B Biomed Sci Appl,* 729(1-2). 191-8, 1999.

[13] Nierenberg, D.W., *et al.*, Effects of 4 y of oral supplementation with beta-carotene on serum concentrations of retinol, tocopherol, and five carotenoids. *Am J Clin Nutr,* 66(2). 315-9, 1997.

[14] Higueret, P. and H. Garcin, Peripheral metabolism of thyroid hormones in vitamin A-deficient rats. *Ann Nutr Metab,* 26(3). 191-200, 1982.

[15] Curran-Celentano, J., *et al.*, Alterations in vitamin A and thyroid hormone status in anorexia nervosa and associated disorders. *Am J Clin Nutr,* 42(6). 1183-91, 1985.

[16] Goswami, U.C. and S. Choudhury, The status of retinoids in women suffering from hyper- and hypothyroidism: interrelationship between vitamin A, beta-carotene and thyroid hormones. *Int J Vitam Nutr Res,* 69(2). 132-5, 1999.

[17] Block, G., *et al.*, Nutrient sources in the American diet: quantitative data from the NHANES II survey. II. Macronutrients and fats. *Am J Epidemiol,* 122(1). 27-40, 1985.

[18] Scriver, C. and L. Rosenberg, *Amino Acid Metabolism and It's Disorders.* Major problems in clinical pediatrics, ed. A. Schaffer. Vol. X. 1973, Philadelphia: W. B. Saunders. 491.

[19] Lonsdale, D., Hypothesis and case reports: possible thiamin deficiency. *J Am Coll Nutr,* 9(1). 13-7, 1990.

[20] Chuang, D. and V. Shih, Disorders of branched chain amino acid and keto acid metabolism, in *The metabolic and molecular bases of inherited disease,* C. Scriver, *et al.*, Editors. 1995, Mcgraw-Hill: New York. p. 1239-1277.

[21] Thompson, R., The value of blood pyruvate determinations in the diagnosis of thiamine deficiency., in *Thiamine deficiency: Biochemical lesions and their clinical significance,* G. Wolstenholme, Editor. 1967, Churchill: London.

[22] Chong, Y.H. and G.S. Ho, Erythrocyte transketolase activity. *Am J Clin Nutr,* 23(3). 261-6, 1970.

[23] Lossy, F., G. Goldsmith, and H. Surett, A study of test dose excretion of five B complex vitamins in man. *J Nutr,* 45. 213, 1951.

[24] *Requirements of vitamin A, thiamine, riboflavine and niacin.* Report of a Joint FAO/WHO Expert Group. FAO Nutrition Meeting Report No. 41. 1967, Rome: Food and Agriculturer Organization of the United Nations.

[25] Sperl, W., *et al.*, Stridor as the major presenting symptom in riboflavin-responsive multiple acyl-CoA dehydrogenation deficiency. *Eur J Pediatr,* 156(10). 800-2, 1997.

[26] Scholte, H.R., *et al.*, Riboflavin-responsive complex I deficiency. *Biochim Biophys Acta,* 1271(1). 75-83, 1995.

[27] Dawson, D.B., *et al.*, Transient organic aciduria and persistent lacticacidemia in a patient with short-chain acyl-coenzyme A dehydrogenase deficiency. *J Pediatr,* 126(1). 69-71, 1995.

[28] Oduho, G.W., Y. Han, and D.H. Baker, Iron deficiency reduces the efficacy of tryptophan as a niacin precursor. *J Nutr,* 124(3). 444-50, 1994.

[29] Fujii, T., *et al.*, Complex I (reduced nicotinamide-

adenine dinucleotide-coenzyme Q reductase) deficiency in two patients with probable Leigh syndrome. *J Pediatr*, 116(1). 84-7, 1990.

[30] Victoria, C.R. and U.G. Meneghelli, [Fecal excretion of fats and nitrogen in rats with niacin deficiency]. *Arq Gastroenterol*, 27(2). 62-6, 1990.

[31] Victoria, C.R. and U.G. Meneghelli, [Morphometry and kinetics of jejunal epithelium in rats with niacin deficiency]. *Arq Gastroenterol*, 26(4). 111-5, 1989.

[32] Brody, T., *Nutritional Biochemistry*. 2 ed. 1999, San Diego: Academic Press. 1006.

[33] Gries, C.L. and M.L. Scott, The pathology of thiamin, riboflavin, pantothenic acid and niacin deficiencies in the chick. *J Nutr*, 102(10). 1269-85, 1972.

[34] Moiseenok, A.G., V.M. Sheibak, and V.A. Gurinovich, Hepatic CoA, S-acyl-CoA, biosynthetic precursors of the coenzyme and pantothenate-protein complexes in dietary pantothenic acid deficiency. *Int J Vitam Nutr Res*, 57(1). 71-7, 1987.

[35] Fry, P.C., H.M. Fox, and H.G. Tao, Metabolic response to a pantothenic acid deficient diet in humans. *J Nutr Sci Vitaminol*, 22(4). 339-46, 1976.

[36] Bergmark, C., *et al.*, Hyperhomocysteinemia in patients operated for lower extremity ischaemia below the age of 50--effect of smoking and extent of disease. *Eur J Vasc Surg*, 7(4). 391-6, 1993.

[37] Wilcken, D.E., N.P. Dudman, and P.A. Tyrrell, Homocystinuria due to cystathionine beta-synthase deficiency--the effects of betaine treatment in pyridoxine-responsive patients. *Metabolism*, 34(12). 1115-21, 1985.

[38] Tobias, S.L., *et al.*, Alcohol intakes and deficiencies in thiamine and vitamin B_6 in black patients with cardiac failure. *S Afr Med J*, 76(7). 299-302, 1989.

[39] Leinert, J., I. Simon, and D. Hotzel, [Evaluation of methods to determine the vitamin B_6 status of humans. 1. alpha-EGOT: methods and validation]. *Int J Vitam Nutr Res*, 51(2). 145-54, 1981.

[40] Tsuge, H., N. Hotta, and T. Hayakawa, Effects of vitamin B_6 on (n-3) polyunsaturated fatty acid metabolism. *J Nutr*, 130(2S Suppl). 333S-334S., 2000.

[41] Norman, E.J., Urinary methylmalonic acid/creatinine ratio: a gold standard test for tissue vitamin B_{12} deficiency [letter; comment]. J Am *Geriatr Soc*, 47(9). 1158-9, 1999.

[42] Michaud, J.L., *et al.*, Nutritional vitamin B_{12} deficiency: two cases detected by routine newborn urinary screening. *Eur J Pediatr*, 151(3). 218-20, 1992.

[43] Toyoshima, S., *et al.*, Accumulation of methylmalonic acid caused by vitamin B_{12}-deficiency disrupts normal cellular metabolism in rat liver. *Br J Nutr*, 75(6). 929-38, 1996.

[44] Rosenberg, I.H. and J. Dyer, The prevalence and causes of folic acid deficiency in the United States. In: *Folic acid in neurology, psychiatry, and internal medicine*. New York: Raven Press,. 19-22, 1979.

[45] Allen, W.P., Folic acid in the prevention of birth defects. *Curr Opin Pediatr*, 8(6). 630-4, 1996.

[46] Roon-Djordjevic, B.v. and S. Cerfontain-van, Urinary excretion of histidine metabolites as an indication for folic acid and vitamin B 12 deficiency. *Clin Chim Acta*, 41. 55-65, 1972.

[47] Scholtissek, J., *et al.*, Biotin supply by large bowel bacteria in minipigs: evidence from intracaecal avidin. *Br J Nutr*, 64(3). 715-20, 1990.

[48] Wolf, B., Disorders of biotin metabolism, in *The Metabolic and Molecular Bases of Inherited Disease*, C. Scriver, *et al.*, Editors. 1995, McGraw-Hill, Inc.: New York. p. 3151-3177.

[49] Liu, Y.Y., *et al.*, Abnormal fatty acid composition of lymphocytes of biotin-deficient rats. *J Nutr Sci Vitaminol (Tokyo)*, 40(3). 283-8, 1994.

[50] Mock, N., *et al.*, Increased urinary excretion of 3-hydroxyisovaleric acid and decreased urinary excretion of biotin are sensitive early indicators of decreased biotin status in experimental biotin deficiency. *Am J Clin Nutr*, 65. 951-8, 1997.

[51] Bermond, P., Letter: Clinical symptoms of malnutrition and plasma ascorbic acid levels. *Am J Clin Nutr*, 29(5). 493, 1976.

[52] Levine, M., *et al.*, Vitamin C pharmacokinetics in healthy volunteers: evidence for a recommended dietary allowance. *Proc Natl Acad Sci U S A*, 93(8). 3704-9, 1996.

[53] Cheraskin, E. and W.M. Ringsdorf, Jr., A lingual vitamin C test: 3. Relationship to plasma ascorbic acid level. *Int Z Vitaminforsch*, 38(1). 120-2, 1968.

[54] Ringsdorf, W.M., Jr. and E. Cheraskin, The lingual ascorbic acid test. *Quintessence Int*, 9(12). 81-5, 1978.

[55] Inouye, L., S.A. Miller, and M.C. Alfano, Lingual ascorbic acid test: poor correlation with blood and tissue levels of ascorbic acid in guinea pigs. *J Dent Res*, 54(6). 1180-2, 1975.

[56] Leggott, P.J., *et al.*, Response of lingual ascorbic acid test and salivary ascorbate levels to changes in ascorbic acid intake. *J Dent Res*, 65(2). 131-4, 1986.

[57] Price, P.A., J.G. Parthemore, and L.J. Deftos, New biochemical marker for bone metabolism. Measurement by radioimmunoassay of bone GLA protein in the plasma of normal subjects and patients with bone disease. *J Clin Invest*, 66(5). 878-83, 1980.

[58] Asami, T., T. Kawasaki, and M. Uchiyama, Unique form of rickets with low serum 25-hydroxyvitamin D in two normally nourished children. *Acta Paediatr Jpn*, 37(2). 182-8, 1995.

[59] Heldenberg, D., G. Tenenbaum, and Y. Weisman, Effect of iron on serum 25-hydroxyvitamin D and

24,25-dihydroxyvitamin D concentrations. *Am J Clin Nutr*, 56(3). 533-6, 1992.

[60] Lind, C., J. Chen, and I. Byrjalsen, Enzyme immunoassay for measuring 25-hydroxyvitamin D3 in serum. *Clin Chem*, 43(6 Pt 1). 943-9, 1997.

[61] Franck, P.F., G.A. de Maaker, and A.J. Moolenaar, 25-Hydroxyvitamin D and 1,25-dihydroxyvitamin D determined in serum by "SPE-Octadecyl (C18)" column extraction and radioassay. *Clin Chem*, 35(9). 1995, 1989.

[62] Naber, T.H., *et al.*, Serum alkaline phosphatase activity during zinc deficiency and long- term inflammatory stress. *Clin Chim Acta*, 249(1-2). 109-27, 1996.

[63] Mahmoodian, F., A. Gosiewska, and B. Peterkofsky, Regulation and properties of bone alkaline phosphatase during vitamin C deficiency in guinea pigs. *Arch Biochem Biophys*, 336(1). 86-96, 1996.

[64] Kaddam, I.M., *et al.*, Comparison of serum osteocalcin with total and bone specific alkaline phosphatase and urinary hydroxyproline: creatinine ratio in patients with Paget's disease of bone. *Ann Clin Biochem*, 31(Pt 4). 327-30, 1994.

[65] Pietschmann, P., W. Woloszczuk, and H. Pietschmann, Increased serum osteocalcin levels in elderly females with vitamin D deficiency. *Exp Clin Endocrinol*, 95(2). 275-8, 1990.

[66] Jono, S., *et al.*, 1,25-Dihydroxyvitamin D3 increases in vitro vascular calcification by modulating secretion of endogenous parathyroid hormone-related peptide. *Circulation*, 98(13). 1302-6, 1998.

[67] Kamal-Eldin, A. and L.A. Appelqvist, The chemistry and antioxidant properties of tocopherols and tocotrienols. *Lipids*, 31(7). 671-701, 1996.

[68] Ghatak, A., *et al.*, Oxy free radical system in heart failure and therapeutic role of oral vitamin E. *Int J Cardiol*, 57(2). 119-27, 1996.

[69] Lang, J.K. and L. Packer, Quantitative determination of vitamin E and oxidized and reduced coenzyme Q by high-performance liquid chromatography with in-line ultraviolet and electrochemical detection. *J Chromatogr*, 385. 109-17, 1987.

[70] McClean, S.W., *et al.*, Liquid-chromatographic assay for retinol (vitamin A) and retinol analogs in therapeutic trials. *Clin Chem*, 28(4 Pt 1). 693-6, 1982.

[71] Einhorn, T.A., *et al.*, Fracture healing and osteocalcin metabolism in vitamin K deficiency. *Clin Orthop*, (237). 219-25, 1988.

[72] Hodges, S.J., *et al.*, Circulating levels of vitamins K1 and K2 decreased in elderly women with hip fracture. *J Bone Miner Res*, 8(10). 1241-5, 1993.

[73] Weber, P., The role of vitamins in the prevention of osteoporosis--a brief status report. *Int J Vitam Nutr Res*, 69(3). 194-7, 1999.

[74] Binkley, N.C. and J.W. Suttie, Vitamin K nutrition and osteoporosis. *J Nutr*, 125(7). 1812-21, 1995.

[75] Sokoll, L.J., *et al.*, Changes in serum osteocalcin, plasma phylloquinone, and urinary gamma-carboxyglutamic acid in response to altered intakes of dietary phylloquinone in human subjects [see comments]. *Am J Clin Nutr*, 65(3). 779-84, 1997.

[76] Berner, Y.N., *et al.*, Low plasma carnitine in patients on prolonged total parenteral nutrition: association with low plasma lysine. *JPEN J Parenter Enteral Nutr*, 14(3). 255-8, 1990.

[77] Bernsen, P.L., *et al.*, Successful treatment of pure myopathy, associated with complex I deficiency, with riboflavin and carnitine [see comments]. *Arch Neurol*, 48(3). 334-8, 1991.

[78] Mandel, H., *et al.*, The importance of recognizing secondary carnitine deficiency in organic acidaemias: case report in glutaric acidaemia type II. *J Inherit Metab Dis*, 11(4). 397-402, 1988.

[79] McCarty, M.F., Maturity-onset diabetes mellitus--toward a physiological appropriate management. *Med Hypotheses*, 7(10). 1265-85, 1981.

[80] Matsumoto, M., *et al.*, Organic acid and acylcarnitine profiles of glutaric aciduria type I. *Acta Paediatr Jpn*, 32(1). 76-82, 1990.

[81] Elpeleg, O.N., *et al.*, Recurrent, familial Reye-like syndrome with a new complex amino and organic aciduria. *Eur J Pediatr*, 149(10). 709-12, 1990.

[82] Feller, A.G. and D. Rudman, Role of carnitine in human nutrition. *J Nutr*, 118(5). 541-547, 1988.

[83] Beyer, R.E., *et al.*, Elevation of tissue coenzyme Q (ubiquinone) and cytochrome c concentrations by endurance exercise in the rat. *Arch Biochem Biophys*, 234(2). 323-9, 1984.

[84] Beyer, R.E., The relative essentiality of the antioxidative function of coenzyme Q-- the interactive role of DT-diaphorase. *Mol Aspects Med*, 15 Suppl. s117-29, 1994.

[85] Bendahan, D., *et al.*, 31P NMR spectroscopy and ergometer exercise test as evidence for muscle oxidative performance improvement with coenzyme Q in mitochondrial myopathies [see comments]. *Neurology*, 42(6). 1203-8, 1992.

[86] Bresolin, N., *et al.*, Clinical and biochemical correlations in mitochondrial myopathies treated with coenzyme Q_{10}. *Neurology*, 38(6). 892-9, 1988.

[87] Boitier, E., *et al.*, A case of mitochondrial encephalomyopathy associated with a muscle coenzyme Q_{10} deficiency. *J Neurol Sci*, 156(1). 41-6, 1998.

Chapter 3

Minerals

TABLE 3-1. Summary of Essential Mineral Repletion, Deficiency, and Toxicity

				REPLETION			DEFICIENCY		TOXICITY
Element	*Primary function*	*Specimen*	*Positive*	*Child 2-12yrs†*	*Adult Male†*	*Adult Female*†*	*Causes*	*Signs*	*Signs*
Calcium (Ca)	Bone and tooth formation, blood clotting, nerve and muscle contraction	Urine Hair	Low High	800-1200 mg	800-1200 mg	800-1200 mg*	Decreased intake, malabsorption, renal failure	Tetany, osteomalacia, arrhythmia	Anorexia, nausea, constipation, vomiting, confusion
Chromium (Cr)	Insulin receptor binding factor	RBC Hair	Low Low	50-150 mcg	200-300 mcg	200-300 mcg	Decreased intake malabsorption, total parenteral nutrition (TPN)	Glucose insensitivity	None
Copper (Cu)	Enzyme cofactor (erythrocyte superoxide dismutase)	Urine RBC Hair	Low Low Low	1.0-10 mg	2.0-20 mg	2.0-20 mg	Decreased intake malabsorption, prematurity, malnutrition	Anemia, neutropenia	Vomiting, hepatic necrosis
Iodine (I)	Thyroid hormone	Serum T4, etc.	Low	90-150 mcg	150-300 mcg	150 mcg*	Decreased intake, impaired conversion	Neurologic, developmental, reproductive	None (see hyperthyroid)
Iron (Fe)	Cofactor for enzymes, cytochromes, RBC production, heme function	Serum TIBC %Sat Ferritin	Low Low Low Low	10-20 mg	10-30 mg	18-50mg*	Decreased intake, blood loss, increased requirement due to pregnancy	Anemia	Cirrhosis, heart failure, skin pigmentation, diabetes, arthropathy
Magnesium (Mg)	Muscle, nerve action, coenzyme for carbohydrate, and protein metabolism	RBC Serum Hair	Low Low High	200-600 mg	350-750 mg	300-700 mg*	Decreased intake malabsorption, alcoholism, renal tubular leak	Muscular twitching and weakness, arrhythmias, irritability, convulsions, poor growth	Nausea, vomiting, hypotension, weakness, hyporeflexia, confusion

TABLE 3-1 Cont. Summary of Essential Mineral Repletion, Deficiency, and Toxicity

Manganese (Mn)	Cofactor for enzymes	RBC Hair	Low Low	1-10 mg	5-13 mg	5-13 mg	Decreased intake malabsorption TPN	Hypercholesterolemia, weight loss	Hallucinations, neural damage
Potassium (K)	Intracellular fluid control, nerve & muscle contraction, energy metabolism	RBC Hair	Low Low	50-150 mg	150-300 mg	150-300 mg	Decreased intake, losses from vomiting, diarrhea, renal losses, diuretics	Muscle weakness, arrhythmia	Paresthesia, confusion, cardiac depression, weakness
Selenium (Se)	Cofactor for glutathione peroxidase, thyroxine oxidase	RBC Hair	Low Low	50-150 mcg	50-250 mcg	50-250 mcg	Decreased intake malabsorption TPN	Cardiomyopathy	Hair loss, fatigue, peripheral neuropathy
Zinc (Zn)	Cofactor for many enzymes	RBC Hair Plasma LA to DGLA ratio	Low Low High	8-30 mg	16-65 mg	15-65 mg*	Decreased intake, malabsorption, acrodermatitis enteropathica, diarrhea, sickle cell disease	Rashes, anorexia, lethargy, growth, retardation, alopecia, impaired immunity, slow wound healing, altered taste	Cu deficiency, impaired immunity

†Amounts shown are daily oral doses for overcoming deficiency. Source for low end of ranges: Recommended Dietary Allowances, National Academy of Sciences, National Research Council.

*Pregnant or Lactating – Calcium 1200 mg, Iodine 200 mcg, Magnesium 450 mg, Zn 25 mg. Iron requirements during pregnancy may be 30-60 mg but revert to non-pregnant amounts 2-3 months after delivery, even if lactating.

Introduction

Health risks linked to inadequate intakes of nutritionally essential minerals and to increased exposure to toxic elements in industrialized societies are documented in the scientific literature [1-11]. Groups particularly vulnerable to mineral deficiencies include children [12], pregnant women [13], the elderly [14], and those with infections [15]. In addition, depleted soil [16] and poor food choices lead to mineral deficiencies that relate directly to heart disease, hypertension [17], and cancer [18]. Chronic exposure to toxic elements can lead to disparate clinical disorders with manifestations vastly different from the classic symptoms of acute toxicity [19].

Laboratory testing has the potential to reveal mineral deficiencies and metal toxicity. Practitioners new to the field frequently raise questions regarding the best specimen for screening versus confirmatory evidence of deficiency or toxicity and the clinical significance of abnormalities detected. The first section of this chapter focuses on the key concepts of mineral metabolism, while the second section highlights the advantages and limitations surrounding specimen and test choice. Finally, specific discussion is given for each essential mineral and toxic metal. After reading the concepts, the reader is encouraged to peruse the last section for the mineral or toxic metal that may be most interesting.

Biochemistry and Toxicology of Minerals

Concepts

Thirteen trace elements are currently known to be nutritionally essential for human health. Present at concentrations of micrograms per gram tissue or less, these elements -- frequently referred to as minerals -- are iron, copper, zinc, iodine, selenium, boron, cobalt, chromium, molybdenum, manganese, vanadium, silicon, and nickel [20]. Sometimes referred to as macro-minerals, the other essential minerals present at significantly larger concentrations are calcium, magnesium, and phosphorus. Other macro-minerals like sodium, potassium, and chloride are classified as electrolytes.

Plants and animals have evolved highly efficient mechanisms for extracting nutritionally important elements that are often present in soil and food in minute amounts. Furthermore, plants and animals have developed homeostatic systems to preserve the required tissue levels of essential elements so that survival is assured. A defining characteristic of a trace element is that only a very small amount is necessary for proper function of the whole organism. Yet a trace element deficiency can disproportionately cause biochemical abnormalities in enzyme or hormone systems that regulate the metabolism of substrates present at far greater concentrations (Fig. 3-1). If these substrates, in turn, exert regulatory control within the body, negative effects of insufficient amounts of a trace element of

FIGURE 3-1. Magnification Giving Major Effects of Trace Element Deficiency

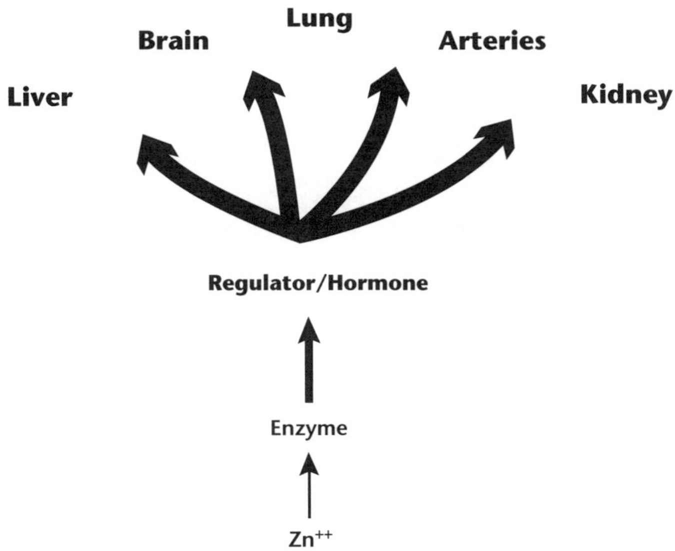

nutritional importance are greatly amplified [20].

Trace elements such as iron, copper, or molybdenum have biological oxidation-reduction functions due to their ability to easily release or accept electrons (*i.e.*, change valence states). Other trace elements, such as zinc and nickel, are stable in one particular valence state and, therefore, play a structural role, primarily as coordination sites. For example, the role of zinc can be seen maintaining the structure of the digestive enzyme, carboxypeptidase [21]. The two zinc atoms occupy a central position in the protein structure, conferring considerable stability (Figure 3-2). Nonmetal elements (selenium, iodine) and metalloids (silicon, boron) form covalent bonds with carbon and oxygen and are primarily found as part of a structural protein or enzyme (*i.e.*, selenium as selenocysteine in enzymes and silicon in connective tissue). Magnesium and calcium are found in three states: bound to proteins, bound in the crystalline matrix of bone, and as free ions in all body fluids. In their free ionic forms, calcium and magnesium are regulators or modifiers of enzyme activity.

FIGURE 3-2. Carboxypeptidase-Zn

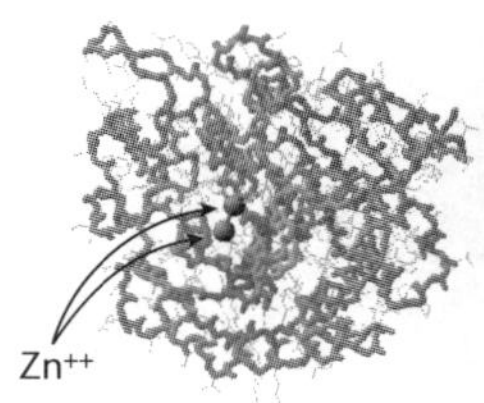

All elements are characterized by bell-shaped dose-response curves (Figure 3-3). The curve represents overall functional response to varying intakes of a given element. Low intake results in a deficiency state, while high intake results in toxicity symptoms. The optimal amount of trace element intake is in the middle of the curve where function is maximized and toxicity is minimized. Any element can become harmful if the body's intake exceeds its ability to utilize, store, or eliminate the element. Nutrient elements are no exception to this rule. For some minerals, the difference between intakes leading to deficiency or toxicity spans a 100-fold range [22]. For others, the range is only 10-fold. For example, signs of toxicity manifest from daily selenium intakes of 5 mg/day. This exceeds the upper US Recommended Daily Allowance (RDA) by only a factor of 25 [23]. Of course, some elements are characteristically toxic and cause harm even in small amounts. Lead, cadmium, and mercury rank highest on this list. Modern life in an industrialized society has resulted in increased exposure because of the enormous amounts of toxic metals released into the environment.

FIGURE 3-3. Typical Nutrient Optimal Intake Range Variation

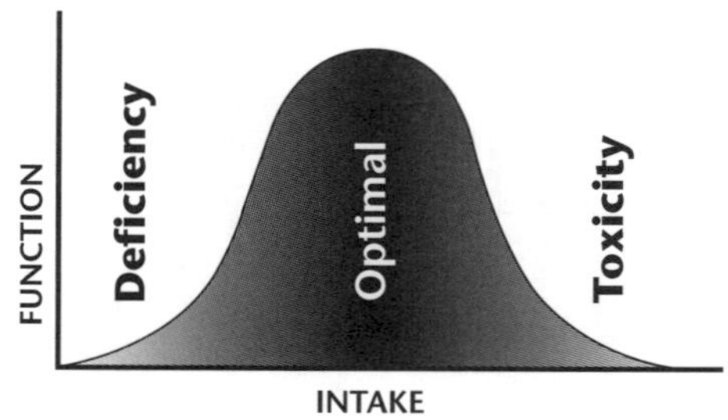

Nutrient Interactions

The deficiency or excess of one element can interfere or otherwise affect the metabolic utilization of other elements. For instance, the excessive intake of zinc antagonizes intestinal copper absorption to the point that a copper deficiency can result despite proper copper intake. Copper deficiency exacerbates iron deficiency anemia. Iron deficiency exacerbates the toxic effects of lead and cadmium. Beneficial interactions also exist. For example, selenium mitigates the toxic effects of cadmium and mercury, and calcium is protective against lead toxicity. Examples of vitamins effecting mineral status are the promotion of intestinal absorption of iron by vitamin C and the intestinal calcium transport stimulated by vitamin D.

Mineral deficiencies produce a myriad of disorders that can affect almost any organ and tissue of the body. Overt deficiencies of some minerals cause diseases producing pathognomonic symptoms such as iron deficiency anemia. Chronic sub-optimal essential mineral status coupled with toxic metal poisoning results in many symptoms that are easily misdiagnosed (Table 3-1).

Intervention Options

When minerals contained in food pass through the normal digestive process, they undergo several changes in chemical bonding as various metal complexes are formed. The low pH of the stomach serves to weaken the native protein-bound forms in food and cause increased free ionic concentrations. When the pH is raised by the action of bicarbonate-rich pancreatic fluid (in response to secretin hormone action), the divalent ions form new complexes with negatively charged amino and organic acids. Thus, the mixture that passes down the small intestine contains a variety of mineral salts like calcium aspartate, magnesium oxalate, vanadium citrate, and numerous other complexed forms. Before they are transported across the intestinal mucosa, most elements undergo at least an additional transition in complexed form. The calcium in milk is thought to undergo multiple transitions, one of the most available forms being calcium orotate. Digestive factors that can dramatically affect element absorption include inadequate (or diluted) stomach acid, low alkaline pancreatic output, or low pancreatic digestive enzymes.

Meat, organ meats, shellfish, dark green leafy vegetables, nuts, seeds, and legumes are generally good sources of minerals and trace elements, while fruits and bread contain variable amounts. The milling of grain results in the loss of many elements. Bananas, oranges, and potatoes are excellent sources of potassium. Dairy products are rich sources of calcium and magnesium, but are not good sources of many trace elements. The typical American diet is likely to provide less than optimal intake of essential minerals and trace elements.

Dietary supplements of minerals are available in many forms. Because simple forms like magnesium oxide or magnesium sulfate (Epsom Salt) are very poorly absorbed, complexed or chelated forms are always recommended. The most common of these are the lactate, gluconate, citrate, picolinate, or amino acid chelates (especially aspartate). Chelation processing methods are important considerations when choosing vendors of supplements. Simply mixing an elemental oxide with an amino acid does little good to promote absorption of the element. The formation of organic ligands with metal ions requires careful adjustment of pH with introduction of the new complexing agent at the proper time as described above for the normal digestive process. Oral supplementation ranges for properly complexed forms of elements are shown in Table 3-1.

Mineral Status Assessment

The total body content of elements spans several orders of magnitude; the optimal concentration depends on the element and the tissue. For example, calcium and magnesium comprise sizable portions of our skeletal mass -- up to a quarter of our body weight -- while total body content of manganese is less than 20 mg, one millionth that of calcium. Tissue distribution of most elements is determined by the existence of specific protein factors that bind, transport, and store each element under hormonal or other type of homeostatic control. Likewise, toxic heavy metals preferentially accumulate in different body organs. All these factors must be considered in choosing a representative specimen for testing.

Apparent mineral status can vary depending upon the specimen type and when it is sampled (Figure 3-4). Dietary intake of an element like zinc, for example, can vary greatly from day to day. Serum concentrations of the element change more slowly but respond to average daily intake, while levels in cells (*e.g.*, hepatocytes) and in tissues that act as stores for the element (*e.g.*, kidney) change on an even slower time scale. The pertinent clinical question of when and what type of intervention and supplementation will guide specimen choice for each element. Multiple element profiles on a single specimen (*e.g.*, erythrocytes) provide helpful clues about individual needs, but data from other types of testing may be needed to provide definitive answers.

FIGURE 3-4. Body Status Variations with Time

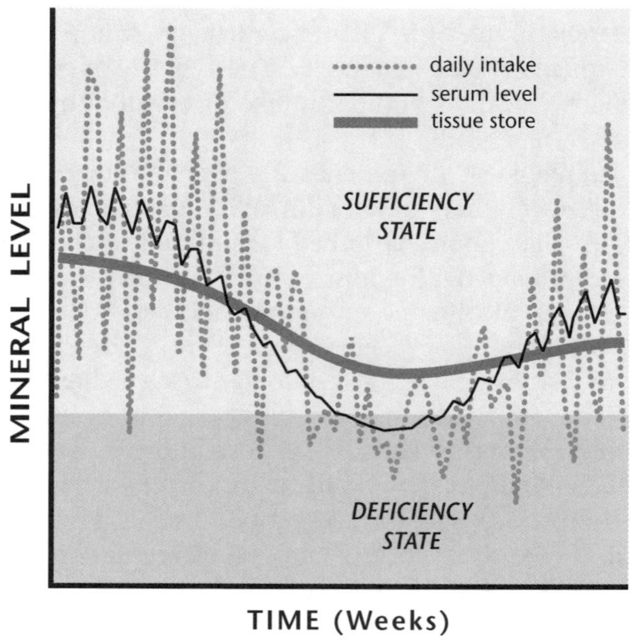

An ideal biopsy tissue for mineral status measurement is one whose mineral stores are exchangeable with other major body pools, the rate of exchange being sufficient to provide the best sensitivity with respect to changes in dietary mineral intake. Too slow an exchange rate results in insufficient sensitivity, meaning that the parameter being measured -- whether it be mineral concentration, mineral-mineral-dependent enzyme activity or physiological function -- is unresponsive to changes in mineral pools outside the tissue. A tissue with low sensitivity makes a poor indicator of change in whole body mineral status. For instance, in human subjects fed a zinc-deficient diet for 7 weeks, erythrocyte zinc levels remained unchanged despite clinical signs of zinc deficiency [24]. The reason for lack of sensitivity of erythrocytes to zinc deficiency is, presumably, because the rate of erythrocyte turnover (about four months) is slower than the turnover rate of other zinc-sensitive tissues. When low erythrocyte zinc is found, there has been a history of many weeks of zinc deficiency. On the other hand, too much sensitivity will only serve to reveal day-to-day variations that follow the daily fluctuations of mineral intake. For example, urinary levels of potassium are of little use for this reason. Such values are not meaningful in terms of showing long-term mineral status unless serial measurements are averaged over a period of weeks or months. The best biopsy tissue for essential mineral status determination is likely to be metabolically active tissue containing mineral-dependent proteins that 1) exhibit rapid rates of turnover, or 2) do not bind the mineral so tightly that exchange with the extracellular fluid is prevented.

Choosing the Best Specimen for Mineral Testing

Advances in laboratory techniques have opened avenues for detecting trace amounts of toxic metals in small amounts of samples such as whole blood, serum, plasma, urine, or hair. While there is no single best specimen for simultaneously assessing the status for both essential elements and toxic heavy metals, the following discussion will present the merits and limitations of the specimens commonly tested (*i.e.*, serum or plasma, whole blood and its cellular constituents, urine, and hair). A flowchart based on initial decisions regarding the need for toxic metal assessment is shown in Figure 3-5.

Serum or Plasma

In the scientific literature, the most popular specimens for assessing element status are serum and plasma. This is despite disagreement among many leading nutritional scientists concerning the meaning of serum or plasma mineral concentrations in terms of representing whole body mineral status. Nonetheless, there are more available data on serum and plasma mineral levels than on any other specimen. If a comparison with data published in the older literature is of paramount importance, such as for meta-analysis, then serum is the best choice.

Whole Blood

Whole blood is a particularly valuable emergency room tool for the assessment and follow-up of acute toxic metal poisoning. The obvious reasons are because blood levels parallel poisoning severity, and blood can be obtained quickly from even the unconscious patient. Even in cases of chronic exposure, whole blood is useful. For example, the routine measurement of lead in spot blood samples is the procedure recommended by the Centers for Disease Control and Prevention for screening lead exposure. The use of whole blood for essential mineral status is limited to the electrolytes, sodium, and potassium. While magnesium status can be assessed using whole blood, other types of specimens can provide more clinically useful information.

Blood cells

Many elements are considered to correlate best with nutritional status when intracellular (*e.g.*, erythrocytes or mononuclear leukocytes) levels are measured. This is because blood cells are metabolically active, and they accumulate minerals against a concentration gradient to maintain cellular function.

Hence, if levels in these cells are low, then it is likely that other body tissues are depleted in the same mineral(s). This is truer for isolated mononuclear leukocytes than for washed red blood cells (mostly erythrocytes, which lack a nucleus and mitochondria). Nutritional intervention studies indicate that even leukocyte mineral levels do not always vary as a function of mineral intake. Red blood cell mineral levels serve to give an approximate picture of an individual's mineral status and this type of specimen remains popular and useful.

Hair

Hair is often the best specimen for the routine screening for toxic metals in the body. Owing to its high metabolic activity, high sulfhydryl group content, and, consequently, its great affinity for multivalent metals, hair concentrates toxic heavy metals several hundred-fold above concentrations found in blood. Hence, toxic metals become detectable in hair before other tissues or body fluids. For assessing a history of toxic metal exposure, hair serves the purpose of being a chronological "recorder." For these reasons, when exogenous metal contamination is excluded, hair is described in toxicology textbooks as a very useful specimen for toxic metal testing [25]. The analysis of hair for nutrient element status assessment can also be useful. Nutritional deficiencies of calcium and magnesium can lead to paradoxically high levels of these minerals in hair. This is because hair calcium and magnesium levels change in response to the same homeostatic factors that control total body calcium and magnesium losses. Further details are given below as each element is discussed.

Because of the ease of sampling and the apparent rich information content in a profile of 25 or more elements, hair element testing has become a favorite tool among some practitioners. Unfortunately, some laboratories have greatly overstated the diagnostic potential of hair element analysis. The calculation of long lists of meaningless ratios and the inclusion of detailed, misleading metabolic interpretations have led to investigations that paint a poor picture of the testing. Most recently, analysis of a split hair sample sent to six commercial laboratories was reported [26]. Some of the results showed poor correlation between laboratories, and the interpretive comments varied widely, leading the authors to recommend private restraint and regulatory restrictions on all laboratories offering elemental testing of hair. While some restrictions may become necessary to avoid misguided interventions, the conclusions of such reports should be evaluated in context of the approximately 3,000 reported studies of hair elemental analysis that have demonstrated its strong potential as a useful clinical tool. Many of these reports have been the subject of independent reviews [27-29].

Urine

Urine presents a different aspect of mineral balance. For individuals whose day-to-day dietary intake is relatively constant, normal levels of urinary zinc and chromium can reflect nutritional adequacy for these minerals since urine is the main route of excretion when mineral intake is plentiful. On the other hand, for individuals with large daily variations in diet, urinary levels are more difficult to interpret.

Chelation Challenge ("Provocation") Tests

Modern laboratory techniques still have certain limitations. Sometimes it is difficult to demonstrate the presence of a toxic metal in static samples of hair, blood, and urine despite clinically significant body burdens for these metals. This is because, over a long period of time, toxic metals tend to accumulate in relatively non-exchangeable pools within specific tissues (*e.g.*, bone, brain, liver, kidneys, and adipose tissue). For acute exposures, the more time that elapses between exposure and testing, the more likely the result will be low concentrations of toxic metals in blood, urine, and hair despite a significant total body burden. Thus, it is reasonable for a clinician to want to corroborate an initial "normal" or "borderline" toxic metal finding with adjunctive testing. This is justifiable if a patient's complaints and history are consistent with one or more hallmark symptoms of metal toxicity. Since organ biopsy (*e.g.*, liver, brain, or bone) is rarely an option, the next best course of action is to disturb the body's stores of toxic metals so that a certain quantity will redistribute into the blood as a stable complex and be eliminated in the urine, where it can be measured. Low molecular weight chelating agents have been found to work best for this purpose due to the fact that they specifically bind and mobilize toxic metals from tissues into circulation. Typically, a challenge or provocation consisting of an oral or intravenous dose of the chelating agent is administered. Urine is collected for the next 4 to 24 hours and analyzed for toxic metals. This result may be compared with the results of pre-challenge urine. An elevation above reference levels in one or more toxic metals in the post-challenge urine confirms the diagnosis of elevated toxic metal body burden with potential toxic consequences. Two chelation test protocols are described below.

DMSA Challenge Procedure

- Empty the bladder and discard the urine.
- Take one 500 mg DMSA capsule with a glass of water between meals (at least one hour before or after eating).
- Take a second 500 mg capsule four hours after the first.
- Collect total urine for 24 hours after taking the first DMSA capsule. Empty the bladder for the last time at the same time of day that the first DMSA capsule was taken on the previous day.
- Mix and transfer a portion of urine to a specimen tube for analysis.

DMSA Challenge Test

DMSA (meso-2,3-dimercaptosuccinic acid) is an FDA-approved chelating agent that mobilizes heavy metals, such as lead and mercury from human tissues [30-35]. DMSA transiently diffuses into and effectively competes with the tissue binding sites to release toxic metals that are normally sequestered in bone, nerve, liver, and other body tissues. Effectiveness in mobilization of cadmium, lead, and mercury has been demonstrated, and the safety of DMSA use in children with elevated blood lead has been validated [36]. In addition to its heavy metal binding capacity, DMSA has also been found helpful in reducing toxicity of organic compounds [37], and in restoring immune function in lead intoxication [38].

DMSA may be administered orally in doses of 300 to 1500 mg. The total output of urine over the 24 hours following the initial dose is collected, and the daily output of metals is determined. (Since almost all of the DMSA is cleared by the kidneys in about four hours, some prefer a more convenient four or six hour collection [39].) Elevated levels demonstrate the presence of a potentially toxic body burden of heavy metal.

Typical IV EDTA Procedure

- Evaluate renal capacity by creatinine clearance
- Prepare IV solution
- 500cc bag of sterile water
- EDTA, 1-3gm according to renal capacity
- Multi-trace mineral solution containing Zn, Cu, Cr, Co Mn, Se, Mo (may be added to last 1/3 of infusion)
- Magnesium chloride solution, 2gm
- Ascorbic acid solution, 6-7gm
- B-vitamin solutions, including 1000mcg B12, 100mg pyridoxine, 250mg dexpanthenol
- Administer at 166cc/hr

* Variable components include heparin, lidocaine, and calcium gluconate

EDTA Challenge Test

Intravenous ethylenediamine tetraacetic acid (EDTA) is widely used as a therapy for atherosclerosis and other age-associated diseases [40]. The substance is used in an intravenous drip solution in the form of the disodium magnesium or disodium calcium salt, which binds with variable avidity to most divalent metal ions. Clinical efficacy has been demonstrated for EDTA chelating therapy in a case of occupational parkinsonism due to manganese exposure, including MRI confirmation of the evolution of improved clinical pattern due to reduction of heavy metal deposition in basal ganglia. Occupational exposure to manganese compounds in this case had resulted in high blood and urinary levels of the metal [41].

The responses of multiple trace elements in plasma to varying doses of EDTA have demonstrated the mobilization of toxic heavy metals [42]. Some concern has arisen concerning the potential for single dose EDTA causing redistribution of endogenous toxic metals to tissues like brain. One study, designed to demonstrate such a phenomenon in rats using radioactive tracer lead monitoring, failed to find any evidence for transient increase in brain lead with EDTA treatment [43]. Combination of DMSA and EDTA treatments may offer superior methods for reduction of toxic metal body burdens. An earlier treatment with EDTA may increase lead excretion after a subsequent dose of DMSA [39]

Since essential minerals are lost from the body during the therapy, it is desirable to determine the extent of losses of both toxic and essential elements by performing a pre- and post-chelation analysis. For this purpose, it is satisfactory to use single catches of urine and to report the data as amount of mineral per mg creatinine. The inset text box outlines a typical procedure for IV EDTA therapy [44]. Extensive data has been reported showing the changes in excretion of chemical elements with challenge by intravenous EDTA [45].

Summing Up Specimen Choices

Most practicing clinicians want a reliable, no nonsense approach to mineral testing to help complete the nutritional picture of their patient. First, reduce the clinical questions to the most relevant elements on a case-by-case basis. This simplifies the process of specimen choice (Figure 3-5). Erythrocyte or red blood cell (RBC) testing is highly recommended for routine screening of essential elements. Hair is a very useful specimen for toxic metal screening. Hair also provides unique information about calcium status and reflects long-term body status of most essential trace elements. Use urine to obtain information about habits of dietary element intake or to measure toxic metal excretion after a chelation challenge. Compare the results of multiple elements in analyses from different types of specimens to provide a more complete picture of mineral competency and more clinically useful information. An essential element does not function in isolation but synergistically with other elements and nutrients. Multiple testing methods will reveal abnormalities in the mobilization, utilization, and excretion of minerals – information that can be extremely useful in difficult cases. You must balance the cost of multiple assays with clinical imperatives dictated by a patient's history.

FIGURE 3-5. Practical Considerations for Choosing the Best Specimen

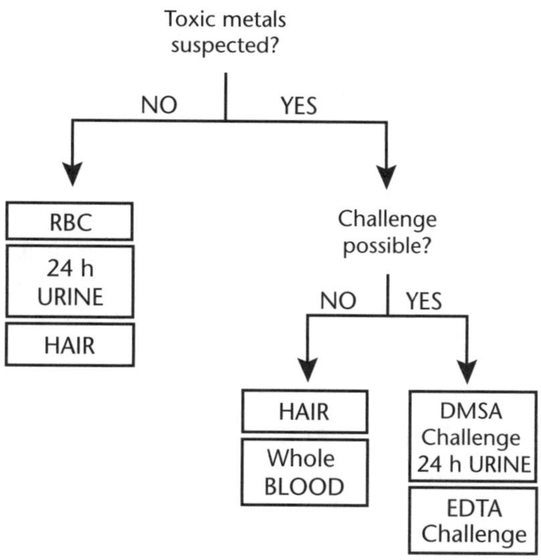

The Essential Minerals

Magnesium

Magnesium is often cited as the element most deficient in modern diets. Its role in forming a stable complex with adenosine diphosphate (ADP) places magnesium in the position of most widely required for metabolic function in all cells. Because of this ubiquity, early signs of magnesium depletion are often vague and are confused with other disorders. Using a cutoff of 0.74 mmol/L, 47% of 1033 patients submitted for serum electrolyte analyses were found to be hypomagnesemic [46].

Serum magnesium is influenced by many factors. Whereas a low serum magnesium level might indicate a deficiency, higher levels serum of magnesium correlate poorly with total body magnesium stores. A better index of whole body magnesium status is the concentration of magnesium inside cells. There is little debate that magnesium stores in muscle -- accounting for approximately 27% of whole-body magnesium -- represents a significant pool of biologically active magnesium [47]. Since measuring muscle magnesium concentrations requires biopsy, it is unlikely this method will find routine application outside of research. There are a number of reports associating magnesium levels in mononuclear blood cells (MBCs) with magnesium status [48-52]. A significant correlation exists between the magnesium concentration in mononuclear blood cells (MBCs) and muscle. Like MBCs, red blood cells (RBCs) contain appreciable amounts of magnesium. The cellular composition of blood is 99% RBCs whereas MBCs comprise less than 1% of the total number of cells found in circulation. Thus, the magnesium pool in RBCs is significantly greater than that in MBCs. It has been shown that RBC magnesium correlates better with bone magnesium, a major storage site of body magnesium, than muscle magnesium [53]. A disadvantage of using RBCs for magnesium determination is that the magnesium content declines with the age of the RBCs [54]. Nonetheless, RBCs provide a useful means of testing for magnesium status.

With attention to caveats regarding the interpretation of results, hair can be a useful specimen for magnesium testing for the following reasons: 1) high concentrations of magnesium are present in hair, 2) hair magnesium is responsive to parathyroid hormone, the same hormone that determines the release of magnesium from bone, a major storage site for magnesium, and 3) its content is representative of long term dietary habits. Therefore, high hair magnesium is indicative of magnesium mobilization from bone to replenish depleted magnesium stores in metabolically active tissues. Certainly, low hair magnesium is not indicative of magnesium deficiency.

An important mechanism for magnesium conservation is kidney tubular re-absorption [55]. For this reason, during periods of magnesium depletion, kidney magnesium excretion is markedly reduced. Therefore, in the presence of normal renal function, reduced urine magnesium is indicative of low magnesium intake. It should be kept in mind, however, that this test can reflect a recent decrease in dietary magnesium despite normal total body stores of this element. This limitation can be overcome by conducting a magnesium load-retention test. As originally described [1], this functional test consists of introducing magnesium sulfate intravenously (2 mL of a 50% solution) and collecting pre- and post-injection 24 hour urines, which are assayed for magnesium and creatinine. Retention of magnesium greater than 25% of the load indicates magnesium deficiency. The increase of post- over pre-challenge urine magnesium should be greater than 180 mg/day. However, due to time requirements and specimen collection challenges, this test is impractical for most clinics. A modification of this test consists of an equivalent oral dose of magnesium in place of the intravenous route. Although adjusting for variations in gastrointestinal absorption of magnesium between patients does present a challenge, this form of the test is considerably easier to conduct and deserves further study.

Calcium

Serum total calcium is so tightly regulated by the parathyroid hormone (PTH) and calcitonin that its measurement is of value only in cases of serious hormone or bone disorders. Measurements of urinary calcium are useful for monitoring therapy with patients known to be in negative calcium balance [56].

Hair analysis promises to be a practical screening test for calcium balance. Unlike serum, hair calcium levels vary over wide ranges. Two studies have reported a relationship of hair calcium to coronary artery disease [57, 58]. Another group showed that a high dietary phosphorus to calcium intake (a pattern known to induce negative calcium balance) resulted in as much as three times more calcium in the hair compared to normal [59], indicating that elevated hair calcium reflects negative calcium balance. A direct correlation has been reported between high hair calcium and abnormal glucose tolerance [60]. Females have higher calcium in hair than males, but both sexes show a parabolic relationship of hair calcium with age, with maxima around 40-45 years (Fig. 3-6). The data shown here represent total population calcium and should not be mistaken for a normal, healthy population with minimal bone loss. Values for physically fit, well-nourished females should not exceed those for

FIGURE 3-6. Hair Calcium Variations with Age

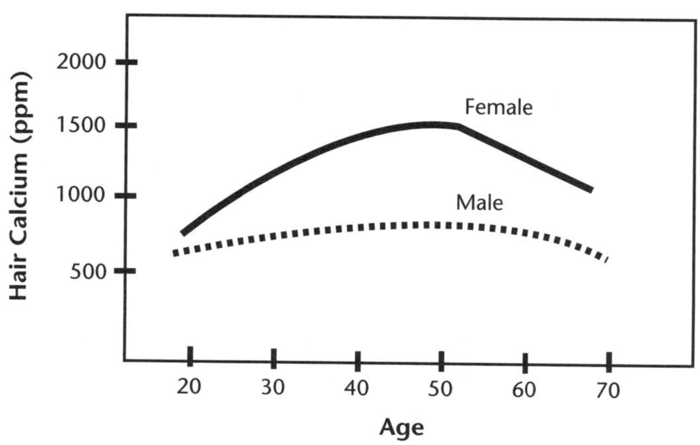

males. Males in optimal health should exhibit a rise during the middle years to values only slightly higher than those found at 20–25 years old. Because of shared hormonal control mechanisms, levels of calcium, magnesium, and strontium are highly correlated in hair.

Iron

Iron is the element with the longest history of clear association with a deficiency syndrome. Early techniques could accurately report iron in serum because the levels were high enough to be measured. The recognition of iron's biochemical role in association with hemoglobin and its association with anemia was one of the early achievements of molecular medicine. In addition to clinical concern over iron deficiency, the question of excessive iron has received attention because of the realization that free iron greatly increases non-specific oxidation reactions that are involved in free radical degenerative effects [61]. The longstanding practice of adding iron in the production of bread and other foods adds to this concern. Several measures of iron status are available that allow the detection of iron excess or iron deficiency (ID).

Serum iron concentration is decreased in many but not all patients with iron deficiency anemia, chronic inflammatory disorders, and myocardial infarction. Elevated serum iron occurs in hemochromatosis, acute hepatitis, and excessive dietary or supplemental iron ingestion. A common measurement of iron status is the Total Iron-Binding Capacity (TIBC). TIBC is the maximum concentration of iron that serum proteins, principally transferrin, can bind. Transferrin saturation, expressed as a percentage, is the ratio of serum iron concentration to TIBC. Measurements of serum iron, TIBC, and transferrin saturation are unreliable for the diagnosis of depleted total body iron pools, though they are useful tests for screening iron overload disorders [62].

Decreased levels in serum ferritin serves as an early sign of declining stores of body iron. As iron deficiency progresses to anemia, decreased serum iron and elevated TIBC values will follow, but the physician should not wait for hypochromia and microcytosis before providing therapy for iron deficiency. Overall, serum ferritin usually provides a reliable index of total body iron status. Exceptions to this rule involve cases of cancer, infections, inflammation, and liver disease [63, 64].

A positive hemoglobin response to iron supplementation is a reliable way to diagnose iron deficiency [65]. For this assessment, hemoglobin must be measured before and several weeks after iron supplementation. Agreement about the best single measurement method for detecting ID is still elusive but recent reports have served to clarify the issue. In one study, premenopausal women were depleted in iron and then fed iron-supplemented diets. The results are summarized in Table 3-2. Using the time of appearance in changes for each measurement, the relative sensitivities for detecting depletion of iron were, in decreasing order, ferritin, % transferrin saturation, plasma iron, hemoglobin, hematocrit, and erythrocyte zinc protoporphyrin (ZnPP). For patients already depleted in iron, however, the earliest sign of repletion was heme synthesis via protoporphyrin levels [66]. This study also confirmed other observations of the efficacy of ascorbic acid supplementation in promoting the assimilation and repletion of iron.

Serum ferritin and ZnPP both were found to identify hemodialysis patients who would show an increased hematocrit when intravenous iron dextran was administered [67]. In another study, pertinent to the question of which test most sensitively reveals a need for iron, the issue is further clarified by associating testing with phases of ID [68]. Ferritin reflects the iron stores, while ZnPP indicates whether the ID in a given patient is clinically relevant or not. The measured ZnPP, hemoglobin concentration, and red cell indices can reveal the extent of a clinically relevant ID. It is valuable to note that misleading conclusions about the value of testing for ID may be drawn from reports of simple correlations between the measurements [69, 70]. While it is difficult to ascribe to any single test the distinction of "best," the different tests efficiently complement each other by detecting different stages of depletion or deficiency and showing the clinical extent of ID. The pertinent clinical question is whether a patient's iron-critical functions will respond to supplemental iron.

TABLE 3-2. Testing for Phases of Iron Deficiency

Phase of Iron Deficiency	Ferritin	% Saturation	ZnPP	Hemoglobin or Hct
Iron replete	**N**	>15	N	N
Body stores reduced (ID Grade I)	**Low**	>15	N	N
Functional impairment Iron deficient erythropoiesis (ID Grade II)	Low	**<=15**	**High**	N
End-stage effects Iron deficiency anemia (ID Grade III)	Low	<=15	High	**Low**

N = Normal

Zinc

Blood accounts for a mere 0.5% of the total body content of zinc. Blood zinc is partitioned between plasma (12-22%), leukocytes and platelets (3%), and erythrocytes (75-88%). Many zinc nutrition studies have relied on measurements of plasma zinc levels [28]. While plasma zinc concentration is clinically useful, the result can be complicated by conditions of acute stress and illness such as those following myocardial infarction [71]. Most plasma zinc is bound to its major carrier protein, albumin. Thus, any treatment that alters albumin levels will alter plasma zinc concentration. For example, corticosteroids, oral contraceptives, and endogenous gonadal hormones of pregnancy will lower plasma zinc [72]. Plasma zinc also exhibits diurnal variation; it declines after meals and rises with short-term fasting. These conditions make difficult the task of extrapolating from plasma zinc levels to whole body zinc status.

Mononuclear blood cells (MBCs) and red blood cells (RBCs) containing zinc dependent proteins make an attractive alternative for determining whole body zinc status. For instance, RBCs contain about 10-times more zinc than plasma due to their content of carbonic anhydrase and other zinc-containing enzymes. Additionally, since RBCs slowly turn over about every 120 days, zinc levels in these tissues are insensitive to recent changes in whole body zinc stores [24]. Like RBCs, MBCs have zinc levels that are indicative of whole body zinc status [73]. MBCs have considerably faster turn over rates (~10 days) and, therefore, exhibit intermediate sensitivity to changes in zinc nutriture compared to RBCs.

Other measures of zinc status can be obtained from the activities of various zinc-containing enzymes found in blood: alkaline phosphatase, carbonic anhydrase, nucleoside phosphorylase, and ribonuclease [9]. Of these enzymes, alkaline phosphatase has been given the most study. Zinc-deficient patients who received zinc supplementation showed increases in serum alkaline phosphatase activity that paralleled the degree of zinc repletion [74]. Similar studies that measured alkaline phosphatase activity in erythrocytes have shown this membrane enzyme to be responsive to zinc depletion even when other markers, including plasma, erythrocyte and platelet zinc concentrations, and plasma alkaline phosphatase activity failed to change with zinc deficiency [24, 75]. Another zinc dependent marker with promise as a sensitive indicator of mild zinc deficiency is the plasma enzyme, 5′-nucleotidase [76].

Urinary excretion of zinc will decrease as a result of zinc deficiency. Measurement of zinc in a 24-hour urine can be clinically useful provided that chronic renal disease, liver cirrhosis, sickle cell anemia, and other conditions known to cause urinary hyperexcretion of zinc are ruled out. The zinc-loading test, in which an oral dose of zinc is given to the patient and the percent increase in urinary excretion is calculated, may be helpful in certain cases [77].

Hair zinc has been reported by a number of studies to reflect the dietary zinc intake in both animals and humans. Low levels of zinc in the hair of children were shown to correlate with poor growth in the Near East [78], and with poor growth, anorexia, and hypogeusia in the U.S. [10]. Such findings have been corroborated in zinc deficient children with Crohn's disease [79]. However, the use of hair for the diagnosis of zinc deficiency is not without caveats. Hair zinc can be in the normal or high range, even in patients with clinical evidence of zinc deficiency [78]. Such false negative results arise when sudden zinc deficiency causes the arrest of hair growth (induction of telogen phase). In such situations, the normal concentration of zinc in hair is reflective of the previous nutritional state before the deficiency occurred. This is more of a problem in cases of periodic zinc deficiency whereas prolonged zinc deficiency will most likely result in consistently low hair zinc levels.

Copper

Although the assessment of copper status in humans is not perfected, a few biochemical indices are useful as aids for the diagnosis of copper deficiency. Plasma or serum copper are easy tests for the routine clinical assessment of copper status; low concentrations may indicate depleted copper stores [80]. However, due to homeostatic mechanisms that regulate changes in circulating ceruloplasmin, the main copper binding protein to which 95% of plasma copper is bound, plasma copper levels do not change unless copper stores are severely depleted [71]. Like zinc, plasma copper levels are prone to physiological factors not directly related to copper status. Plasma copper levels tend to be higher in women than men [81]. Infection, inflammation, and estrogen hormones increase plasma copper levels, while corticosteroid and adrenocorticotropic hormones and Wilson's disease lower copper concentrations [82]. An evaluation of copper status based on plasma levels requires careful consideration of these conditions.

Erythrocyte copper is useful as a marker for copper deficiency because of the tight relationship with the cuproenzyme, superoxide dismutase (SOD). Most of the copper in erythrocytes is associated with this enzyme. Direct measurement of SOD in erythrocytes provides meaningful insight into copper status and vice versa [83]. The activity of superoxide dismutase or cytochrome C oxidase in platelets or leukocytes is reduced in animals and human volunteers fed copper deficient diets. Another promising marker of copper deficiency is ceruloplasmin, which can be assayed immunochemically or by its oxidase activity. Enzymatically measured ceruloplasmin has been shown to be a sensitive indicator of copper status in animals and humans fed diets low in copper [84]. A good index of copper status is the ratio of enzymatic to immunoreactive ceruloplasmin, which may be more sensitive for early copper deficiency than either plasma copper or erythrocyte SOD. Serum ceruloplasmin is commonly used to confirm Wilson's disease, a hereditary copper storage disorder.

The copper content in hair is a good indicator of copper status [85, 86]. However, two situations can occur that will produce misleading results: 1) copper loading diseases, especially Wilson's disease, will produce results not reflective of total body copper stores, and 2) contamination due to hair treatments or swimming pool water (containing Cu algaecides). Urinary copper is useful to monitor copper poisoning; otherwise, copper levels in the urine are not indicative of the body's copper status due to the inability of ceruloplasmin-bound copper to be filtered by the kidney [87].

Chromium

Early efforts to measure chromium in biological tissues were hampered by poor sensitivity until advances in analytical instrumentation allowed for quantitative analysis with acceptable accuracy. Additionally, past measurements grossly overestimated tissue chromium levels due to difficulties with contamination.

Plasma chromium concentration is often less than one part per million. Although measurable with advanced instrumentation, plasma chromium is not in equilibrium with tissue chromium stores [88]. This makes plasma chromium an undesirable choice for assessing chromium nutriture. Since the major route of chromium elimination is via renal excretion, urinary chromium may be the most valid indicator of chromium status when dietary intake is >40 µg/day [89]. Hair concentrates chromium over 100 times above those levels found in blood. For this reason, hair chromium has found use as an indicator of chromium status [90]. Erythrocytes may be a useful specimen for assessing chromium status because they contain about 20-fold greater concentrations of chromium than serum. Chromium concentration is higher in cells that actively transport glucose than in those that don't. Chromium is involved in the formation of insulin receptors in the membranes of glucose-dependent cells.

Selenium

Erythrocytes are a good tissue for selenium assessment because almost all of the selenium is associated with the enzyme, glutathione peroxidase (GP). GP serves a primary antioxidant defense function through its ability to transfer free radical electrons from fatty acid peroxides to glutathione. Selenium supplementation resulted in increased selenium in erythrocytes in children with Down's syndrome [91].

Erythrocyte selenium-dependent GP activities were significantly lower in cystic fibrosis patients than healthy subjects. When selenium intake was increased with graded doses of pancreatic enzymes containing selenium in both groups, erythrocyte GP activities and plasma selenium concentrations increased. Changes in GP activities during the initial 8 months correlated with those in selenium intake [92]. Hair selenium correlates well with intake and may present the most accurate way to screen for selenium deficiency [93].

Potassium

If renal failure is not present, urinary potassium excretion accurately reflects daily potassium intake. Red cell potassium reflects the potassium content of other tissues and is a good indicator of total body potassium [94]. Whole blood potassium is about as useful as

RBCs in monitoring tissue levels since it reflects the potassium contained in RBCs. One study, which measured the electrocardiograms of an elderly population, showed correlations of RBC potassium levels with alterations in the potassium-dependent repolarization phase. On the other hand, the correlation with serum potassium levels was poor [94, 95].

Manganese, Molybdenum, and Vanadium

Manganese status assessment is best done using whole blood or erythrocytes. Hair manganese is a valid indicator of toxicity in cases of manganese excess, but there is controversy over its use for deficiency states [96].

Molybdenum is present in the body at extremely small concentrations and is difficult to measure. Nonetheless, molybdenum is concentrated several-fold in hair owing to its affinity for the large number of sulfhydryl groups in hair. Therefore, hair is the preferred specimen for molybdenum analysis. Vanadium body stores are assessed best using urine or hair [8].

The Toxic Metals

Our environment contains both toxic organic compounds and toxic metals. From an environmental standpoint, an important difference between toxic compounds and toxic metals is that while organic compounds can be chemically destroyed, toxic metals are inherently indestructible. The only recourse for the disposal of toxic metals is to bind them to ligands that alter their toxicologic and environmental behavior and then sequester them where they cannot reenter the environment. A poignant point that must be realized is that toxic metals tend to accumulate undetected in the body over time. As biological levels of a toxic metal increase, a point is reached when subclinical effects become clinically relevant. In a population whose mean exposure to a toxic metal has reached this threshold, a small increase in intake, for example 20%, can lead to a doubling in the number of symptomatic persons (Figure 3-7). Though a relatively small increase in exposure to a toxic metal may seem trivial, the chance that the metal is causing health problems increases significantly.

The toxicity of an element depends on its physiochemical properties. Those properties that increase the absorption and transport of a toxic metal (*i.e.*, its bioavailability) will determine its ability to reach one or more target sites within the body. Targets include sites such as membrane or structural proteins, enzymes, membrane lipids, or DNA molecules. Once at the target site in sufficient concentration, the metal may substitute for or otherwise displace a mineral of physiological importance from its binding site. This substitution is called "molecular mimicry" [97]. Yet, because it does not share the exact same chemical properties of the displaced

FIGURE 3-7. Toxic Mineral Symptom Threshold
A small increase in exposure to a toxic metal can lead to a disproportionate increase in the number of individuals who will experience symptoms of toxicity. For example, a 20% increase in the exposure rate shifts the frequency distribution curve (*i.e.* the number of individuals found to have toxic metal at each concentration interval plotted for all possible toxic metal concentrations) to the right as indicated at "a." The toxic metal concentration at which symptoms appear ("symptom threshold") remains unchanged. Therefore, the area under the new distribution curve to the right of the symptom threshold increases from "b" to "b+c". Plainly stated, the number of individuals experiencing toxicity symptoms approximately doubles [83].

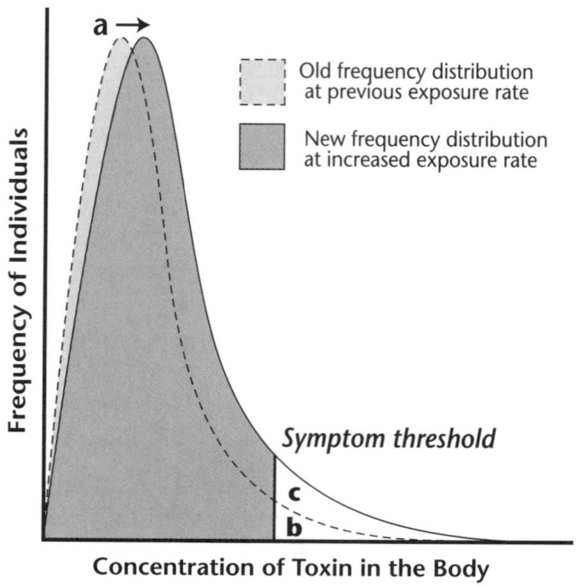

mineral, the toxic metal fails to substitute for its function. This "foreign-coin-jams-the-juke-box" paradigm deserves a more sophisticated discussion. Nonetheless, it provides a conceptual framework for understanding an important underlying mechanism for metal toxicity.

Three other mechanisms of toxicity deserve special attention, particularly in the context of metals as neurotoxins: 1) inorganic metal transformation to an organic species, 2) sulfhydryl group binding, and 3) free radical generation. The first mechanism determines the solubility of the metal within the various compartments in the body and is an important factor in determining bioavailability. Biological systems have the ability to add methyl groups to metals, which serve to make the metals more soluble in lipid environments. A lipid-soluble metal is many times more neurotoxic than aqueous forms because it can penetrate lipid membranes and concentrate in lipid-rich tissue such as the brain. The second mechanism revolves around structure-function changes that occur separately from molecular mimicry. Many toxic metals (in particular, lead, mercury, and arsenic) avidly form covalent bonds with the sulfhydryl groups of cysteine amino acids belonging to proteins. The formation of complexes can lead to structural changes to the detriment of the protein's function. The third mechanism, free radical generation, involves the formation of highly reactive oxygen species having unpaired electrons. Unchecked formation of free radicals leads to lipid peroxidation, cellular membrane degeneration, and ultimately to cell death (see chapter on Aging and Oxidant Stress). Many metals are potent catalysts of free radical reactions. Though cells have evolved defensive mechanisms to guard against free radical formation or to repair the damage caused by them, metals present in sufficient concentrations will promote formation of enough free radicals to saturate these defenses.

Assessing the impact of toxic metal exposure is difficult unless clinical symptoms are noticeable (Table 3-3). Even if symptoms are obvious, it is judicious to confirm the presence of the metal in a biological tissue sampled from the patient. Proper interpretation of the toxic metal concentration in a particular tissue requires an understanding of the relationship between exposure and levels in the body. It is also important to understand the distribution pattern of the metal, particularly regarding the organ most affected by the toxicity and the routes of elimination. In the past, the possibilities of assessing toxic metal exposure were limited by the performance of the analytical instruments. This situation has dramatically improved with the introduction of modern techniques utilizing inductively-coupled plasma/mass spectroscopy (ICP-MS) and neutron activation analysis (NAA). In most cases, the tissues available for toxic metal analysis are blood, urine, and hair. Often, the specimen choice can make the difference between obtaining a positive or inconclusive result, which is another reason to consider multiple testing methods, especially with serious or confusing clinical conditions.

There is another mode of metals eliciting toxic responses that is quite different. A metal atom that can become attached to a large molecule (*e.g.*, a protein) has the potential for becoming antigenic.

TABLE 3-3. Toxic Metal Exposure Symptoms

Metal	Symptoms of Toxicity	Protective Effect	
Aluminum	Abnormal speech, myoclonic jerks, osteomalacia, progressive encephalopathy, Alzheimer's disease	**Phosphorus**	Lowers intestinal absorption of aluminum
Lead	Microcytic hypochromic anemia, renal dysfunction, hypertension, anorexia, muscle discomfort, constipation, metallic taste, low IQ (children)	**Calcium**	Lowers intestinal absorption of lead
Mercury	Mental symptoms (irritability, insomnia, fatigue, poor short-term memory), tremor, stomatitis, gingivitis, GI and renal disturbances, decreased immunity	**Selenium**	Protects against cellular toxic effects of mercury
Cadmium	Femoral pain, lumbago, osteopenia, renal dysfunction, hypertension, vascular disease	**Zinc**	Competes for cadmium binding sites
Arsenic	Peripheral arteriosclerosis ("blackfoot disease"), "rice-water" stools, proteinuria, hyperkeratosis, "milk and roses" hyperpigmentation, garlic breath odor, stomatitis	**DMSA, DMPS, DMPA, Emblica officinalis**	Removal of arsenic by competitive binding

The immune system can mount responses to such products that result in the full range of symptoms as for any other allergic response. Such responses involve cell-mediated immune mechanisms that may be detected by testing for lymphocyte elevations when the patient is challenged with dilutions of metal-bound protein. The term, "metal hypersensitivity responses," is used to differentiate this type of toxicity. Because of its propensity to form stable bonds with organic compounds, mercury is particularly suspected for producing such responses, but any reactive element can initiate metal hypersensitivity reactions, because it is not the metabolic toxicity of the element but the response to antigenically modified proteins that initiate them. Molybdenum used in alloys to retain prosthetic devices, for example, has been found to cause metal hypersensitivity that was alleviated only when the patient was treated with immunologic desensitization methods [98].

Lead

Lead, a known toxin since antiquity, is widely distributed in the environment largely due to human activity. Water transported through old lead pipes or pipes joined by lead-containing solders can carry significant amounts of lead. Lead is used in the making of crystal glassware and some imported ceramic products from which wine and fruit juices can leach lead out. Some traditional home medicines have been found to contain lead. Lead was added to gasoline as an anti-knock agent until it was phased out in the early 1980s. As a result, lead is still found in the soil around highways. Lead was also a common ingredient in paints until its use was banned in 1972. Nonetheless, lead is still found in paint products intended for non-domestic use and in artist pigments. The soil around many older homes painted with lead-based paints is still contaminated with lead; this remains a significant source of lead exposure among children living in poorer neighborhoods to this day, especially since lead ingested by mothers can be passed on to infants via breast milk [99].

Exposure to lead can occur by ingestion, inhalation, and even by dermal contact. As much as 300 µg of lead per day may be ingested in a normal diet. Well-nourished adults absorb only about 1 to 10% of ingested lead, while children can absorb as much as 50% of their dietary intake. The absorption of lead is increased in patients with compromised gastrointestinal integrity and low dietary intakes of calcium [100], magnesium [101], iron, and vitamins C and D [102], so supplementation of these nutrients may prevent symptoms. Absorbed lead is rapidly incorporated into bone and erythrocytes, though eventually, lead will distribute into all tissues. When not deposited in tissue, lead is excreted into bile or urine. The kidney is susceptible to damage by lead when exposed to high lead concentrations during clearance. Renal competence should be assured by demonstration of normal creatinine clearance before beginning any intervention to detoxify concentrations of lead (or other heavy metal contaminants) from the tissue.

Lead toxicity is expressed by several mechanisms. As an electrophile, lead readily forms covalent bonds with the sulfhydryl group of cysteine. The tertiary structures of proteins depend on free sulfhydryl groups that are adversely affected by lead's affinity for a protein's cysteine residues. Loss of structure of some proteins invariably leads to their instability and loss of biological activity. Proteins contained within astrocytes of the nervous system [103] and renal tubular cells are particularly susceptible to the structural changes imposed by lead. Lead interferes in the heme biosynthetic pathway. Delta-aminolevulinic acid dehydratase (ALAD) and ferrochelatase, the second and last enzymes of the committed pathway of heme biosynthesis, are inhibited by lead. A significant hallmark of lead toxicity is anemia due to the lack of heme. Inhibition of heme synthesis by lead results in a rise in urinary coproporphyrin and erythrocyte protoporphyrin, which at one time were the only reliable markers of lead toxicity [104].

Only in recent years has it been realized that the nervous system is particularly sensitive even to low levels of lead. Chronic lead exposure reduces nerve conduction velocity in peripheral nerves in adult subjects without obvious symptoms or signs of toxicity. Lead, which readily passes the blood-brain barrier [105], can cause symptoms and signs involving the central nervous system (CNS). Lead interferes with a number of calcium-dependent ion channels, receptors, and enzymes, including calmodulin [106], protein kinase C, L-type calcium- and nicotinic acetylcholine-gated channels [107], and magnesium-dependent cGMP phosphodiesterase. The fact that lead alters calcium-mediated cellular processes in laboratory experiments may have clinical significance. In children without obvious signs of lead intoxication, subjective and nonspecific symptoms (*e.g.*, fatigue, impaired concentration, short-term memory deficits, insomnia, anxiety, and irritability) may occur along with reduced performance on visual intelligence and visual-motor coordination tests [108]. In a landmark study of lead exposure in children, tooth lead concentrations were shown to be significantly and positively correlated with teacher-rated learning disabilities [109].

Hair keratin has a high affinity for lead owing to its high fraction of cysteine residues relative to its other amino acids. Thus, high lead in hair is a good marker for lead exposure. Normally, hair lead content is <5 μg/g; lead levels >25 μg/g indicate severe lead exposure. Serum or plasma is not a very useful specimen for lead screening except in cases of very recent or current chronic exposure. This is because serum lead returns to normal levels within 3-5 days of last exposure. Urinary lead concentrations increase with lead poisoning, although urinary elimination of lead is a process that occurs for many days after a single exposure. Treatment with EDTA or dimercaprol (also used for chelating arsenic, gold, and mercury) accelerate the urine excretion rate of lead; thus, this therapy is useful for revealing lead exposure.

Whole blood concentrates lead about 75-fold greater than serum or plasma, and it has the highest correlation with toxicity. For this reason, whole blood lead is defined by the Centers for Disease Control as the preferred test for detecting lead exposure. The World Health Organization has defined whole blood lead levels >200 µg/L in adults to be elevated, while in children levels >100 µg/L are considered high. Whole blood lead concentrations above these limits indicate significant lead exposure, and appropriate measures should be taken to reduce exposure.

Mercury

Mercury as a neurotoxin has an intriguing history. The phrase "mad as a hatter" has its origins with the 17th and 18th century hat makers who suffered from mercurialism due to their use of liquid mercury in the manufacture of the popular felt-brimmed hats. Sir Issac Newton, the famous seventeenth century physicist, experienced a year of dark moods and marked personality changes that puzzled friends and close associates. Posthumous analysis of archived samples of Newton's hair revealed highly elevated concentrations of mercury, which is evidence that supports the historical hypothesis that Issac Newton's "madness" was a result of his exposure to the toxic metal while he was conducting experiments to study its properties [110].

The human population is exposed daily to naturally occurring mercury. The earth's crust releases approximately 30,000 tons of mercury per year as a product of natural out-gassing from rock. However, man-made sources of mercury are significant too -- approximately 6000 tons/year of mercury are used in the manufacture of electrical switches, for electrolysis, and as a fungicide. Ninety tons of mercury are used each year for making dental amalgams.

In its elemental form, mercury (Hg^0) is nontoxic. However, once chemically or enzymatically altered to the ionized, inorganic form (Hg_2^+), it becomes toxic. Bioconversion of mercury to organic alkyl form produces highly toxic compounds with avidity for the nervous system [111]. Microorganisms in the environment and in the human intestinal tract can bioconvert relatively nontoxic elemental mercury to its toxic forms, inorganic Hg_2^+ and organic mercurous alkyl compounds. As organic mercury passes up the food chain, it accumulates in deep-sea fishes (*e.g.*, albacore and tuna) and man. With chemical properties similar to lead, mercury is toxic in three ways: 1) it reacts with sulfhydryl groups and impairs the activity of enzymes, 2) with changes in tertiary structure, some proteins become immunogenic, eliciting a proliferation of B-lymphocytes that generate immunoglobins to bind the new antigen (collagen is particularly susceptible to this effect) and 3) alkyl mercury compounds are very lipophilic and have a high affinity for lipid-rich tissue of the nervous system.

Mercury intoxication produces a triad of symptoms: 1) mental changes including erethism (excessive irritability, excitability, or sensitivity to stimulation), depression, poor short-term memory, difficulty concentrating, insomnia, and fatigue, 2) spontaneous tremor and deficits in psychomotor performance, and 3) stomatitis and gingivitis. Signs and symptoms of mercury intoxication involving other organ systems include renal and gastrointestinal disturbances, pain in joints and limbs, weight loss, metallic taste in the mouth, and increased susceptibility to infections.

There is a valid concern that the mercury released from dental amalgams, which are composed of as much as 50% mercury, can have a negative impact on an individual's health. Although mercury-containing amalgams have been in use for over 100 years, their use was intensely debated at the turn of the century and again in the 1930s. Small yet measurable amounts of mercury (up to 20 µg/d) are continuously released from the amalgam surface; the rate of release is accelerated by hot liquids and chewing. Normal bacterial flora converts a fraction of the released elemental mercury to its toxic forms, Hg_2^+ and alkyl mercury. A portion of the elemental mercury released from amalgams is unavoidably inhaled into the lungs where it can be biotransformed to its toxic forms. Studies have suggested that chronic mercury exposure in amounts released by amalgams provokes an increase in both mercury- and antibiotic-resistant strains of bacteria in the oral and intestinal flora.

The World Health Organization has set the safe standard of mercury exposure in food and water to 45 µg/d. If the amount of mercury absorbed from dental amalgams is combined with all other sources of mercury (*e.g.*, fish, environmental, occupational, medicinal), the daily exposure to mercury can be expected to exceed the WHO limit for some individuals. Based on a study of normal, presumably healthly populations, whole blood mercury concentration was typically found to be <5 µg/L. About 1% of this population had whole blood levels of mercury greater than 5 µg/L. Individuals with occupational exposure to mercury, such as dentists and dental technicians, may routinely have whole blood mercury up to 15 µg/L. Significant exposure is evident when the whole blood level of mercury is >50 µg/L if exposure is due to alkyl mercury, or >200 µg/L if exposure is due to Hg_2^+. The level of mercury in urine is the most reliable way to assess exposure to inorganic mercury. Daily urinary levels greater than 50 µg/d indicate a Hg_2^+ overload. Hair levels of mercury greater than 1 µg/g also indicate mercury toxicity. The quantity of mercury assayed in blood and hair, but not urine, correlates with the severity of toxicity symptoms.

Removing the source should be the first treatment for mercury toxicity. More aggressive treatment calls for chelation therapy. British Anti-Lewisite (BAL) – containing dimercaprol, penicillamine, ethylenediamine tetracetic acid (EDTA), 2,3 dimercaptosuccinic acid (DMSA), and 2,3-dimercapto-1-propanesulfonic acid (DMPS) will all mobilize mercury and cause a rise in the daily urinary mercury excretion rate. The preferred chelation agents, based on their affinity for mercury and low toxicity, are DMSA or DMPS.

Arsenic

Arsenic exists in toxic and non-toxic forms. The toxic forms are the inorganic species of arsenic (As-III and As-V). As-III is more toxic than As-V. Preserved "pressure-treated" wood contains appreciable amounts of the toxic forms of arsenic, as do many insecticides. The organoarsines, arsenobetaine and arsenocholine, are non-toxic forms of arsenic that are present in many foods, especially shellfish and other predators in the seafood chain (*e.g.*, cod and haddock). Arsenic is partially detoxified by methylation to mono- and dimethylarsine.

Urinary inorganic arsenic peaks at about 10 hours and returns to normal 20-30 hours after ingestion. Non-toxic organic arsenic is completely excreted in 24-48 hours after ingestion. Arsenic expresses its toxicity by: 1) binding with any hydrated sulfhydryl group on protein, which causes distortion of tertiary structure and loss of activity; 2) tightly binding dihydrolipoic acid, a necessary cofactor for pyruvate dehydrogenase (binding of dihydrolipoic acid inhibits the conversion of pyruvate to acetyl CoA, the first step in the citric acid cycle); and 3) competing with phosphate for binding to ATP during its mitochondrial synthesis, which causes the formation of lower energy ADP .

A majority of the arsenic normally found in urine is non-toxic, probably due to the ingestion of organic arsenic naturally found in seafood. There is a need to rule out ingestion of seafood in cases of borderline or moderate arsenic elevation. Consumption of seafood before collection of a urine specimen for arsenic testing is likely to result in elevated urinary arsenic, which can be clinically misleading. Specialized laboratory methods of separation are required to differentiate non-toxic organic forms from the toxic inorganic forms of arsenic. The concentration of arsenic in hair is typically greater than in other tissues because of its high affinity for keratin found in hair. Keratin, which is rich in cysteine residues, is the major structural protein in hair. Arsenic binds keratin in hair, trapping it at the time of exposure. Hair arsenic greater than 1 µg/g dry weight indicates excessive exposure. Arsenic combines readily with sulfhydryl groups of tissue proteins; thus, it is rapidly cleared from the blood. Therefore, elevated arsenic in the serum rapidly becomes normal within 6-10 hours. For this reason, serum is not a good specimen for arsenic screening purposes.

British Anti-Lewisite, an effective antidote for treating arsenic poisoning, has been widely used but has received criticism because of:

1. Its low therapeutic index
2. Its tendency to redistribute arsenic to brain and testes
3. The need for (painful) intramuscular injection
4. Its unpleasant odor

Newer antidotes, such as DMPS and DMSA, have been favored because they feature low toxicity and high therapeutic indices [112]. Other candidates, including crude extract of Emblica officinalis [113] and N-(2,3-dimercaptopropyl)phthalamidic acid (DMPA)[114], also are reported to be effective.

Cadmium

Cadmium, obtained as a byproduct of zinc and lead smelting, is used in industry for electroplating, in the production of rechargeable batteries, and as a paint pigment. The spray painting of organic-based paints without use of protective breathing apparatus is the most common cause of chronic cadmium overexposure. Another source of cadmium exposure is tobacco smoke. Auto mechanics also are susceptible to cadmium toxicity from automobile exhaust emissions.

Cadmium competes with zinc at all cellular binding sites. The displacement of zinc from an enzyme active site results in loss of enzyme activity. The process is depicted in Figure 3-8 where the metabolic reaction converting "Substrate" to "Product" is impaired. This type of toxicity is sometimes referred to as enzyme poisoning. The target organs of cadmium toxicity are primarily the kidneys. Chronic exposure typically results in the slow onset (over a period of years) of renal dysfunction with proteinuria. Breathing aerosols containing cadmium can lead to pulmonary congestion resembling chronic emphysema. Excretion of cadmium is proportional to creatinine except when renal damage has occurred. Renal damage due to cadmium exposure can be detected by increased cadmium excretion relative to creatinine. Monitoring urinary output of cadmium is an excellent means of assessing exposure. Daily output of cadmium >3µg per 24 hour urine indicates toxicity.

FIGURE 3-8. **Zinc-enzyme Poisoning by Cadmium**

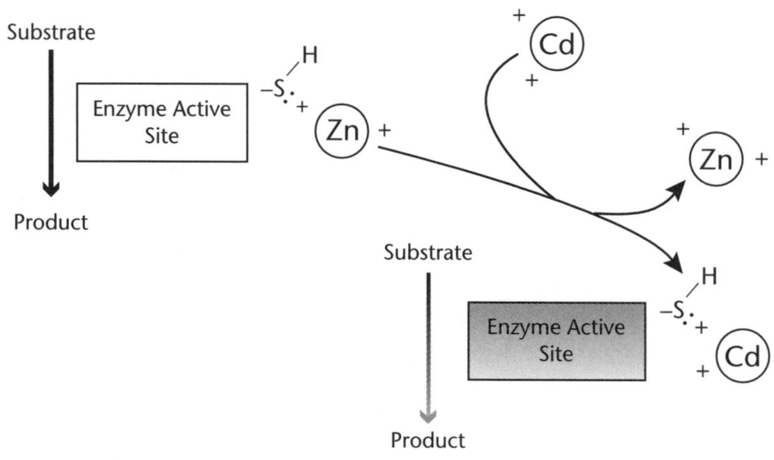

Aluminum

After oxygen and silicon, aluminum is the most abundant element in our earth's crust and is ubiquitous in our environment. Aluminum is commonly used in many consumer products such as cat litter, antiperspirants, infant formula, baking soda, foils, and others. Daily ingestion of aluminum from water and food sources is 5-10 mg/d. This amount of metal far exceeds the safety dose of lead, mercury, cadmium, and arsenic, so it might come as a surprise that, though humans apparently tolerate aluminum well, this metal is a potent neurotoxin. Aluminum is commonly used in manufacturing and contained in medicinal products. Acid rain pollution is also increasing the rate of liberation of aluminum and other metals from otherwise poorly soluble minerals, like feldspar and kaolinite, into ground water supplies. This has important implications for the aging population who are exposed to increasing levels of aluminum. Increased exposure levels also have strong etiological implications for several neurological disorders, particularly Alzheimer's disease. Although it is not proven that aluminum is the causative agent, it is striking that the neurofibrillary tangles of neurons that characterize the brains of Alzheimer's patients (as well as in patients with amyotrophic lateral sclerosis, and Parkinson's and Huntington's diseases) accumulate aluminum.

Although trace levels of aluminum may not be of concern for adults, because of its tendency to accumulate over time, aluminum exposure during infancy and adolescence is a different matter. Levels of aluminum are 2-3 fold higher in cow milk than human milk, and some infant formulas have up to ten times as much [115]. The effect is seen as higher plasma aluminum in infants fed soy formula compared to breast fed infants [116].

While only a small portion of ingested aluminum is absorbed, in patients with renal failure the absorbed aluminum is not eliminated, and it accumulates in the blood. Aluminum avidly binds blood proteins such as albumin, which rapidly distribute aluminum throughout the body. Aluminum overload results in its accumulation in brain and bone; this is particularly true in patients with renal failure who lose the ability to filter and eliminate aluminum, and for whom the dialysis procedure is not efficient at removing aluminum. Encephalopathy develops as a consequence of prolonged hemodialysis due to exposure to aluminum-laden dialysis water and to the aluminum contained in oral medications that contain phosphate binders. Aluminum toxicity is characterized by abnormal speech, myoclonic jerks, convulsions, and a predisposition to osteomalacic fractures. In bone, aluminum replaces calcium, which disrupts normal osteoid formation and mineralization. Aluminum also interrupts calcium exchange, preventing resorption of calcium back into the blood and restoration of circulating concentrations of calcium. Aluminum effectively interferes with the normal physiologic control of parathyroid hormone.

The normal reference range for urinary aluminum is 0-32 µg/24 hr urine. Biliary excretion is normally the major pathway of aluminum elimination; nonetheless, urinary excretion of aluminum is increased in patients with nephrotic syndrome due to leakage through the kidney. Acute aluminum toxicity can be identified and treated with desferrioxamine, a chelator of aluminum and iron. Desferrioxamine will considerably increase the urinary excretion of aluminum; hence, it is useful for the treatment of acute aluminum overload. Chelation treatment with desferrioxamine should be handled conservatively due to the risk of it inadvertently mobilizing large amounts of aluminum to the brain, which may enhance encephalopathy rather than improving it.

Special evacuated blood collection tubes are required for testing aluminum in blood. Blood collected in the standard evacuated blood tubes will be contaminated by as much as 0.5 µg of aluminum. This is due to the fact that standard evacuated collection tubes have rubber stoppers comprised of aluminum-silicate.

What can be done about heavy metal toxicity?

1. Avoid exposure.

Review sources of exposure (Table 3-4) that may be adding to your body burden. In many cases it is difficult to identify specific sources. Since the Table 3-4 does not show all possible sources, you may need to check contents of other materials to which you have frequent exposure. Everyone is exposed to certain levels of heavy metals from foods, beverages, and environmental sources. This is just the first, necessary step in reducing your body burden.

2. Reduce intestinal absorption.

The absorption of ingested heavy metals is usually very low, but the rate can be variable from one individual to the next. One of the food components that is most helpful in reducing the percentage of toxics absorbed is dietary fiber. A diet high in a variety of sources of fiber has many health benefits. By including foods from the following list regularly, you can assure that your fiber intake remains high enough to help restrict heavy metal absorption.

- Beans, all kinds
- Cooked vegetables
- Whole grain breads
- Whole grain cereals, especially oatmeal
- Fresh fruits, especially apples

Products are also available as dietary supplements that contain naturally occurring agents that bind heavy metals, helping reduce the amount that is absorbed. Bentonite and algal extracts are examples of such ingredients.

3. Increase elimination.

Natural chelating and complexing substances may be used as dietary supplements (as shown below) to increase the removal of heavy metals by forming soluble complexes that can be cleared by the kidney. More aggressive chelation therapy using either prescription oral or intravenous compounds may by used by physicians trained in treating toxic metal exposures.

- Methionine, 3000 mg/day*
- Ascorbic acid, 3000 mg

•Lipoic acid, 100 mg TID

** Assure adequate vitamin B12 and folate to prevent homocysteine elevation.*

4. Add competing nutrient elements.

Another way to deal with the presence of toxic heavy metals is by a generous supply of other elements that compete with and displace the toxic metals. Increasing intake of the nutrient elements may help overcome the effects of toxicity [117].

Elevated Metal	*Competing Nutrient*	*Usage level*
Mercury	Selenium	50 mcg
Cadmium	Zinc	50 mg
Lead	Calcium	1000 mg
Lead	Magnesium	500 mg

Elements of Potential Toxicity

There is a class of chemical elements that meets neither the definition of those currently known to be essential for humans nor of those currently thought to be highly toxic to humans. Many of these elements are frequently found at detectable levels in human tissues (especially hair, where they are concentrated along with the other elements described above). Their presence, however, even when found above the "Normal" cutoff values is not necessarily associated with clinical toxic effects. At sufficiently high levels, they constitute a toxic threat -- therefore, the term "potentially toxic elements" has been applied.

Elements of this type most often included on clinical laboratory reports of mineral profiles are barium, beryllium, bismuth, boron, lithium, nickel, and strontium. Some of these are present at relatively high concentrations in pharmaceutical or diagnostic solutions. Concentrated solutions of barium salts are used for x-ray contrast studies and bismuth compounds are present in over-the-counter anti-acid formulas, for example. Serum lithium is routinely monitored in patients using lithium salts for depression [118].

The stable, non-radioactive isotopes of barium and strontium are interesting because of their chemical properties and food source occurrences that lead to physiological and nutritional behaviors paralleling calcium. These elements are frequently present in hair at concentrations roughly equal to or greater than those for the essential elements, cobalt, manganese, and vanadium. They tend to rise and fall in population data along with calcium and magnesium in hair, displaying variations with age similar to those shown in Figure 3-6. The relationship between calcium and strontium is quite strong. Concentrations of strontium incorporated into bone are correlated with age. The strontium/calcium ratio in bone is elevated in renal osteodystrophy, and significantly higher strontium/calcium ratios are present in end-stage renal failure patients [119]. Although it does not compete with calcium in the ionic cellular control functions of calcium, strontium intestinal uptake and bone

TABLE 3-4. Common Sources of Toxic Heavy Metals

Aluminum	Cadmium	Lead	Mercury	Arsenic
Aluminum cooking utensils	Drinking water	Automobile exhaust (not as much since Pb-free fuels)	Dental amalgams	Rat poisons
Baking powder (Al sulfate)	Soft water (from galvanized pipes)	Leaded house paint	Broken thermometers and barometers	Insecticide residues on fruits and vegetables
Antacids (certain brands, see labels)	Soft drinks from dispensing devices with galvanized plumbing	Drinking water from lead plumbing	Grain seeds treated with methylmercury fungicide	Drinking water, well water, seawater
Antiperspirants	Refined wheat flour	Vegetables grown in Pb-contaminated soil	Predator fish, certain lake fish	Automobile exhaust
Aluminum cans	Canned evaporated milk	Canned fruit and juices	Mercuric chloride	Wine (if pesticides used containing arsenic)
Drinking water (alum used as bactericide)	Processed foods	Canned evaporated milk	Calomel (body powder, talc, laxatives)	Household detergents
Milk and milk products (from equipment)	Oysters, kidney, liver	Milk from animals grazing on Pb-contaminated land	Cosmetics (check labels if possible)	Colored chalk
Alum in pickled foods (check labels)	Cigarette smoke, tobacco products	Bone meal	Latex and solvent-thinned paints	Sewage disposal
Nasal spray	Superphosphate fertilizers	Organ meats such as liver	Hemorrhoid suppositories	Wood preservatives
Toothpaste	Dental appliances	Lead-arsenate pesticides	Mercurochrome, merthiolate	Wallpaper dye and plaster
Ceramics (made from AL 203 clay)	Ceramic glazes	Leaded caps on wine bottles	Fabric softeners	
Dental amalgams	Paint pigments	Rainwater and snow	Floor waxes and polishes	
Cigarette filters and tobacco smoke	Electroplating	Improperly glazed pottery	Air conditioner filters	
Automotive exhausts	Silver polish	Painted glassware	Wood preservatives	
Pesticides	Polyvinyl plastics	Painted pencils	Certain batteries	
FD&C color additives	Rubber carpet backing	Toothpaste	Fungicides for lawns and shrub	
Vanilla powder	Nickel-Cd batteries	Newsprint	Leather tanning products	
Table salt, seasonings	Rust-proofing materials	Colored printed materials	Felt	
Bleached flour		Eating utensils	Adhesives	
American cheese		Curtain weights	Skin-lightening creams	
Medications containing kaolin (Al silicate)		Putty	Certain ointments to treat psoriasis	
Sutures with wound-healing coatings		Car batteries	Photoengraving	
Rat poisons		Cigarette ash, tobacco	Tattooing	
		Lead shot, firing ranges	Sewage sludge used as fertilizer	

deposition and resorption are apparently under the control of the same hormonal influences and nutritional dependencies as calcium [120]. There is some evidence that strontium intake may have positive effects on bone formation [121]. Ethanol intake had negative effects on bone content of both barium and strontium [121], but there is little data to corroborate other relationships for barium.

Other, more exotic elements are included by some laboratories because they can be measured at little additional costs, although one is hard pressed to find any significant clinical associations. These include antimony, gold, palladium, platinum, rubidium, silicon, thallium, titanium, tungsten, and uranium.

Summary

For all of those chemical elements known to be essential for human function, the technology for measurement allows accurate concentration determinations in body fluids and tissues. Deriving clear answers to the pertinent clinical questions of mineral status, however, is complicated by the nature of mineral physiology and the variety in characteristics of individual elements. Levels of elements in any single body fluid or tissue provide partial information about total body content and toxic load. The continued growth in popularity of multi-element profiles is evidence of the clinical utility for data that aids detection and treatment of toxic metal exposure and nutrient mineral deficiency. Erythrocyte, urine, and hair are specimens that can reveal useful aspects of both toxic and nutrient elements. Whole blood continues to be useful in demonstrating toxic metal exposure, although sensitivity to chronic low-level exposure may be low. Methods utilizing markers of biochemical impact for deficiency or toxic effects are proving to be helpful for certain elements.

References

[1] Seelig, M.S., *Magnesium Deficiency in the Pathogenesis of Disease*. 1980, New York: Plenum Medical Books.

[2] Klevay, L.M., S.J. Reck, and D.F. Barcome, Evidence of dietary copper and zinc deficiencies. *JAMA*, 241(18). 1916-8, 1979

[3] Guigoz, Y., Recommended dietary allowances (RDA) for the free living elderly, in *Facts and Research in Gerontology (Supplement on Nutrition and Aging)*. 1994, Springer Publishing Co: New York. p. 113-143.

[4] Anderson, R.A., N.A. Bryden, and M.M. Polansky, Serum chromium of human subjects: effects of chromium supplementation and glucose. Am J *Clin Nutr*, 41(3). 571-7, 1985.

[5] Moser-Veillon, P.B., Zinc: consumption patterns and dietary recommendations. *J Am Diet Assoc*, 90(8). 1089-93, 1990.

[6] Garry, P.J., *et al.*, Nutritional status in a healthy elderly population: dietary and supplemental intakes. *Am J Clin Nutr*, 36(2). 319-331, 1982.

[7] Pennington, J.A., *et al.*, Selected minerals in foods surveys, 1974 to 1981/82. *J Am Diet Assoc*, 84(7). 771-80, 1984.

[8] Nielsen, F.H., Other Trace Elements, in *Present Knowledge in Nutrition*, E. Ziegler, Editor. 1996, International Life Sciences Institute Press: Washington DC. p. 353-377.

[9] Prasad, A.S., Clinical manifestations of zinc deficiency. *Annu Rev Nutr*, 5. 341-63, 1985.

[10] Hambidge, K.M., *et al.*, Low levels of zinc in hair, anorexia, poor growth, and hypogeusia in children. *Pediatr Res*, 6(12). 868-74, 1972.

[11] Matkovic, V., Calcium intake and peak bone mass. *N Engl J Med*, 327(2). 119-20, 1992.

[12] Dwyer, J.T., Adolescence, in Present Knowledge in Nutrition, E.E. Ziegler and L.J. Filer, Jr, Editors. 1996, ILSI: Washington DC. p. 404-413.

[13] Picciano, M.F., Pregnancy and lactation, in *Present Knowledge in Nutrition*, E.E. Ziegler and L.J. Filer, Jr, Editors. 1996, ILSI: Washington DC. p. 384-395.

[14] Heird, W.C., Nutritional requirements during infancy, in *Present Knowledge in Nutrition*, E.E. Ziegler and L.J. Filer, Jr, Editors. 1996, ILSI: Washington DC. p. 396-403.

[15] Keusch, G.T., Nutrition and infection, in *Modern Nutrition in Health and Disease*, M.E. Shils, J.A. Olson, and M. Shike, Editors. 1994, Lea & Febiger: Philadelphia. p. 1241-1258.

[16] Marier, J.R., L.C. Neri, and T.W. Anderson, *Water hardness, human health and the importance of magnesium*. 1979, National Research Council of Canada: Ottawa, Canada.

[17] Seelig, M.S., Consequences of magnesium deficiency on the enhancement of stress reactions; preventive and therapeutic implications (a review). *J Am Coll Nutr*, 13(5). 429-46, 1994.

[18] Yang, G., *et al.*, Selenium-related endemic diseases and the daily selenium requirement of humans. *World Rev Nutr Diet*, 55. 98-152, 1988.

[19] National Research Council Committee on Neurotoxicology and Models for Assessing Risk, in *Environmental Neurotoxicology*. 1992, National Academy Press: Washington, D.C.

[20] Mertz, W., *The essential trace elements*. Science, 213(4514). 1332-8, 1981.

[21] information, N.c.f.b., http://www.ncbi.nlm.nih.gov/cgi-bin/Entrez/Structure/mmdbsrv?form=6&db=t&Dopt=s&uid=12018. 2000, National Library of Medicine.

[22] Frieden, E., A survey of the essential biochemical elements. *Biochemistry of the Essential Ultra-trace Elements*, ed. E. Frieden. 1984, New York: Plenum Press. 1-15.

[23] Yang, G., *et al.*, Studies of safe maximal daily dietary Se-intake in a seleniferous area in China. Part II: Relation between Se-intake and the manifestation of clinical signs and certain biochemical alterations in blood and urine. *J Trace Elem Electrolytes Health Dis*, 3(3). 123-130, 1989.

[24] Ruz, M., *et al.*, Erythrocytes, erythrocyte membranes, neutrophils and platelets as biopsy materials for the assessment of zinc status in humans. *Br J Nutr*, 68(2). 515-27, 1992.

[25] Dreisbach, R.H., ed. *Handbook of Poisoning: Diagnosis and Treatment*. Vol. 7th. 1971, Lange Medical Publications: Los Altos, CA.

[26] Seidel, S., *et al.*, Assessment of commercial laboratories performing hair mineral analysis [In Process Citation]. *Jama*, 285(1). 67-72, 2001.

[27] Valkovic, V., *Trace elements in human hair*. 1977, New York: Garlang STPM Press. 194.

[28] Mertz, W., ed. *Trace Elements in Human and Animal Nutrition*. Fifth ed. 1986, Academic Press: Boston. 979.

[29] Chatt, A. and S. Katz, Hair analysis. *Applications in the biomedical and environmental sciences*. 1988, New York: VCH Publishers. 134.

[30] Chen, S., *et al.*, Persistent effect of in utero meso-2,3-dimercaptosuccinic acid (DMSA) on immune function and lead-induced immunotoxicity. *Toxicology*, 132(1). 67-79, 1999.

[31] de la Torre, A., *et al.*, Comparison of the effectiveness of 2,3-dimercaptopropanol (BAL) and meso-2,3-dimercaptosuccinic acid (DMSA) as protective agents against mercuric chloride-induced nephrotoxicity in rats. Biol Trace Elem

Res, 63(1). 1-10, 1998.

[32] Miller, A.L., Dimercaptosuccinic acid (DMSA), a non-toxic, water-soluble treatment for heavy metal toxicity. *Altern Med Rev*, 3(3). 199-207, 1998.

[33] Ren, M. and R. Yang, Clinical curative effects of dimercaptosuccinic acid on hepatolenticular degeneration and the impact of DMSA on biliary trace elements. *Chin Med J (Engl)*, 110(9). 694-7, 1997.

[34] Schwartz, B.S. and W.F. Stewart, Different associations of blood lead, meso 2,3-dimercapto-succinic acid (DMSA)-chelatable lead, and tibial lead levels with blood pressure in 543 former organolead manufacturing workers. *Arch Environ Health*, 55(2). 85-92, 2000.

[35] O'Connor, M.E. and D. Rich, Children with moderately elevated lead levels: is chelation with DMSA helpful? *Clin Pediatr (Phila)*, 38(6). 325-31, 1999.

[36] Chisolm, J.J., Jr., Safety and efficacy of meso-2,3-dimercaptosuccinic acid (DMSA) in children with elevated blood lead concentrations. *J Toxicol Clin Toxicol*, 38(4). 365-75, 2000.

[37] Merkord, J., *et al.*, Antidotal effects of 2,3-dimercaptopropane-1-sulfonic acid (DMPS) and meso-2,3-dimercaptosuccinic acid (DMSA) on the organotoxicity of dibutyltin dichloride (DBTC) in rats. *Hum Exp Toxicol*, 19(2). 132-7, 2000.

[38] Blanusa, M., *et al.*, Mobilisation of cadmium by meso- and racemic-2,3-dimercaptosuccinic acid (DMSA) in rats [In Process Citation]. *Pharmacol Toxicol*, 87(4). 179-81, 2000.

[39] Lee, B.K., *et al.*, Provocative chelation with DMSA and EDTA: evidence for differential access to lead storage sites. *Occup Environ Med*, 52(1). 13-9, 1995.

[40] Olszewer, E. and J.P. Carter, EDTA chelation therapy in chronic degenerative disease. *Med Hypotheses*, 27(1). 41-9, 1988.

[41] Discalzi, G., *et al.*, Occupational Mn parkinsonism: magnetic resonance imaging and clinical patterns following CaNa2-EDTA chelation. [In Process Citation]. *Neurotoxicology*, 21(5). 863-6, 2000.

[42] Ilback, N.G., T. Stalhandske, and U. Lindh, Effects of EDTA on trace elements and cardiovascular function in the anesthetised rabbit [In Process Citation]. *Biol Trace Elem Res*, 76(2). 133-48, 2000.

[43] Seaton, C.L., J. Lasman, and D.R. Smith, The effects of CaNa(2)EDTA on brain lead mobilization in rodents determined using a stable lead isotope tracer. *Toxicol Appl Pharmacol*, 159(3). 153-60, 1999.

[44] Cranton, E., ed. A textbook on EDTA chelation therapy. *Journal of Advancement in Medicine*. 1989, Human Sciences Press: New York. 416.

[45] Cranton, E., *et al.*, Urinary trace and toxic elements and minerals in untimed urine specimens relative to urine creatinine, Part II: Provoked increase in excretion following intravenous EDTA., in *A textbook on EDTA chelation therapy*, E. Cranton, Editor. 1989, Human Sciences Press: New York. p. 416.

[46] Whang, R. and K.W. Ryder, Frequency of hypomagnesemia and hypermagnesemia. Requested vs routine [see comments]. *Jama*, 263(22). 3063-4, 1990.

[47] Shils, M.E., *Magnesium, in Present Knowledge in Nutrition*, M.L. Brown, Editor. 1990, Nutrition Foundation: Washington DC. p. 224-234.

[48] Elin, R.J., Assessment of magnesium status. *Clin Chem*, 33. 1965, 1987.

[49] Elin, R.J., Status of the mononuclear blood cell magnesium assay. *J Am Coll Nutr*, 6(105)1987.

[50] Geven, W.B., *et al.*, Reference values of magnesium and potassium in mononuclear cells and erythrocytes of children. *Clin Chem*, 36(7). 1323-7, 1990.

[51] Elin, R.J., J.M. Hosseini, and J.R. Gill, Jr., Erythrocyte and mononuclear blood cell magnesium concentrations are normal in hypomagnesemic patients with chronic renal magnesium wasting. *J Am Coll Nutr*, 13(5). 463-6, 1994.

[52] Loun, B., *et al.*, Intracellular magnesium content of mononuclear blood cells and granulocytes isolated from leukemic, infected, and granulocyte colony-stimulating factor-treated patients. *Clin Chem*, 41(12 Pt 1). 1768-72, 1995.

[53] Alfrey, A.C., N.L. Miller, and D. Butkus, Evaluation of body magnesium stores. *J Lab Clin Med*, 84(2). 153-62, 1974.

[54] Elin, R.J., *et al.*, Effect of magnesium deficiency on erythrocyte aging in rats. *Am J Pathol*, 100(3). 765-78, 1980.

[55] Quamme, G.A., Free cadmium activity in renal epithelial cells is enhanced by Mg2+ depletion. Kidney Int, 41(5). 1237-44, 1992.

[56] Albanese, A.A., Osteoporosis. *J Am Pharm Assoc*, 17(4). 252-3, 1977.

[57] Basco, J., On determination of calcium in hair and its use for investigation of coronary heart disease and calcium metabolic rate. *Atomki Kozl*, 26. 2324-5, 1985.

[58] MacPherson, A. and J. Baklint, Beard calcium concentration as a marker for coronary heat disease by supplementation with micro nutrients including selenium. *Analyst*, 120. 871-875, 1995.

[59] Bland, J., Dietary calcium, phosphorus and their relationships to bone formation and parathyroid activity. *J John Bastyr Col Naturopathic Med*, 1. 1, 1979.

[60] Tamari, G. and Z. Rona, Hair mineral levels and their correlation with abnormal glucose

tolerance. *Cytobiological Review*, 1985.
[61] Giordani, A., *et al.*, Contrasting effects of excess ferritin expression on the iron-mediated oxidative stress induced by tert-butyl hydroperoxide or ultraviolet-A in human fibroblasts and keratinocytes. *J Photochem Photobiol B*, 54(1). 43-54, 2000.
[62] Balan, V., *et al.*, Screening for hemochromatosis: a cost-effectiveness study based on 12,258 patients. *Gastroenterology*, 107(2). 453-9, 1994.
[63] Cook, J., Serum ferritin as a measure of iron in normal subjects. *Am J Clin Nutr*, 27. 9681-9687, 1974.
[64] Frank, P. and S. Wang, Serum iron and total iron binding capacity compared with serum ferritin in assessment of iron deficiency. *Clin Chem*, 27. 276-9, 1981.
[65] Freire, W.B., Hemoglobin as a predictor of response to iron therapy and its use in screening and prevalence estimates. *Am J Clin Nutr*, 50(6). 1442-9, 1989.
[66] Milne, D.B., S.K. Gallagher, and F.H. Nielsen, Response of various indices of iron status to acute iron depletion produced in menstruating women by low iron intake and phlebotomy. *Clin Chem*, 36(3). 487-91, 1990.
[67] Fishbane, S. and R.I. Lynn, The utility of zinc protoporphyrin for predicting the need for intravenous iron therapy in hemodialysis patients. *Am J Kidney Dis*, 25(3). 426-32, 1995.
[68] Hastka, J., *et al.*, Laboratory tests of iron status: correlation or common sense? *Clin Chem*, 42(5). 718-24, 1996.
[69] Gambino, R., *et al.*, The relation between chemically measured total iron-binding capacity concentrations and immunologically measured transferrin concentrations in human serum. *Clin Chem*, 43(12). 2408-12, 1997.
[70] Gambino, R., Serum transferrin (total iron-binding capacity) in evaluation of iron status *Clin Chem*, 42(12). 2053, 1996.
[71] Solomons, N.W., On the assessment of zinc and copper nutriture in man. *Am J Clin Nutr*, 32. 856, 1979.
[72] Davies, S., Assessment of zinc status. *Intl Clin Nutr Rev*, 4. 122-129, 1984.
[73] Whitehouse, R.C., *et al.*, Zinc in plasma, neutrophils, lymphocytes, and erythrocytes as determined by flameless atomic absorption spectrophotometry. *Clin Chem*, 28. 475-480, 1982.
[74] Weismann, K. and H. Hoyer, Serum alkaline phosphatase and serum zinc levels in the diagnosis and exclusion of zinc deficiency in man. *American Journal of Clinical Nutrition*, 41. 1214-1219, 1985.
[75] Samman, S., *et al.*, Is erythrocyte alkaline phosphatase activity a marker of zinc status in humans? *Biol Trace Elem Res*, 51(3). 285-91, 1996.
[76] Bales, C.W., *et al.*, Marginal zinc deficiency in older adults: responsiveness of zinc status indicators. *J Am Coll Nutr*, 13(5). 455-62, 1994.
[77] Linder, N., *et al.*, An oral zinc loading test in psoriasis. *Metabolism,* 37(9). 807-9, 1988.
[78] Strain, W.H., *et al.*, Analysis of zinc levels in hair for the diagnosis of zinc deficiency in man. *Journal of Laboratory Clinical Medicine*, 68. 244, 1966.
[79] Solomons, N.W., *et al.*, Growth retardation and zinc nutrition. *Pediatr Res*, 10(11). 923-7, 1976.
[80] Solomons, N., Biochemical, metabolic and clinical role of copper in human nutrition. *J Am Coll Nutr*, 4. 83-105, 1985.
[81] Milne, D.B. and F.H. Nielsen, Effects of a diet low in copper on copper-status indicators in postmenopausal women. *Am J Clin Nutr*, 63(3). 358-64, 1996.
[82] Peretz, A., J. Neve, and J.P. Famaey, Effects of chronic and acute corticosteroid therapy on zinc and copper status in rheumatoid arthritis patients. *J Trace Elem Electrolytes Health Dis*, 3(2). 103-8, 1989.
[83] Fisberg, R.M., *et al.*, Plasma zinc, copper, and erythrocyte superoxide dismutase in children with phenylketonuria. *Nutrition*, 15(6). 449-52, 1999.
[84] Blakley, B.R. and D.L. Hamilton, Ceruloplasmin as an indicator of copper status in cattle and sheep. *Can J Comp Med*, 49(4). 405-8, 1985.
[85] Piccinni, L. and *et al.*, A case control study on selenium, zinc, and copper in plasma and hair of subjects affected by breast and lung cancer. *Biol Trace Elements Res*, 24. 105-109, 1990.
[86] Donma, M., O. Donma, and *et al.*, Hair zinc and copper concentrations and zinc/copper ratios in pediatric malignancies and health children from southeastern Turkey. *Biol Trace Elements Res*, 36. 51-58, 1993.
[87] Danks, D.M., Copper deficiency in humans. *Annu Rev Nutr*, 8. 235-57, 1988.
[88] Anderson, R.A., *et al.*, Effects of supplemental chromium on patients with symptoms of reactive hypoglycermia. *Metabolism*, 36. 351, 1987.
[89] Anderson, R.A. and A.S. Kozlovsky, Chromium intake, absorption and excretion of subjects consuming self-selected diets. *Am J Clin Nutr*, 41(6). 1177-83, 1985.
[90] Aharoni, A., *et al.*, Hair chromium content of women with gestational diabetes compared with nondiabetic pregnant women. *Am J Clin Nutr*, 55(1). 104-7, 1992.
[91] Anneren, G., M. Gebre-Medhin, and K.H. Gustavson, Increased plasma and erythrocyte selenium concentrations but decreased

erythrocyte glutathione peroxidase activity after selenium supplementation in children with Down syndrome. *Acta Paediatr Scand*, 78(6). 879-84, 1989.

[92] Winklhofer-Roob, B.M., *et al.*, Effects of pancreatic enzyme preparations on erythrocyte glutathione peroxidase activities and plasma selenium concentrations in cystic fibrosis. *Free Radic Biol Med*, 25(2). 242-9, 1998.

[93] Lane, H.W., *et al.*, Blood selenium and glutathione peroxidase levels and dietary selenium of free living and institutionalized elderly subjects. *Proc Soc Exp Biol Med*, 173. 87-95, 1985.

[94] Bahemuka, M. and K. Hodkinson, Red blood cell potassium as a practical index of potassium status in elderly patients. *Age and Aging*, 5. 24-29, 1976.

[95] Lans, K., I.F. Stein, and K.K.I. A., The relationship of serum potassium to erythrocyte potassium in normal subjects and patients with potassium deficiency. *Am J Med Sci*, 223. 65-74, 1955.

[96] Keen, C., Whole blood manganese as an indicator of body manganese. *N Eng J Med*, 308. 1230, 1983.

[97] Clarkson, T.W., Molecular and ionic mimicry of toxic metals. *Annu Rev Pharmacol Toxicol*, 33. 545-71, 1993.

[98] Rea, W., *Chemical Sensitivity*. 1994, Boca Raton: Lewis. 2015.

[99] Watson, G.E., *et al.*, Influence of maternal lead ingestion on caries in rat pups. *Nat Med*, 3(9). 1024-5, 1997.

[100] Mahaffey, K.R. and J.L. Annest, Association of erythrocyte protoporphyrin with blood lead level and iron status in the second National Health and Nutrition Examination Survey, 1976-1980. *Environ Res*, 41(1). 327-38, 1986.

[101] Fine, B.P., *et al.*, Influence of magnesium on the intestinal absorption of lead. *Environ Res*, 12(2). 224-7, 1976.

[102] Cheng, Y., *et al.*, Relation of nutrition to bone lead and blood lead levels in middle-aged to elderly men. The Normative Aging Study. *Am J Epidemiol*, 147(12). 1162-74, 1998.

[103] Holtzman, D., *et al.*, Maturation of resistance to lead encephalopathy: cellular and subcellular mechanisms. *Neurotoxicology*, 5(3). 97-124, 1984.

[104] Simmonds, P.L., C.L. Luckhurst, and J.S. Woods, Quantitative evaluation of heme biosynthetic pathway parameters as biomarkers of low-level lead exposure in rats. *J Toxicol Environ Health*, 44(3). 351-67, 1995.

[105] Bradbury, M.W. and R. Deane, Permeability of the blood-brain barrier to lead. *Neurotoxicology*, 14(2-3). 131-6, 1993.

[106] Sandhir, R. and K.D. Gill, Effect of lead on the biological activity of calmodulin in rat brain. *Exp Mol Pathol*, 61(1). 69-75, 1994.

[107] Oortgiesen, M., *et al.*, Nanomolar concentrations of lead selectively block neuronal nicotinic acetylcholine responses in mouse neuroblastoma cells. *Toxicol Appl Pharmacol*, 103(1). 165-74, 1990.

[108] Skerfving, S., *et al.*, Biological monitoring of inorganic lead. *Scand J Work Environ Health*, 19(Suppl 1). 59-64, 1993.

[109] Needleman, H.L., *et al.*, Deficits in psychologic and classroom performance of children with elevated dentine lead levels [published erratum appears in *N Engl J Med* 1994 Sep 1;331(9):616]. N Engl J Med, 300(13). 689-95, 1979.

[110] Broad, W.J., Sir Isaac Newton: mad as a hatter. *Science*, 213(4514). 1341-2, 1344, 1981.

[111] Hamada, R. and M. Osame, Minamata disease and other mercury syndromes, in *Toxicology of Metals*, L. Chang, Editor. 1996, CRC: Boca Raton. p. 337-351.

[112] Muckter, H., *et al.*, Are we ready to replace dimercaprol (BAL) as an arsenic antidote? *Hum Exp Toxicol*, 16(8). 460-5, 1997.

[113] Biswas, S., G. Talukder, and A. Sharma, Protection against cytotoxic effects of arsenic by dietary supplementation with crude extract of Emblica officinalis fruit. *Phytother Res*, 13(6). 513-6, 1999.

[114] Stine, E.R., *et al.*, N-(2,3-dimercaptopropyl)-phthalamidic acid: protection, in vivo and in vitro, against arsenic intoxication. *Toxicol Appl Pharmacol*, 75(2). 329-36, 1984.

[115] Fernandez-Lorenzo, J.R., *et al.*, Aluminum contents of human milk, cow's milk, and infant formulas. *J Pediatr Gastroenterol Nutr*, 28(3). 270-5, 1999.

[116] Litov, R.E., *et al.*, Plasma aluminum measurements in term infants fed human milk or a soy-based infant formula. *Pediatrics*, 84(6). 1105-7, 1989.

[117] Kostial, K., Cadmium, in *Trace Elements in Human and Animal Nutrition*, M. Mertz, Editor. 1986, Academic Press, Inc.: Orlando. p. 319-345.

[118] Decina, P. and R.R. Fieve, Prophylactic serum lithium levels in recurrent unipolar depression. *J Clin Psychopharmacol*, 1(3). 150-2, 1981.

[119] D'Haese, P.C., *et al.*, Aluminum, iron, lead, cadmium, copper, zinc, chromium, magnesium, strontium, and calcium content in bone of end-stage renal failure patients. *Clin Chem*, 45(9). 1548-56, 1999.

[120] Gonzalez-Reimers, E., *et al.*, Relative and combined effects of ethanol and protein deficiency on strontium and barium bone content and fecal and urinary excretion. *Biol Trace Elem Res*, 68(1). 41-9, 1999.

[121] Grynpas, M.D., *et al.*, Strontium increases

vertebral bone volume in rats at a low dose that does not induce detectable mineralization defect. *Bone*, 18(3). 253-9, 1996.

Chapter 4

Amino Acids

TABLE 4-1. Summary of Nutrients Indicated by Abnormal Amino Acid Results

	Amino Acids	Low Result	High Result
Amino Acids for Polypeptides and Energy	Isoleucine (Ile), Leucine (Leu), Valine (Val)	*	B_6, 100mg; Check for insulin insensitivity
	Histidine (His)	Folate, 800 mcg; His, 500mg TID	
	Lysine (Lys)	Carnitine, 1-2 g	Vitamin C, 1g BID; Niacin, 50 mg; B_6, 100 mg; Iron, 15 mg
	α-Aminoadipic acid (αAAA)		B_6, 100 mg; αKG, 600 mg BID
	Threonine (Thr)	*	B_6, 100 mg; Zinc, 30 mg
	α-Amino-N-butyric acid (αABA)	αKG, 600 mg BID; B_6, 100 mg	
Neurotransmitters and Precursors	Phenylalanine (Phe)	*	Iron, 30 mg; Vitamin C, 1 gm TID; Niacin, 50 mg; Low Phe diet
	Tyrosine (Tyr)	Iron, 30 mg, Tyrosine, 500 mg TID; Vitamin C, 1 gm BID; Niacin 50 mg	Copper, 3 mg; Iron, 30 mg; Vitamin C, 1 gm TID B_6, 100 mg BID
	Tryptophan (Trp)	5-HTP, 50 mg TID	Niacin, 50 mg; B_6, 100 mg BID
	Glutamic Acid (Glu)	B_6, 100 mg; αKG, 600 mg TID	Niacin, 50 mg; B_6, 100 mg BID
	γ-Aminobutyric acid (GABA)		αKG, 600 mg BID; B_6, 50 mg
Glycine, Serine and Related Amino Acids	Glycine (Gly)	Gly, 1000 mg TID	Folate, 800 mcg; B_6, 100 mg; B_2, 50 mg; B_5, 500 mg
	Serine (Ser)	B_6, 100 mg; Folate, 800 mcg, Manganese, 15 mg	*
	Ethanolamine		Magnesium, 200 mg BID
	Phosphoethanolamine (PE)		SAM, 200 mg BID; B_{12}, 1000 mcg; Folate, 800 mcg; Betaine, 1000 mg
	Phosphoserine		Magnesium, 200 mg BID
	Sarcosine		B_2, 50 mg

TABLE 4-1. Continued

Supply and Removal of Amino Groups (Urea Cycle)	Alanine (Ala)	*	B_6, 100 mg
	Arginine (Arg)	* or Arg, 500 mg BID	Manganese, 15 mg
	Aspartic Acid (Asp)	αKG, 600 mg BID; B_6, 100 mg	Magnesium, 200 mg BID; Zinc, 30 mg
	Asparagine (Asn)	Mg, 200 mg BID	
	Glutamine (Gln)		αKG, 600 mg BID; B_6, 100 mg
	Citrulline		Magnesium, 200 mg BID; Aspartic acid, 1 mg BID
	Ornithine (Orn)	Arginine, 500 mg BID	Magnesium, 200 mg BID; αKG, 600 mg BID; B_6, 100 mg
Sulfur-Containing Amino Acids	Methionine (Met)	*	B_6, 100 mg; αKG, 600 mg BID; Magnesium, 200 mg BID; SAM, 200 mg BID
	Cystine (Cys)	NAC, 500 mg BID	B_2, 50 mg
	Homocystine (HCys)		B_6, 100 mg; Folate, 800 mcg; B_{12}, 1000 mcg; Betaine, 1-2 g TID
	Cystathionine		B_6, 100 mg
	Taurine (Tau)	Taurine, 300 mg BID; B_6, 100 mg	Vitamin E, 800 IU; Vitamin C, 1 g TID; β-Carotene, 25,000 IU; Coenzyme Q_{10}, 30 mg; Lipoate, 200 mg
Collagen-Related Amino Acids	Proline (Pro)	αKG, 600 mg BID	Vitamin C, 1000 mg TID; Niacin, 50 mg
	Hydroxyproline (HPro)		Vitamin C, 1000 mg TID; Iron, 15 mg
	Hydroxylysine (HLys)		Vitamin C, 1000 mg TID; Iron, 15 mg
β-Amino Acids and Derivatives	β-Alanine		Lactobacillus and Bifidobacteria; B_6, 100 mg
	Anserine		Zinc, 30 mg
	Carnosine		Zinc, 30 mg
	β-Aminoisobutyric acid		B_6, 50 mg
	1-Methylhistidine		Vitamin E, 400 IU; B_{12}, 1000 mcg; Folate, 800 mcg
Markers of Tissue Injury	3-Methylhistidine		Branched Chain Amino Acids, Vitamin E, 800 IU; Vitamin C, 1 g TID; β-Carotene, 25,000 IU; Coenzyme Q_{10}, 30 mg; Lipoate, 200 mg
	3-Nitrotyrosine		Vitamin E, 800 IU; Vitamin C, 1 g TID; β-Carotene, 25,000 IU; Coenzyme Q_{10}, 30 mg; Lipoate, 200 mg
	Assymetric dimethylarginine		Arginine, 3-6 gm
	Methionine sulfoxide		Vitamin E, 800 IU; Vitamin C, 1 g TID; β-Carotene, 25,000 IU; Coenzyme Q_{10}, 30 mg; Lipoate, 200 mg

* Low levels are generally best normalized with balanced or custom mixtures of essential amino acids

General Concepts in Amino Acid Testing

Amino acids are molecules with intriguing chemical properties. They have proton accepting carboxyl groups and proton-donating amine groups. They have chiral centers so that no two groups can be interchanged about the central carbon without producing a new compound‡. They join in infinite variety to form chains of two to two thousand. They have solubility ranging from extremely soluble to virtually insoluble in water. Several amino acids contain sulfur ether (C-S-C) or sulfhydryl (-SH) groups that are the most electron-rich groups among biological molecules. Because of their proton exchange properties, they act as pH and redox buffers. Because of their chiral centers and varieties of solubility, proteins formed from amino acid chains have unique and stable three-dimensional shapes that can form associations to build increasingly complex structures. The physiological functions of amino acids, either singly or in peptide sequences, include the following:

- Transmission and control of neuronal signals
- Regulation of muscle activity
- Maintenance of redox status
- Hormonal regulation
- Enzymes and blood transport proteins
- Formation of the ligaments, tendons, and bone matrix that holds the human form together
- Supply of energy pathway intermediates

Genetic information is translated into proteins using twenty amino acids. The structural variety, specificity, and stability of proteins are due to the variety of sizes, shapes, polarities, and electronegativities of their amino acid side chains. Dietary sources of protein generally provide all of the amino acids needed by humans since the same twenty amino acids occur in all life forms. There is considerable variety, however, in the amino acid composition of dietary protein and in individual ability to efficiently digest protein and assimilate amino acids. During periods of acute stress protein breakdown increases by as much as 20% with concurrent increased amino acid demands [1]. Any combination of the following factors can lead to amino acid inadequacies:

- Dietary protein restrictions
- Insufficient stomach acid
- Insufficient gastric and pancreatic enzymes
- Inadequate secretin or other hormonal responses
- Increased losses due to stress

A summary of information on amino acid abnormalities and corresponding intervention options that are frequently used is shown in Table 4-1.

---

‡ Because of this property, all amino acids except glycine occur as D- and L- sterioisomers. Because only the L- form is utilized in human intermediary metabolism, the chiral designation, "L", is dropped from all names for amino acids in this book. The D- prefix will be used for any reference to a D-amino acid.

During the life cycle in man, nine amino acids are generally considered essential dietary compounds. (The question of conditional essentiality of several others will be discussed later.) The essential amino acids (EAA) are sometimes referred to as "indispensable" amino acids because they must be obtained from food intake (Table 4-2). Transfers of amino groups to organic acids that are produced in other metabolic pathways can form the other eleven amino acids used for protein synthesis. Additional amino acids, such as hydroxyproline, are formed by reactions that add chemical groups after proteins are synthesized. These amino acids are released into body fluids at cell death or during protein turnover in viable cells. Biochemical transformations in detoxification or neurotransmitter pathways lead to the formation of still other amino acids and simple peptides. All of these classes of compounds may be measured in profiles of physiologically significant amino acids in plasma or urine. Plasma was used to assess the fasting amino acid profiles in orally-fed nursing home men and tube-fed nursing home men with dementia [2]. Such individuals have significantly lower plasma levels of methionine and branched-chain amino acids than healthy controls. The ratio of essential to nonessential amino acids is also significantly lower suggesting that the intake of essential amino acids is inadequate in this group.

TABLE 4-2. Essential Amino Acids

- Arginine
- Isoleucine
- Leucine
- Lysine
- Methionine
- Phenylalanine
- Threonine
- Tryptophan
- Valine

Amino acid nutrition has an obvious relationship to protein nutrition. There are numerous clinical situations, however, in which amino acid needs exceed amounts provided by dietary protein. In some cases, the amino acids that achieve the improvement are not listed as essential. Instances discussed in this section lead to re-examination of the assignment of "essential" to several amino acids. The question of adequate levels of intake has been the subject of numerous scientific reports. Conclusions regarding the amino acid needs for adults have been based primarily on nitrogen balance studies, and those for infants and children are predicted on the smallest amounts compatible with normal growth. Published estimates show a wide range of individual needs, even within a single study [3, 4]. Because they are more difficult to quantify, the more subtle effects of amino acids on neurological and psychological status are largely omitted from these studies. Some common clinical presentations suggest assessment of amino acids (Table 4-3).

TABLE 4-3. Common Clinical Presentations Indicating Need For Amino Acid Assessment

History

Weight control by severe diet restrictions
Uninformed vegetarianism
Antacid use
Any digestive impairment

Symptoms

Failure to thrive
Autism spectrum disorders
Rapid weight loss
Dyspepsia
Behavior disorders
Chronic fatigue
Depression

Utilization of amino acids is highly tissue dependent. The most active participants are liver, skeletal muscle, and brain. Sometimes the optimal function at one site is dependent on normal functions at other sites. For example, the utilization of serine in the brain to form acetylcholine is regulated, in part, by the availability of glycine. The avid binding and metabolism of glycine that occurs in the liver normally keeps availability low. In the presence of liver disease, however, the amino acids are free to remain in circulation among other tissues. Thus, patients with liver disease may have higher circulating glycine and, when they are given doses of serine, the resulting glycine and serine overload can induce psychotic episodes [5].

Amino Acids in Plasma and Urine

Profiling of amino acids in plasma and urine has been used to elucidate a rapidly growing number of aminoacidopathies since the introduction of partition chromatography methods in 1945 [6]. The fact that a large number of metabolic disorders involve amino acids is due to their complex biochemistry, involving dozens of enzyme systems. The 1995 edition of *The Metabolic and Molecular Basis of Inherited Disease* by Scriver, *et. al.* devotes 12 chapters to disorders of amino acids [7].

Is plasma or urine the preferred specimen for amino acid testing? This question is frequently in the mind of the clinician who is considering the best advice for a patient regarding amino acid testing. An informed response must consider what principal clinical answers are sought and which amino acids are being tested (Table 4-4). Numerous sources are available for authoritative information about the clinical evaluation and therapeutic use of amino acids [6, 8-11].

TABLE 4-4. Specimen Preference According to Critical Information

Desired Information	Specimen
Status of essential amino acid supply Leu, Ile Val, Phe, Lys, Thr, Met, Trp	PLASMA
Status of conditionally essential amino acid supply Arg, Tyr, His, Gly, Ser, Tau	PLASMA
Markers of neurological dysfunction Trp, Tyr, Phe, Met, GABA	PLASMA
Markers of vitamin and mineral deiciency (Table 4-1)	BOTH
Markers of bone loss or muscle wasting HyProl, HyLys, Cys, 3-MeHis	URINE
Markers of intestinal dysbiosis β-Ala	URINE

To state categorically that profiling of amino acids is best performed on plasma or urine is to oversimplify. The question of preferred specimen can be answered only when it is addressed to specific amino acids or to the specific type of information desired. One way to judge the relative value of results from two specimen types is to ask which specimen has been most used for scientific studies. Specimens studied in 100 of the sources used in the preparation of this chapter are listed in Table 4-5. It is clear that the majority of published studies have used plasma as the specimen for analysis. This is primarily because most investigations have been concerned with essential amino acid status. Urine is preferred for studies of bone loss markers and gut microbial overgrowth.

Most reports of neonatal inherited disorders of metabolism are omitted from Table 4-5. The aminoacidemias and aminoacidurias associated with such investigations are approximately equally divided. Inherited metabolic disorders generally result in extreme elevations, and the abnormality is easily detected in either specimen type. The branched chain amino acids (BCAAs), for example, are elevated in both plasma and urine in maple syrup urine disease. The newer application of amino acid profiling of older children and

TABLE 4-5. Amino Acid Studies Using Plasma and Urine

#	Plasma	Urine	Amino Acid	Ref.
1	✓		Leu, Phe, Gln	[1]
2	✓		Met, BCAA, etc.	[2]
3		✓	Profile	[3]
4		✓	Profile	[4]
5	✓		Gly, Ser, Tau	[5]
6		✓	Profile	[6]
7	✓		Profile	[11]
8	✓		Profile	[13]
9	✓		Profile	[14]
10	✓		GABA, Tau	[15]
11	✓		Profile	[16]
12	✓		Profile	[17]
13	--	--	BCAA	[18]
14	--	--	BCAA	[19]
15	✓		Leu	[20]
16	✓		His	[21]
17	✓		His	[22]
18	✓		Met, His	[23]
19		✓	His	[24]
20	--	--	Lys	[25]
21	--	--	Protein	[26]
22	✓		Lys	[27]
23	--	--	Lys	[28]
24	✓		Lys	[29]
25	✓		Lys	[30]
26	✓		Lys	[31]
27	✓		Hcys	[32]
28		✓	aAmAdip	[33]
29		✓	Lys	[34]
30	✓		aAmnAdip	[35]
31	✓		Profile	[36]
32	--	--	Ala	[37]
33	✓		Ala,Ile	[38]
34	✓		Excitatory	[39]
35	✓		Profile	[40]
36	✓		GABA	[41]
37	✓		Excitatory	[42]
38	--	--	Glu	[43]
39	✓		Glu,Cys	[44]
40	✓		Trp	[45]
41	✓		Trp	[46]
42	✓		Trp	[47]
43	✓		Trp	[48]
44	✓		Trp/Neutrals	[49]
45	✓		Trp	[50]
46	✓		Trp	[51]
47	--	--	Neurotrans.	[52]
48	✓		Tyr	[53]
49		✓	Tyr/Glu	[54]
50	--	--	Tyr	[55]
51	✓		Phe/Tyr	[56]
52	--	--	Pyroglu, Gly	[57]
53	✓		B6	[58]
54	✓		Hcys, Ser	[59]
55	✓		Ser, Gly, Tau, His	[60]
56	✓		Ser, Gly	[61]
57	✓		HyPro, Gly, Ser	[62]
58			EtAm, PEtAm	[63]
59	✓	✓	PEtAm	[64]
60	--		PEtAm	[65]
61	--	--	PEtAm, EtAm	[66]
62	✓	--	Thr, PEtAm	[67]
63	✓		Neurotrans.	[68]
64	--		Asp	[69]
65	--	--	Arg	[70]
66	✓	--	Arg	[71]
67	✓		Profile	[72]
68			Cys,GSH	[73]
69	✓	✓	Met, Cys, Tau	[74]
70	✓		Met, Cys	[75]
71	✓		Arg, Cys, Gly	[76]
72	✓		Profile	[77]
73			Hcys	[78]
74	--	✓	Bacterial products	[79]
75	--	--	Met – Pb	[80]
76	--	--	Met – Co	[81]
77	--	--	CysThio	[82]
78		--	Choline, Betaine	[83]
79		✓	HCys	[84]
80	✓	✓	Hcys	[85]
81	✓		Tau	[86]
82	--		Tau	[87]
83	--	--	Tau	[88]
84		--	Tau	[89]
85	✓		Tau	[90]
86	✓		Tau	[91]
87	✓		Tau	[92]
88	✓		Pro	[93]
89			Pro, Leu	[94]
90		✓	Pro, HPro	[95]
91		✓	Pro, HPro	[96]
92		✓	HPro	[97]
93		✓	HPro	[98]
94	✓	✓	HPro	[99]
95			HPro	[100]
96		✓	HPro, HLys	[101]
97	--	✓	HPro, HLys	[102]
98	✓	--	Tau, etc.	[103]
99	--		Profile	[104]
100		--	Tau	[105]
Total	64	20		

adults to determine amino acid status in chronic degenerative diseases is more pertinent for this discussion.

Urinary levels of organic acids formed from amino acid catabolism can be extremely useful as markers of functional adequacy of amino acids. This evaluation is discussed in the chapter on organic acids, but it should be considered when answering the question of specimen selection for direct testing of amino acids. The combination of amino acids in plasma or urine with organic acids in urine provides a more complete picture of amino acid abnormalities.

Nine amino acids cannot be synthesized in human infant tissue (see Table 4-3). They are termed "essential" or "indispensable" because they must be supplied by dietary sources. Other amino acids serve essential functions, but some human tissues utilize nitrogen sources such as other amino acids to produce them. Histidine is essential in infants, but adults usually acquire the ability to synthesize it. When the flow of amino acids from digested protein has ceased, fasting plasma amino acid levels represent the homeostatic balance of dynamic flow into and out of blood. The largest flows are usually from skeletal muscle protein breakdown and hepatic uptake for protein synthesis (Table 4-5). Having the patient fast removes dietary intake effects. Amino acid levels in plasma reach their homeostatic balance point making a fasting specimen ideal for repeated measures to monitor progress. The principal factors effecting changes over time are dietary intake, digestive efficiency, hepatic uptake, and the ability of skeletal muscle to maintain sufficient rates of transamination (see Glutamine section below).

In experimental animals where protein intake can be controlled accurately, amino acids in plasma show distinctive patterns. The essential amino acids threonine, methionine, tryptophan, phenylalanine, tyrosine, valine, isoleucine, leucine, and lysine declined when dietary protein was lowered from 21% to 6% of caloric intake [12]. The average fall for the EAA group was 36% (Figure 4-1). All of the BCAAs correlated with dietary protein decline, but the non-essential amino acids (NEAAs) showed an increase as protein intake decreased. This response is due to the increased mobilization of amino acids from muscle to meet other critical demands. In the same

FIGURE 4-1. Plasma Amino Acid Response to Dietary Protein Restriction

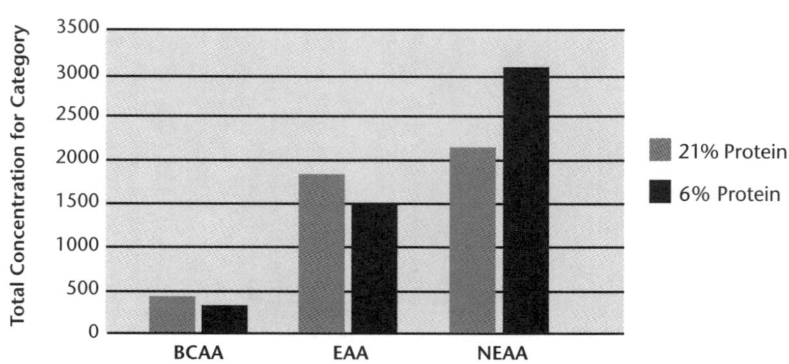

study, insulin and insulin-like growth factor 1 (IGF-1) fell dramatically when plasma amino acids were low (Figure 4-2).

A case will illustrate the ability of plasma amino acid profiles to reflect amino acid status in humans. An 81 year old female with maldigestion and muscle wasting was evaluated with multiple metabolic profiles. A serum chemistry profile revealed frank protein and amino acid deficiency as simultaneous low albumin and total serum protein in the presence of normal BUN/Creatinine ratio (Table 4-6).

FIGURE 4-2. Hormone Responses to Lower Plasma Amino Acids

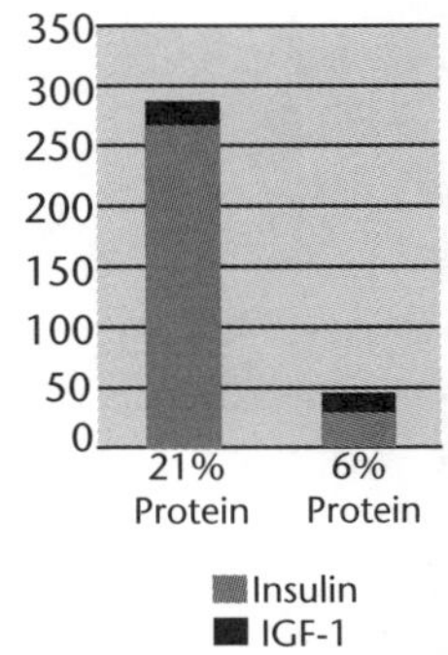

TABLE 4-6. Laboratory Results in Amino Acid Deficiency

Serum Chem-Screen	Result		Reference Range
BUN/Cratinine	15.6 g/dl		12.0 – 20.0
Protein, Total	5.1	L	6.4 – 8.4
Albumin	2.9	L	3.0 – 5.5
Essential Amino Acids	**Result**		**Reference Limits**
Arginine	17 μMole/L	L	50 –160
Histidine	63	L	70 – 140
Isoleucine	28	L	50 – 160
Leucine	86	L	90 – 200
Lysine	57	L	150 – 300
Methionine	15	L	25 – 50
Phenylalaine	31	L	45 – 140
Threonine	41	L	100 – 250
Tryptophan	20	L	35 – 60
Valine	122	L	170 – 420
Essential Amino Acid Derivatives			
Glycine	145	L	200 – 450
Serine	50	L	80 – 200
Taurine	45	L	50 – 250
Tyrosine	59	L	140 – 360

The protein deficiency state was evident in her plasma amino acid profile, which showed abnormally low values for all essential amino acids, as well as the four dispensable amino acids reported. The added dimension of the amino acid profile is seen in the disproportionately low arginine and lysine relative to leucine. Likewise, phenylalanine is moderately low relative to tyrosine, which is severely depleted. These details would be obscured in a 24-hour urine due to retention mechanisms that apply when blood concentrations fall below the renal threshold.

A 24-hour urine amino acid analysis reveals amino acid metabolism throughout the period of metabolic stress of digestion and daily activity. This aspect is of particular value for evaluating those amino acids that primarily reveal tissue degradation. Hydroxylysine and hydroxyproline, for example, are released from collagen of connective tissue and bone. The faithful collection of urine for 24 hours can be a major problem for 24-hour urine amino acid testing. Reports of urinary amino acid studies are virtually always based on data from subjects in a metabolic ward where urine collection is much less subject to error than when performed by free-living individuals. Another source of error is in the reading of the total volume. Individuals can mistake the markings or fail to position the specimen container correctly, resulting in a significant error in the reading. Such errors are expressed directly in the results because of the use of total volume in the data conversions to calculate daily excretions. A properly obtained 24-hour urine, however, provides valuable clinical information and is the preferred specimen for detecting abnormalities of some non-essential amino acid derivatives.

The overall conclusion to be drawn from this analysis is: the great majority of reports documenting clinically useful information from evaluation of essential amino acids have measured plasma levels. We can also say that for most, but not all clinical situations, the greatest array of useful information is derived from the measurement of plasma amino acids. Plasma is especially favored when the prime consideration is the supply of the essential amino acids for optimum balance to maintain or restore health.

Plasma Amino Acid Dynamics

In the fasting state, the primary actions controlling amino acid concentrations in blood plasma are the output from muscle and uptake for hepatic synthesis (Figure 4-3). Plasma from blood drawn at a given moment will reflect the state of the dynamic flux of amino acids leaving sites like skeletal muscle and flowing into sites of utilization in liver, brain, and other tissues. The amount of an essential amino acid in plasma determines the rate of any dependent process in the tissues. For example, low plasma tryptophan results in reduced formation of serotonin in the brain.

When levels of essential amino acids in plasma are low, all dependent functions must decrease. High levels show increased release due to catabolic states (see 3-methyl histidine below) or decreased uptake, as is seen when hepatic activity is inadequate to clear portal amino acid flow [106]. Myriads of genetically determined metabolic factors have bearing on plasma amino acid levels because of impacts on the uptake and release mechanisms for amino acids. For example, lower serum levels of free amino acids due to malabsorption are found in patients with spleen deficiency syndrome [13].

Injury is characterized by a unique pattern of elevated plasma branched chain (leucine, isoleucine, and valine), and aromatic (phenylalanine and tyrosine) amino acids as well as methionine [14]. The nonessential amino acids in muscle tend to decline with injury, glutamine having the most marked change. In hepatic encephalopathy, plasma concentrations of the aromatic amino acids are high, while those of the branched-chain amino acids are low [15].

FIGURE 4-3. Plasma Amino Acid Dynamics

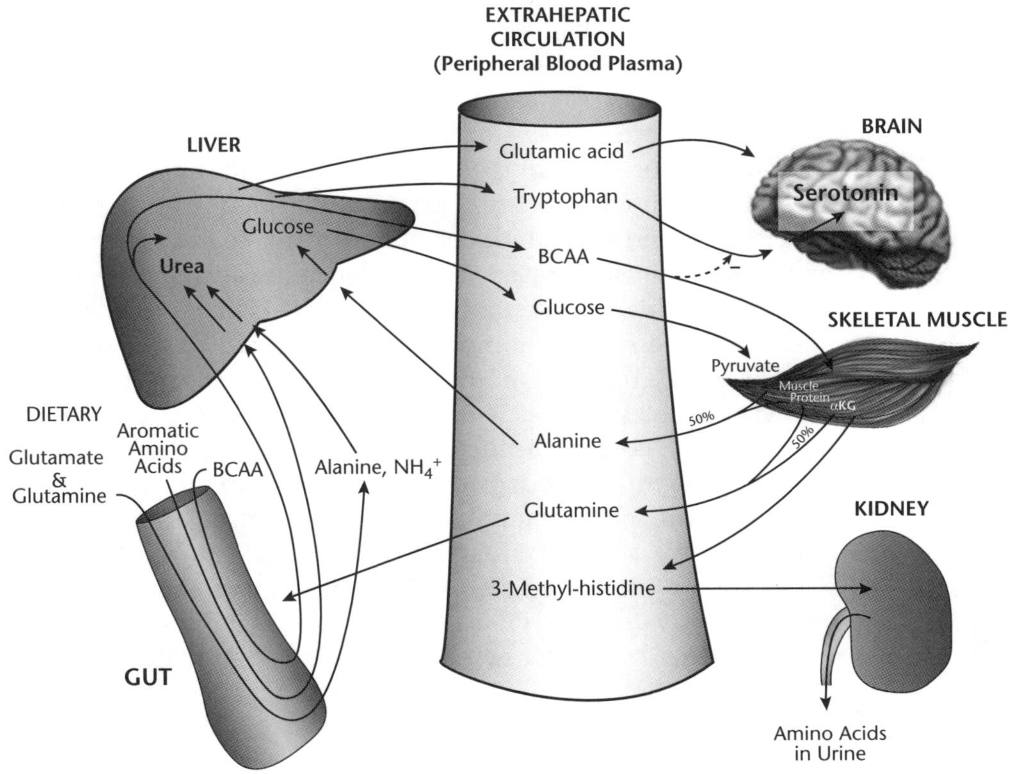

Amino Acid Profiling in Clinical Practice

Neonatal screening and confirmation of genetic abnormalities in amino acid metabolism like Maple Syrup Urine Disease (MSUD) is the most widely accepted application for amino acid testing. Urine is the specimen of choice for newborns, but plasma generally reveals metabolic lesions as well (see above). In MSUD the BCAAs are greatly elevated in both plasma and urine.

Any conditions that manifest symptoms of fatigue, impaired brain function, or muscle weakness are likely to show improvement with optimized amino acid supply. Individually customized formulations may be constructed based on finding low amino acid levels in plasma. An algorithm can be used to adjust relative proportions of each component in a mixture of essential amino acids. Compounding pharmacies can produce the custom formulations from supplies of the free acid form of amino acids. In this form, there is no digestive requirement and absorption is very efficient in most individuals. The delivery of a rich supply of amino acids to the liver can result in significant positive shifts in biosynthetic, energy yielding, and detoxification pathways. The ensuing increase in hepatic output of amino acids provides substrate for neurotransmitter synthesis in brain, actinomyosin replacement in muscle, and cellular control molecules like polypeptide hormones and nitric oxide. Improved supply of arginine as precursor for nitric oxide could explain many of the varied clinical responses to improved delivery of amino acids. Further aspects of this fascinating relationship are discussed at the end of this chapter (under Asymmetrical Dimethylarginine).

It is not necessary to request the full list of about 40 physiological amino acids for developing a custom amino acid supplement because the EAAs are used in the supplemental formula. Given a rich supply of the EAAs and the critical cofactors alpha-ketoglutarate (α-KG) and pyridoxal-5-phosphate (P-5-P), the liver can easily maintain demands for non-essential counterparts. Smaller panels of 10-20 analytes that include the essential amino acids will suffice. Test results and a supplemental free-form amino acid formula based on the results are shown in Tables 4-7 and 4-8. The formula is designed to yield 300 grams of the custom-blended powder. The mixture will last for 30 days at 10 gm/day (2 tsp) dosage. A base mixture is used as the starting point ; extra amounts of individual amino acids are added as indicated by low levels in plasma. In Table 4-8, the "% of Base" column shows the fixed composition of the base. Additional amino acids are added to 253 grams of the base in amounts shown in the "Grams Added" column. A more detailed description of the procedure for free-form amino acid supplementation is given at the end of this chapter.

TABLE 4-7. Example Amino Acid Pattern

Essential Amino Acids	Results		Reference Limits
Arginine	74		50–160
Histidine	72		70–140
Isoleucine	42	L	50–160
Leucine	73	L	90–200
Lysine	204		150–300
Methionine	26		25–50
Phenylalanine	42	L	45–140
Threonine	142		100–250
Tryptophan	40		35–60
Valine	141	L	170–420
Taurine	41	L	50–250

Plasma amino acid profiles measuring the essential amino acids are preferred to provide data for determining oral formulations. Plasma amino acid patterns were used in this way to bring about improvement in patients with chronic fatigue syndrome [16]. The underlying assumption is that the functions of tissues critically dependent on amino acid-supply will benefit from optimal concentrations and balance of the essential amino acids.

Larger profiles of 35-40 amino acids include compounds that reveal tissue-specific abnormalities, especially those related to muscle, bone, and central nervous system. Abnormalities in amino acids specific for muscle turnover and neurotransmitter activities give valuable information about where the emphasis should be placed in nutritional interventions. Since specific vitamins and minerals are required in amino acid metabolic pathways, indications of need for these classes of nutrients are available from specific patterns of abnormalities (Table 4-1). Explanations of the relationships are provided in the discussions of individual amino acids that follow.

TABLE 4-8. Amino Acid Supplement Formula

Base Formula Amount: 253 g	% of Base	Grams Added
Valine	11.10	+ 11 g
Leucine	12.90	+ 14 g
Isoleucine	9.40	+ 9 g
Phenylalanine	12.90	+ 5 g
Trytophan	0.00	+ 0 g
Methionine	7.70	+ 0 g
Threonine	8.10	+ 0 g
Lysine	9.40	+ 0 g
Histidine	10.10	+ 0 g
Arginine	9.40	+ 0 g
Taurine	0.00	+ 8 g
Pyridoxal-5-Phosphate	.30	+ 0 g
Alpha-Ketoglutaric Acid	8.50	+ 0 g

The plasma concentrations of some amino acids have been shown to be age-specific in healthy children [107]. Rapid decreases during the first year of life followed by very gradually increasing concentrations that stabilize at adult levels at 14 –16 years old are found for alanine, arginine, asparagine, methionine, ornithine, phenylalanine, proline, threonine, and tyrosine. A second set of nine amino acids (cystine, glutamine, glycine, histidine, isoleucine, leucine, lysine, tryptophan, and valine) display no drop during the first year but rise steadily from birth to age 16. The overall percentage changes found for most of these amino acids relative to adult levels clustered near 15%. Lysine and tryptophan were exceptions, changing by approximately 30%. Those amino acids not cited in these two groups were found to be stable throughout childhood and adolescence.

Low levels of EAAs in plasma may represent either a poor quality protein diet or inadequate digestion and assimilation of protein. Increased uptake for hepatic protein synthesis or reduced skeletal muscle activity will also lead to low plasma AA levels. Low levels of aromatic amino acids -- tryptophan, phenylalanine, and tyrosine -- may indicate inadequate stomach acid. Hydrochloric acid from the stomach is necessary to activate pepsin-mediated protein digestion. When patterns of multiple low essential amino acids are found in plasma or urine, the individual will benefit from supplementing the diet with a high quality, free-form amino acid formula as described above.

In severe malabsorption states, plasma amino acids are low even when post-prandial testing is performed [108]. Low post-prandial EAAs levels confirm inadequate absorption. Hypochlorhydria significantly inhibits proper digestive activity. Low levels of EAAs may also indicate inadequate pancreatic enzyme activity. Because zinc is required as a cofactor in several digestive enzymes, a deficiency of this element can affect plasma amino acids [109, 110].

Specific vitamins and minerals are required for amino acid metabolism. Abnormal results from amino acid profiling may be due to deficiencies of the nutrients required as cofactors for conversions into other compounds (Table 4-9). These relationships will be discussed under each amino acid below.

TABLE 4-9. Vitamin and Mineral Indications From Amino Acid Reports

Amino Acid or Ratio	Specimen	H/L	Nutrient
His	Plasma	L	Folate
Lys	Plasma	L	Carnitine
Pro/HPro	Urine	H	Vitamin C
Glu/Gln	Plasma	H	B_6 (P_5P)
Sarc/Gly	Plasma	H	B_{12}, Folate
Phe/Tyr	Plasma	H	Iron

The discussions of individual amino acids that follow provide insight into the current information that can aid the clinical interpretation of special cases of abnormalities. It is valuable to keep in mind, however, the many functions that groups of amino acids have in common. Most can serve as carbon source for glucose synthesis and several serve demanding roles in synthesis of critical molecules like DNA, RNA, and glutathione (Table 4-10).

The next section of this chapter deals individually with the amino acids reported on typical laboratory profiles. The groupings used to order our discussion are primarily based on shared functions and relationships. The organization scheme used in Table 4-1 is followed throughout the chapter, but some amino acids have multiple roles and could have been placed in different categories. Several EAAs, for example, are in the first category, "Amino Acids for Polypeptides and Energy," but the primary purpose is to describe functions unique to that group whether the amino acids are essential or not. The grouping of lysine with alpha-ammoadipic acid (αAAA) helps clarify how amino acids function together more than discussing all EAAs together.

TABLE 4-10. Special Metabolic Functions of Amino Acids

The amino acids listed are all required for human tissue protein synthesis, and they are the principal amino acids supplied by dietary protein. The carbon skeletons of most amino acids are able to enter into the biosynthesis of glucose. Those with multiple check marks under "Glucose Synthesis" are most easily transformed into glucogenic compounds, and they play important roles in the maintenance of blood glucose during intervals of fasting. A smaller number of amino acids are involved in each of the other special functions. Multiple involvements are found for aspartic and glutamic acids and glycine. The multiple check marks for glutamine, cysteine, aspartic acid, and asparagine show where supply of these amino acids tend to be a dynamically controlling factor in the biochemical pathways.

Function →	Glucose Synthesis	Urea Cycle	Glutathione	Neuro-transmitters	Creatine	DNA & RNA	Derived Amino Acids
Amino Acid							
Alanine	✓✓						
Arginine	✓	✓			✓		
Aspartic acid	✓	✓✓					
Asparagine	✓					✓✓	
Cysteine	✓		✓✓				
Glutamic acid	✓	✓	✓	✓			Asp, Arg, Pro
Glutamine	✓	✓✓				✓	
Glycine	✓✓		✓		✓	✓	
Histidine	✓			✓			
Methionine	✓		✓				Ser, Cys
Phenylalanine	✓			✓			Try
Serine	✓✓			✓			Gly
Threonine	✓						
Tryptophan	✓			✓			
Tyrosine	✓			✓			
Valine	✓						

Other EAAs occur in other categories where they share functions other than dietary indispensability. In addition to those included in the category "Neurotransmitters and Precursors," glycine, aspartic acid, and taurine are CNS-active compounds. These amino acids have other unique metabolic functions (*e.g.* detoxification) that are highly pertinent in the clinical application of amino acid profiling, so they are placed with other amino acids related to those functions.

Interpretation of Amino Acid Abnormalities

Amino Acids for Polypeptides and Energy

Leucine (Leu), Isoleucine (Ile), and Valine (Val)

The Branched Chain Amino Acids (BCAAs) - isoleucine, leucine, and valine - are essential amino acids that must be obtained from food because they are not synthesized in man. They are used specifically for peptide and protein synthesis and do not serve as precursors to neurotransmitters or bile acids like several other amino acids. Leucine, isoleucine, and valine are ubiquitous among the great variety of proteins that serve as enzymes, transports, and structural components of cells. The BCAAs make up 35% of muscle proteins and comprise almost 50% of the dietary protein amino acids. During strenuous muscle activity, they serve as a source of energy regulation, giving rise to alanine and glutamine for conversion by the liver and kidney to glucose. The carbon skeletons of Ile and Val are catabolized to succinate, a key intermediate in the citric acid cycle. In the brain, the BCAAs appear to play a role in the control of amino acid-derived neurotransmitter metabolism by competing with other large neutral amino acids (*e.g.* tryptophan) for uptake mechanisms. Because of their nitrogen-sparing effects, the BCAAs have been used clinically for recovery from surgery, trauma, and sepsis, and in diabetic regulation. The stress-induced decrease of muscle protein synthesis can be prevented by supplementation with a balanced BCAA formula [17].

Vitamin B6 (pyridoxine) is critical for amino acid metabolism [111]. The cofactor form, pyridoxal-5-phosphate, is required for all transamination reactions. BCAA plasma elevations may result from inadequate B6 because initial metabolism of the BCAAs utilizes a pyridoxal phosphate-dependent enzyme. BCAA utilization is impaired when the cofactor is lacking and BCAA excretion in urine is elevated. Further metabolism of this group of amino acids requires zinc and coenzyme A, a pantothenic acid (B5) derivative. In addition, the enzyme that oxidizes the keto forms of the BCAAs requires thiamin, riboflavin, niacin, and lipoic acid. This enzyme is very active in skeletal muscle, which is the primary site for catabolism of BCAAs [10]. For people with normal dietary protein intake, high urinary or plasma levels of BCAAs indicate deficiencies of these vitamins. Pantothenic acid supplementation has also proven useful in these situations since it conserves the sulfur-containing amino acids. Chronic depletion of BCAAs will eventually show up as low urine and plasma levels with loss of muscle mass and protein synthesis capabilities sometimes experienced as alopecia.

An active insulin response is required for clearing amino acids from plasma. BCAA levels are increased in non-insulin-dependent diabetes mellitus (NIDDM), and the sustained elevated levels may be a causative factor in diabetic microangiopathy [112]. Skeletal muscle is the major destination for plasma BCAAs. Their incorporation into muscle protein is dependent on insulin-like growth factor (IGF-1). Both insulin and IGF-1 are responsible for anabolic effects in skeletal muscle, but for different reasons. IGF-1 stimulates protein synthesis, while insulin inhibits protein breakdown by stimulating amino acid uptake (Figure 4-4). The net effect is that both falling IGF-1 levels and insensitivity to insulin lead to higher plasma BCAA. Conversely, raising BCAA levels in plasma increases the rate of lean body mass increase in response to exercise. Muscle protein synthesis is increased when plasma amino acid levels are raised [113].

Excessive dietary intake can affect plasma levels of these amino acids and will generally cause elevation in urinary amino acid levels. Such excessive intake is common among athletes, especially body builders who may consume large amounts of protein powders in

FIGURE 4-4. Hormonal Effects on Muscle Protein and Plasma Amino Acids

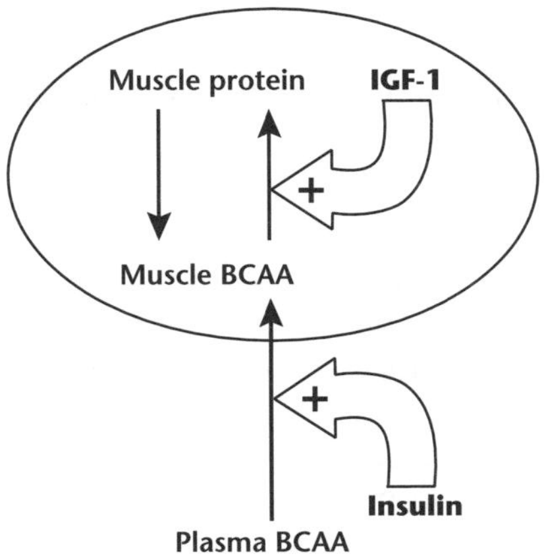

addition to maintaining a high protein diet. Concentrations in plasma will depend on the level of protein intake and the supply of vitamin B6, a vitamin needed in greater amounts to meet the metabolic demands imposed by high amino acid intake. The same can be said for the other amino acids, but the BCAAs are the most abundant in supplemental proteins and most likely to be elevated. An enriched infusion of BCAA increases respiration, decreases sleep apnea in normal adults [18], and stimulates food intake to help the return to normal eating in patients on total parenteral nutrition. The flow of BCAAs into blood suppresses gastric emptying after a meal containing protein. This response is a physiologic adaptation to allow longer residence of protein in the acid milieu of the stomach as required for proper digestion [19].

Leucine (Leu) is among the most abundant amino acids in animal tissues. It has been proposed as an *in vivo* regulator of protein metabolism because it surpasses all other amino acids in inhibiting muscle protein degradation. Because of reduced rates of release from skeletal muscle, plasma essential amino acids tend to fall when leucine is abundant [20]. The finding of such a regulatory role for a single amino acid emphasizes the benefit of balanced intake of all essential amino acids. Heavy use of leucine alone can have undesired effects that would not result if other amino acids were supplied to maintain plasma levels.

Histidine (His)

Elevated plasma or urinary histidine (His) can result from muscle protein breakdown because skeletal muscle is relatively rich in histidine. High 3-methylhistidine (discussed below) may confirm this situation.

Low plasma histidine levels have been associated with rheumatoid arthritis. Supplementation with histidine (around 4000 mg. per day) raises levels in blood and reduces arthritic symptoms in arthritic patients [21]. Folic acid deficiency leads to increased catabolism of histidine [22] [23], and the appearance of degradation products of histidine in urine has been proposed as a functional marker for folate status [24]. Also, consider low dietary intake or malabsorption syndromes, particularly if low levels of other essential amino acids exist. Salicylates and steroids also lower histidine levels.

Lysine (Lys)

Because of its long side chain and positively charged amino group, lysine (Lys) is found on the surface of proteins where it contributes to the total charge of the macromolecule. Indications for therapeutic use of lysine include herpes simplex infections, cardiovascular disease, and osteoporosis [25]. The effect on herpes simplex, the most extensively studied application, is apparently ascribed to competitive inhibition with lysine residues on the surface of the viral particles [26]. The effect makes lysine a useful prophylactic agent for cases of recurrent herpes simplex labialis if serum lysine levels can be maintained at adequate concentrations [27]. Care should be exercised in the long-term oral dosing with lysine, since excessive use of lysine may lead to kidney pathology [28].

The clinical use of lysine clearly calls for monitoring of plasma concentrations of the amino acid. Plasma levels above 165 µmol/L are required to obtain the herpes-inhibiting effect. This level is in the upper-normal range. These findings cause one to focus on the meaning of "normal." Individuals may range from those with low plasma lysine, indicating that availability for metabolic function is impaired, to those in a state of immune incompetence against viral growth where lysine is needed at levels of intake that support upper "normal" plasma levels. Many other individuals fall in the lower "normal" range (near or just below reference limit), where critical biochemical function may be insufficient to meet metabolic challenges even though plasma levels are "normal." Prolonged supplementation at levels over 3 gm per day can result in elevated plasma and urinary lysine concentrations. Such levels generally signal excessive intake that will only add to the total body nitrogen load, challenging ammonia clearance capacity.

Collagen, the connective tissue protein, is stabilized and strengthened because it contains a large percentage of lysine that is used to form the unique cross-linkages. The lysine of collagen is frequently modified to hydroxylysine, which provides a marker for bone loss due to the turnover of collagen during bone resorption (see below). Lysine is also the precursor in the synthesis of carnitine [29]. Serum triglycerides are high in individuals on a lysine-restricted diet [30] because carnitine is essential in the transport of fatty acids. A low lysine level, therefore, is an indication of need for both carnitine and lysine supplementation, particularly if there are signs of muscle weakness, easy fatigability, or high serum triglycerides [31]. Low lysine may suggest poor dietary intake of lysine or excessive arginine intake.

High levels of lysine in fasting plasma or urine are rare, and, if the cause is not identified as lysine supplementation, may indicate impaired metabolism lysine. Proper metabolism of lysine utilizes vitamins B3 and B6. Elevated lysine in urine may also be a result of urea cycle abnormalities resulting in elevated plasma arginine (see Urea Cycle below). The high arginine seems to cause excessive lysine spilling into urine by competition for a shared transport protein active in amino acid resorption in renal tubules.

Alpha-Aminoadipic Acid (αAAA)

Alpha-aminoadipic acid (αAAA) is an intermediate in the normal catabolic breakdown of lysine. αAAA conversion to α-ketoadipic acid involves a transamination requiring B6 and α-ketoglutaric acid. Elevations suggest reduced tissue capability of amine group transfer and inhibited lysine catabolism or excessive lysine intake. Elevated αAAA is a risk factor for heart disease. The high positive correlation between homocysteine and αAAA level explains the reason for accumulation of these 2 compounds in the etiology of atherosclerosis and myocardial infarction [32]. Unlike homocysteine, αAAA is not vitamin B12 dependent, and thus its elevation indicates a specific need for vitamin B6.

After 3 months of vitamin B6 supplementation, the increase of αAAA in blood and urine during an oral lysine loading test was reduced in a mentally retarded girl with a number of dysmorphic features, Raynaud's phenomenon, hypotonia, and petit mal seizures [33]. Another indication of the importance of normal lysine catabolism is the marked pipecolic aciduria in the cerebro-hepato-renal syndrome of Zellweger [34]. The formation of pipecolic acid from lysine (Figure 4-5) is specific to brain tissue. Since pipecolic acid is eventually converted to αAAA by hepatic peroxisomes [35], elevated αAAA can be found in this syndrome.

Threonine (Thr)

Plasma threonine is uniquely resistant to the essential amino acid-lowering effect of dietary protein deficiency [12]. The finding that non-essential amino acids rise in plasma in response to dietary restriction reveals a stimulation of the flow of amino acids from non-critical tissue, mainly skeletal muscle. Threonine is spared from such mobilization. This relationship of threonine to other EAAs allows discriminating the origins of abnormal amino acid patterns in plasma. A pattern of low EAA levels with the exception of threonine indicates inadequate assimilation of dietary protein. This may be due to low protein intake or poor digestive factors. The opposite pattern of low threonine with normal levels of other EAAs is due to factors other than dietary inadequacies. Such factors are generally catabolic, since they are likely to involve increased cortisol and decreased anabolic factors such as insulin-like growth factor and testosterone.

Oral doses of 7.5 g/day will raise plasma threonine levels to upper range of normal [114]. Although threonine may be used to supply

Figure 4-5. Relationships of Lysine, αAAA, and Pipecolic Acid

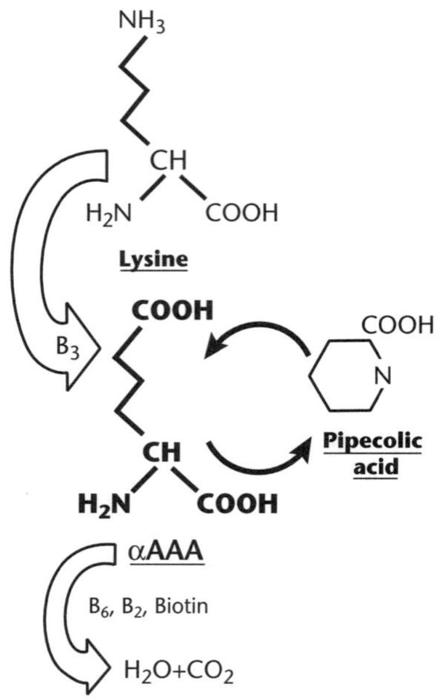

glycine, this amino acid does not rise in plasma with threonine supplementation. Metabolic pools of glycine and serine tend to be drained for numerous biosynthetic pathways. Since threonine is a potential source of glycine and serine, a pattern of concurrent low levels of these three amino acids in plasma is a sign of impaired detoxification, tissue repair, and gluconeogenesis. Chronic low levels can lead to hypoglycemic symptoms of fatigue, headache, anxiety, and shakiness between meals. In cases where threonine is found low, check for potential low or low-normal levels of glycine or serine to confirm the presence of this problem. Supplementation of threonine, as well as other BCAAs between meals, will obviate this condition. In multiple sclerosis, supplementation of threonine is a potential non-sedating, nontoxic approach to the management of spasticity. The effect is due to postsynaptic inhibition of the motor reflex arc in the spinal cord [114].

Alpha-Amino-N-Butyric Acid (AANB)

Alpha-Amino-N-Butyric Acid (AANB) is a metabolite of threonine requiring vitamin B6 and α-ketoglutaric acid for complete oxidation. Low levels, therefore, may reflect increased need for these enzyme cofactors, as well as the essential amino acid threonine. AANB is also one of the amino acids found decreased in cerebrospinal fluid in patients with Parkinson's disease [36]. High levels of AANB have been associated with chronic alcohol intake, and the ratio of AANB to leucine has been proposed as a marker for alcoholism [35]. By calculating the ratio to leucine, variations caused by high protein intake are normalized.

Neurotransmitters and Precursors

The amino acids described in this section are precursors to the "Biogenic Amines," which have wide-ranging functions throughout the body. The aromatic amino acids – phenylalanine, tyrosine, and tryptophan – are converted to catecholamines and serotonin by enzymes in adrenal, intestinal, and nervous tissue. Gamma-amino butyric acid (GABA) and glutamic acid exert CNS-active neurotransmission effects without any modification of their chemical structures. Plasma levels of these amino acids are known to influence CNS concentrations of the respective neurotransmitters. Schizophrenia treatments (and etiologic mechanisms) have been linked to the glutamatergic and dopaminergic excitatory amino acid systems [39]. Alterations in plasma levels of aspartate, glutamate, glycine, and taurine have been suggested as neurochemical markers of epilepsy [40].

Phenylalanine (Phe)

Phenylalanine is an EAA and is the precursor to tyrosine, dopamine, and the catecholamines epinephrine and norepinephrine. Low levels may reflect a potential chronic deficiency state leading to deficits in these important biogenic amines. If tyrosine is low, then conversion of phenylalanine to the catecholamines and/or thyroid hormone may be compromised. Possible symptoms of this would be hypothyroidism, chronic fatigue, learning, memory or behavioral disorders, depression, and autonomic dysfunction [6]. High levels of stress lead to depletion of phenylalanine [52].

The inherited metabolic disorder of phenylketonuria results in greatly elevated phenylalanine in plasma and urine. Excessive protein intake or a metabolic block in the conversion of phenylalanine to tyrosine can also elevate phenylalanine (Figure 4-6). Iron is a cofactor for the enzyme required. Supplementation of iron to an iron-deficient patient may specifically enhance the activity of this enzyme and improve the conversion of phenylalanine to tyrosine.

Tyrosine (Tyr)

Tyrosine is not classified as an essential amino acid because it can be produced by hydroxylation of phenylalanine in normal human metabolism. It serves as a precursor for thyroid hormone, the catecholamines, and melanin (Figure 4-6). Assuming the patient has not been supplementing with tyrosine, high levels may indicate a failure of one or more of these pathways.

The conversion of phenylalanine to tyrosine requires the enzyme phenylalanine hydroxylase and the cofactors iron, tetrahydrobiopterin, and NAD. Deficiency of any of these cofactors creates a functional metabolic block at this step resulting in low tyrosine or a high ratio of phenylalanine to tyrosine in plasma [56]. Supplemental doses of the cofactor-related nutrients may be useful in the optimal metabolic conversion of tyrosine. Because it enhances intestinal absorption of iron, vitamin C can help restore tyrosine formation

FIGURE 4-6. Neurotransmitters and Hormones From Tyrosine

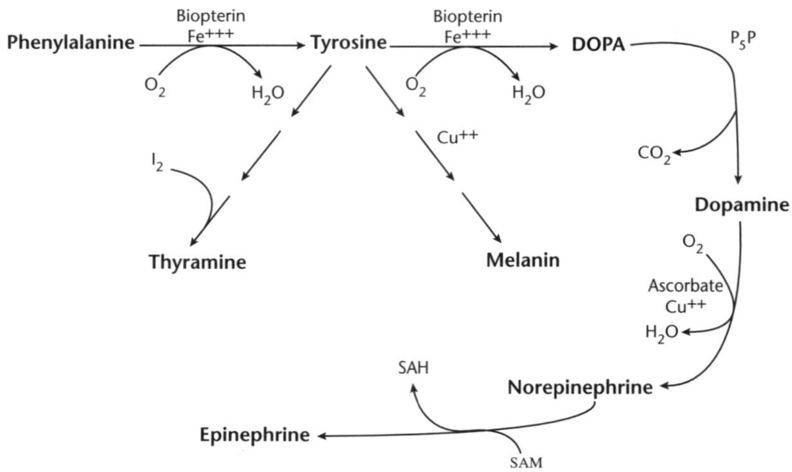

when there is a deficiency of iron. Plasma tyrosine has been proposed as a useful assessment of thyroid function. Low plasma levels of tyrosine have been associated with hypothyroidism [53, 54]. Tyrosine has been used as a treatment for depression and blood pressure modulation [55].

Because the carbon skeletons of both amino acids are catabolized to fumarate, either tyrosine or phenylalanine may be used to maintain adequate levels of citric acid cycle intermediates.

Tryptophan (Trp)

Tryptophan serves as precursor to serotonin and niacin. Approximately 10% of total dietary tryptophan is normally used for serotonin synthesis. Numerous studies have demonstrated that plasma tryptophan is an indirect marker of changes in brain serotonin synthesis [45]. The neurotransmitter is produced in the brain and in the numerous chromaffin cells in the small intestine where it plays a large role in regulating motility.

Tryptophan has been shown to help induce sleep in insomniacs due to increased serotonin production in the brain stem. Plasma tryptophan levels are increased with sleep deprivation because of decreased utilization [46-48]. Tryptophan has been used extensively to treat depression, and it may potentiate the therapeutic action of tricyclic antidepressants. Low plasma levels of tryptophan have been reported in depressed patients [49] and are correlated with the degree of depression [50]. Used alone or with amitryptyline, the amino acid is effective against depression in general practice [51]. Levels in plasma may not rise with supplementation in depressed patients, presumably due to high utilization rates.

High levels may indicate either excessive intake of tryptophan or poor metabolic utilization. Nutrients involved in tryptophan metabolism are vitamin B6 and B3. In the pathway from tryptophan to serotonin, there is only one intermediate – 5-hydroxytryptophan (5-HTP) (Figure 4-7). 5-HTP is available as a nutritional supplement. It is useful because it effectively spares body tryptophan for numerous other functions since 5-HTP is used only in the synthesis of serotonin. The principal catabolic product of serotonin that appears in urine is 5-hydroxyindoleacetate (5-HIA). Low urine levels of 5-HIA indicate lowered production of the neurotransmitter (see Chapter 6 Organic Acids). Oral supplementation of tryptophan or the metabolic equivalent 5-hydroxytryptophan is effective in maintaining adequate tissue supply of the nutrient.

FIGURE 4-7. Metabolism of Tryptophan to Serotonin

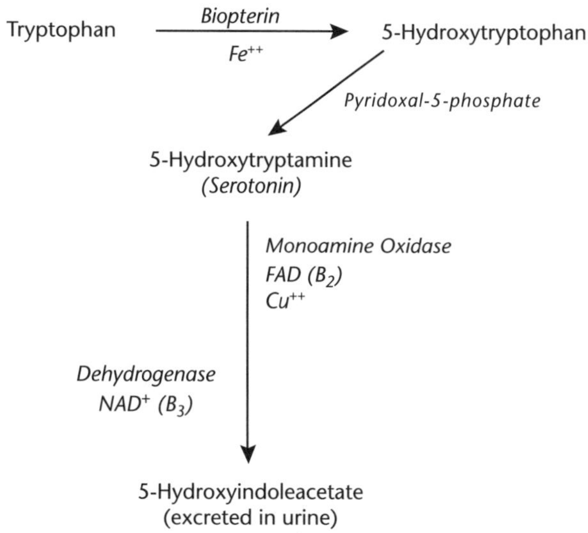

Functions of Glutamate

- *Protein synthesis*
- *Nitrogen transport in muscle and brain*
- *Neurotransmitter*
- *Cell signaling (as polyglutamate)*
- *Bone formation (γ-carboxyGlu)*
- *Glutathione synthesis*
- *Energy source (mucosa)*

Glutamic Acid (Glu)

The importance of glutamate in human physiology is indicated by recent international symposia where this single amino acid has been the theme [115]. From the perspective of nutritional biochemistry, glutamate has played a pivotal role in the design of life processes, especially in human metabolism. Animal proteins contain 11 to 22% glutamate, by weight and some plant proteins contain as much as 40% glutamate. Through the related compounds, α-KG and glutamine, glutamate is involved in the control of a large portion of human metabolism.

Glutamate carries nitrogen from exercising muscle because the various amino acids from degraded muscle proteins transfer their amino groups to α-KG (Figure 4-8). The first step in degradation of excess amino acids in the liver also leads to production of glutamic acid due to the same transaminase enzymes. The nitrogen (or ammonia) from the original amino acids is then transformed into urea in the liver using the enzymes of the urea cycle. The glutamic

FIGURE 4-8. Glutamate in Amino Group Cycling
Glutamate is formed by transfer of amino groups from Asp, Ala, and other amino acids via the enzymes aspartate and alanine transaminase (AST and ALT). α-KG is reformed by deamination under the action of glutamate degydrogenase (GDH).

$$\text{Asp} + \alpha\text{KG} \xrightarrow[\text{P}_5\text{P}]{\text{AST}} \text{Oxaloacetate} + \text{Glu}$$

$$\text{Ala} + \alpha\text{KG} \xrightarrow[\text{P}_5\text{P}]{\text{AST}} \text{Pyruvate} + \text{Glu}$$

$$\text{Glu} \xrightarrow[\text{NAD}]{\text{GDH}} \alpha\text{KG} + \text{NH}_4$$

acid becomes α-KG, which can be used as an energy source by entry into the Citric Acid Cycle.

Glutamate is the principal excitatory neurotransmitter in the brain. Animal studies suggest that the excitatory function of glutamate plays a key role in controlling gastric function, with high glutamate causing a depression of gastric motility [42]. Stimulation of endogenous synthesis of glutamate may contribute to chronic neurodegeneration in such disorders as amyotrophic lateral sclerosis and Huntington's chorea [116]. The latter effect is unrelated to dietary intake of glutamate. Low plasma levels, on the other hand, can impact a large class of neurons in the central nervous system (CNS) that utilize glutamic acid as an excitatory neurotransmitter.

Plasma glutamate levels decrease rather than increase with increased dietary protein intake. This phenomenon apparently is due to the actions controlling fluxes of glutamate synthesis and utilization, and the fact that little or no dietary glutamate or glutamine reaches the portal circulation [117]. Glutamate is actively oxidized and used in glutathione formation in the intestinal mucosa [118].

Glutamic acid is also important in ammonia detoxification in the brain where it combines with ammonia to form glutamine (Figure 4-11). Glutamic acid is elevated in amyotrophic lateral sclerosis patients due to lowered catalytic rates of glutamate dehydrogenase (GD). Treatment with BCAAs showed a significant benefit due to the stimulation of GD [43]. The immune system is influenced by moderate variations of plasma glutamate, and high glutamate seems to be a causative factor in immunopathology of different diseases, including AIDS [44].

Mild hyperammonemia conditions are often seen as low glutamic acid levels and high glutamine levels. Symptoms include headache, irritability, fatigue, mental confusion, poor concentration, and food intolerance reactions, particularly to high protein foods. A low protein, high complex carbohydrate diet with supplementation of alpha-ketoglutaric acid, B_6, and branched chain amino acids will be useful in correcting this condition, particularly if glutamine is high.

Gamma-aminobutyric Acid (GABA)

GABA is an important inhibitory neurotransmitter in the CNS. The concentration in plasma reflects CNS levels, and low levels in plasma is characteristic of one subset of patients with depression [41]. The neurodegenerative condition, Huntington's disease, manifests as lowered levels of GABA as neuron loss proceeds. The synthesis of GABA in muscle tissues from glutamic acid and ornithine is the most likely source of plasma GABA. Higher plasma levels probably reflect low conversion to succinic acid for utilization by the Citric Acid Cycle and energy generation. Therefore, high plasma levels indicate potential inadequate energy generation in muscle and possibly other tissues. α-KG and B6 are cofactors in this GABA metabolic pathway and can be used to alleviate this metabolic impairment and optimize cellular energy generation.

Glycine, Serine, and Related Amino Acids

Glycine (Gly)

The simplest amino acid, glycine, is also one of the most versatile (Figure 4-9). It is not classified as essential because of the ease of formation from other amino acids, but clinical deficiency effects have been long known. Thus, glycine is classified as conditionally essential. A consideration of the magnitude of the demands placed on glycine supply leads to understanding how patients can become glycine-depleted.

Glycine occurs about every third amino acid in collagen, the most abundant protein in the body. It can be used directly as an energy source or for synthesis of glucose to supply blood sugar. It is required for porphyrin synthesis to initiate the pathway leading to heme formation, and every cell function involving formation of RNA or DNA is glycine-dependent because of its role in purine biosynthesis. To these demanding roles, we must add the synthesis of glutathione, bile acids, and detoxification Phase II conjugation reactions.

Because of the demand for detoxification, accumulating evidence indicates that glycine is a conditionally essential amino acid in man, especially during episodes of exposure to several classes of chemicals. For example, glycine is used for direct conjugation to benzoic acid. Benzoic acid is a food component, a food additive, and a product of intestinal bacterial action on unassimilated phenylalanine (see Chapter 6 – Organic Acids). Because glycine is also one of the precursor amino acids for glutathione, individuals experimentally stressed with benzoate show inhibition of glutathione synthesis [57]. These reactions are part of Phase II hepatic detoxification. Plasma glycine is lower in patients with aspirin overdose than in healthy individuals, suggesting depletion of available glycine, which is used for formation of the excretory product, salicyluric acid [119].

Because of its role in gluconeogenesis, low levels of glycine may contribute to hypoglycemic-like symptoms in individuals with impaired insulin responses. The importance of glycine in brain

FIGURE 4-9. Metabolism of Glycine and Serine

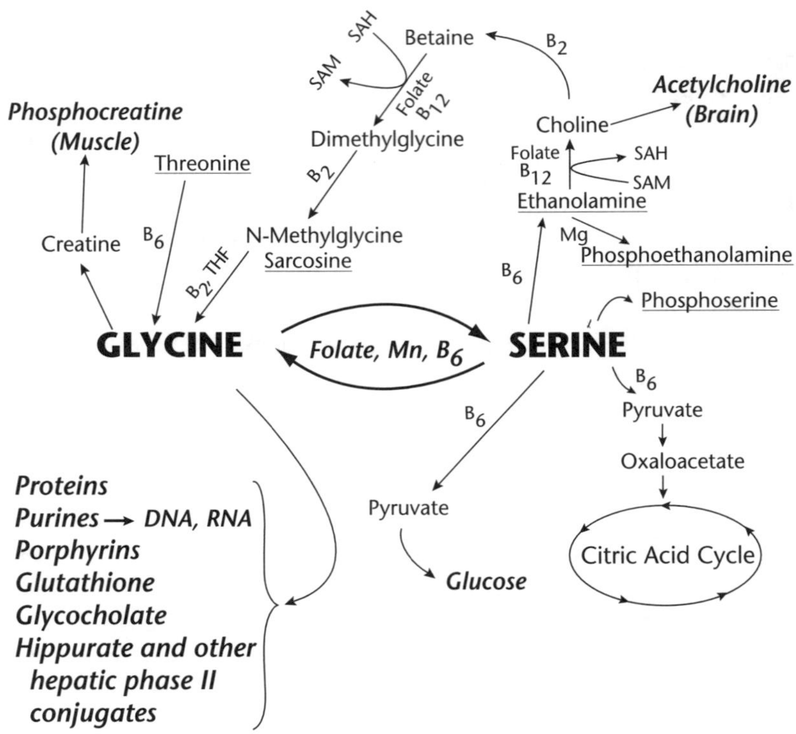

function is illustrated by the finding that non-ketotic hyperglycinemia is associated with early myoclonic encephalopathy which may respond to vitamin B6 [58]. This instance emphasizes the involvement of glycine in the supply of serine for the membrane components phosphoethanolamine and phosphoserine.

Serine (Ser)

Low serine is associated with disordered methionine metabolism and methionine intolerance. Individuals with elevated levels of the heart disease risk factor homocysteine also have low levels of plasma serine, and both factors show normalization with vitamin B6, B12, and folate treatments [59]. Serine combines with homocysteine via a B6-dependent step to produce cystathionine, which is then metabolized to cysteine and taurine (Figure 4-12). During periods of elevated total plasma homocysteine increased amounts of serine may be diverted to lowering plasma homocysteine. Serine is also a critical component in the biosynthesis of acetylcholine, an important CNS neurotransmitter used in memory function and mediator of parasympathetic activity. Patients suffering from episodic acute psychosis display a disturbance of serine-glycine metabolism [5], and a higher serine/glycine ratio is observed in depressed individuals [60].

Plasma serine levels are lower than normal in homocysteinuria patients on folate therapy, compared with healthy adults [59]. Renal transplant recipients with mild elevations of serum creatinine also have lower than normal serine levels. Treatment of renal transplant patients with pyridoxine, folic acid, and vitamin B12 (cofactors required for homocysteine metabolism) caused declines in plasma homocysteine levels with a concurrent rise in plasma serine levels. Treatment of both pyridoxine responsive and non-responsive homocysteinemic patients with betaine lowers plasma homocysteine and normalizes plasma serine levels.

Plasma serine and glycine concentrations are under significant genetic control via a single major gene locus. The serine and glycine-metabolizing enzyme, serine hydroxymethyltransferase, is suggested as the most likely candidate for this single major gene locus [61]. Erythropoietin therapy leads to a normalization of amino acid metabolism with reduction of elevated levels of serine, glycine, and proline in hemodialysis patients [62].

Ethanolamine

Ethanolamine is a precursor in acetylcholine synthesis. High levels may reflect reduced synthesis of this neurotransmitter. Ethanolamine is converted to phosphoethanolamine via a magnesium-dependent reaction. High levels of ethanolamine combined with low levels of phosphoethanolamine may reflect functional magnesium deficiency. Ethanolamine exerts significant effects on DNA synthesis independent of phosphoethanolamine utilization in membrane lipid formation [120]. At high concentrations, ethanolamine exerts inhibition of insulin-induced DNA synthesis. These roles suggest clinical effects of poor wound healing and inadequate maintenance of tissues that have high turnover rates (*e.g.*, small intestine) in cases where plasma ethanolamine is consistently high.

Phosphoethanolamine (PE)

High phosphoethanolamine (PE) may indicate functional impairment of acetylcholine synthesis. PE is methylated with S-adenosylmethionine in the process of acetylcholine synthesis. Functional B12 or folic acid deficiency, or disordered methionine metabolism (*i.e.,* inhibition of methyl group transfer) may account for high PE levels. Increased plasma and urinary PE is found in individuals with osteopenia associated with "adult" hypophosphatasia, along with subnormal circulating total alkaline phosphatase activity [64]. PE, a metabolite of phospholipid metabolism, is decreased in the brains of Alzheimer's disease patients and is structurally very similar to the inhibitory neurotransmitter GABA. The efficient exclusion of PE from GABA binding sites may be an important physiologic mechanism in the control of inhibitory neurotransmission [65]. Individuals with high plasma PE may experience depression or other effects on brain function from the loss of such control because GABA binding is prevented. In addition, cytosolic phosphoethanolamine and ethanolamine have been shown to provide prognostic factors in breast cancer. PE is elevated in patients with high rates of breast tumor growth [66]. As the disease progresses, the values for phosphoethanolamine plus ethanolamine become higher.

Phosphoethanolamine is a precursor to acetaldehyde [67], which can exert metabolic poisoning effects by irreversibly binding to coenzyme A, especially in an individual with restricted intake of pantothenic acid.

Phosphoserine

Parkinson's disease patients have elevated plasma phosphoserine and the CSF/plasma ratio of many amino acids are significantly lower in patients than those of controls. In contrast, Alzheimer's disease patients had lower CSF levels of phosphoserine than controls [68]. Both of these disorders may have origins in the dysfunction of transport of neutral and basic amino acids across the blood-brain barrier [36].

Sarcosine

Sarcosine is N-methylglycine. Sarcosinemia can result from severe folate deficiency because of the folate requirement for the conversion of sarcosine to glycine. This reaction is the final methyl transfer in the sequence of reactions that forms glycine from betaine or choline (Figure 4-10). Sarcosinemia is also found in individuals with defects in the enzyme sarcosine dehydrogenase that catalyzes the formation of glycine from sarcosine [7]. Sarcosine is one of the amino acids found elevated in plasma for patients with Parkinson's disease [36]. Since several amio acids show CSF/plasma ratios lower than controls, the disorder involves a disruption of amino acid transport.

The reverse reaction in which glycine is methylated to form sarcosine is a highly unique part of methyl group biochemistry. The reaction, catalyzed by the enzyme glycine-N-methyl transferase and requiring S-adenosylmethionine (SAM) conversion to S-adenosylhomocysteine (SAH) has been called the only 'unimportant' methylation reaction because it results in no key metabolite synthesis. It appears at first to be a metabolic "dead end." The advantage to the cell seems to be in providing a temporary methyl group storage site in sarcosine while freeing up the sulfur of SAH to be used in other biosynthetic transfer reactions, such as the formation of the glycosaminoglycan sulfates of synovial fluid.

Supply and Removal of Amino Groups (Urea Cycle)

The detoxification of ammonia is a critical metabolic process that requires the functioning of the Urea Cycle. Subtle impairments of the liver's urea cycle function can lead to sub-clinical or mild hyperammonemic conditions. Symptoms include chronic fatigue, headache, irritability, occasional diarrhea or nausea, lack of concentration, mental confusion, and intolerance of foods, particularly high protein ones. Blood urea nitrogen (BUN) levels may be low.

Ammonia toxicity problems can deplete α-KG and vitamin B_6. Supplementation with these nutrients combined with a low protein, high complex carbohydrate diet and BCAAs can help normalize urea cycle function. Intestinal bacteria also are a major source of ammonia.

Alanine (Ala)

During strenuous exercise, muscle protein is degraded to amino acids that must be recovered to prevent wasteful loss of energy. Alanine transaminase (ALT) enzymes are highly active in muscle. They provide a mechanism for the removal of ammonia by transferring amine groups from a variety of amino acids to pyruvate, forming alanine (Fig. 4-10). Thus, alanine is the major carrier of amino acid nitrogen from muscle to liver where its carbon skeleton is converted to glucose (via glucogenesis). This carrier role of alanine is significant because the metabolism of amino acids in skeletal muscle is a major contributor to overall protein metabolism.

FIGURE 4-10 The Action of Alanine Transaminase

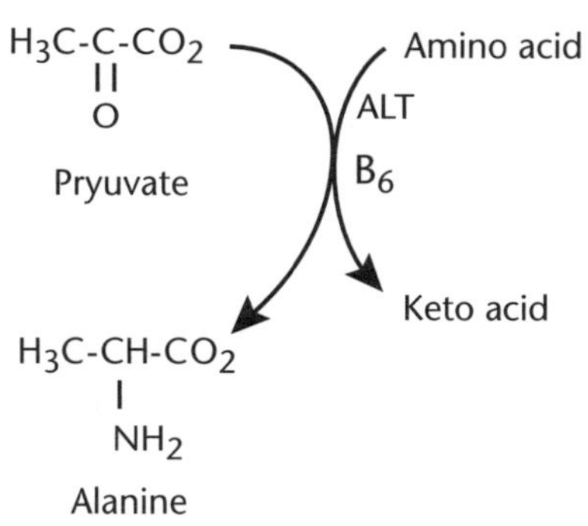

The vitamin B_6-dependent enzyme mentioned above is also known as alanine aminotransferase. It is also one of the most abundant enzymes in hepatic tissue, which has led to its use as a standard marker for liver disease [37]. Hepatic ALT catalyzes the reverse of the muscle reaction so that pyruvate is made available for gluconeogesis and the amine group is passed to glutamic acid, forming glutamine. Glutamine then carries the amine group into the urea cycle for final disposal as urea. When plasma alanine is low, hypoglycemic symptoms may be overcome by additional alanine. Supplementation of alanine and other BCAAs may be indicated if symptoms of easy fatigability between meals are present with a pattern of low amino acids in plasma or urine. A defect in cellular uptake or metabolism of neutral amino acids seems to be a consistent feature of Huntington's disease resulting in low plasma alanine. Genetic polymorphism of the genes that control cell uptake of alanine in non-Huntington's patients may account for variability seen in plasma alanine [38].

Arginine (Arg)

In terms of total quantity passing through metabolic pathways, arginine is most needed for protein and creatinine synthesis. In terms of total physiological impact, arginine status is critical for cell regulation via its precursor requirement for production of nitric oxide and polyamines like spermine [121]. These functions greatly extend (see ADMA below) the most frequently stated function for arginine: a participant in the urea cycle (Figure 4-11).

FIGURE 4-11. Relevant Ammonia Metabolism Pathways

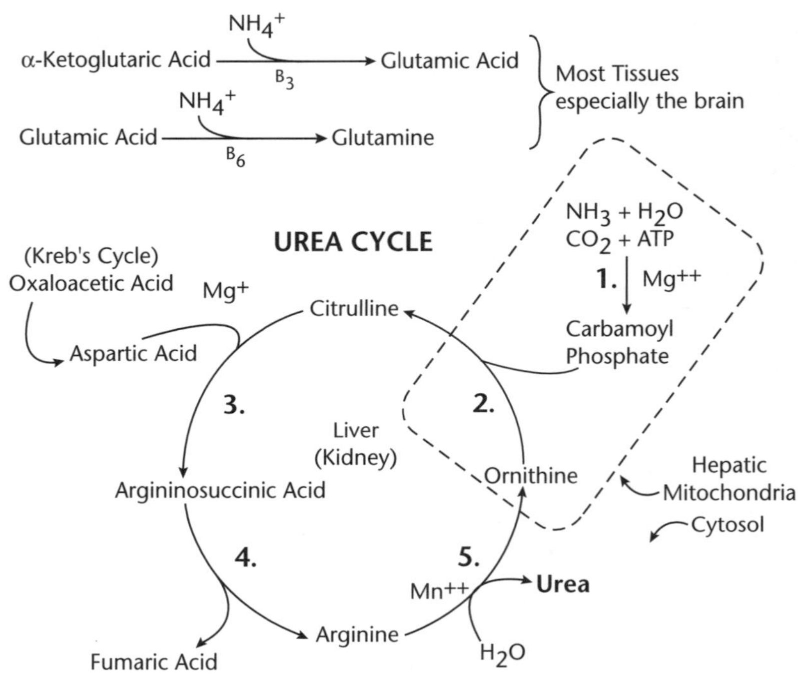

HYPERAMMONEMIA AND:	FAULTY ENZYME:
1. Low Urinary Orotate	Carbamoyl Phosphate Synthetase
2. Oroticaciduria	Ornithine Carbamoyl Transferase
3. Citrullinemia	Argininosuccinate Synthetase
4. Argininosuccinic Aciduria	Argininosuccinate Arginine Lyase
5. Hyperagininemia	Arginase

Even though arginine is created as an intermediate in the urea cycle, the amino acid has been found to be a dietary essential in many metabolic balance studies [122]. Thus, arginine is one of the few compounds that needs to be regularly supplied from dietary sources, even though human tissues contain the enzymes necessary for its synthesis from other carbon and amine precursors. The central question regarding the benefit of dietary intake for a patient is whether their biosynthetic potential can meet the total body demand. Net synthesis is generally negligible for intermediates in dynamic metabolic cycles like the urea cycle.

Arginine released from dietary protein is absorbed in the jejunum and ileum of the small intestine where a specific transport system facilitates the process. This same facilitated transport portal is used for lysine and histidine, which are the other basic amino acids derived from protein digestion. This intestinal transport competition is one reason that heavy supplementation with any one of these amino acids should be done with caution. In addition, intestinal bacteria can decarboxylate arginine to form agmatine, which may increase the virulence of *H. pylori* [123]. Agmatine is also produced in human tissues [124] and may have an important neurotransmitter role [125]. As much as 60% of absorbed arginine is metabolized by enterocytes and never enters portal circulation.

Low levels of arginine in plasma or urine could be reflective of poorly balanced dietary intake because of the American population's preference for high lysine foods, such as meat and dairy products, as opposed to foods high in arginine, such as legumes, whole grains, and nuts [126].

Arginine deficiency quickly leads to hyperammonemia due to failure of ammonia removal. Citric acid and orotic acid appearance in urine are other biochemical markers of this condition (see Chapter 6 – Organic Acids). Arginine also is the direct precursor to the cell regulator nitric oxide. Arginine and nitric oxide are involved in vascular tension. A sequence of events in the endothelial cells results in nitric oxide release. The nitric oxide penetrates into the underlying layer of muscle cells where it elicits release of the final modulator of muscle relaxation, cyclic guanosine monophosphate. This role of arginine and its regulation by the arginine derivative, asymmetrical dimethylarginine, are discussed further in the section on asymmetrical dimethylarginine below. Many of the reported effects of arginine in human health are due to nitric oxide-related cell responses.

There is evidence of a relative deficiency of arginine in diseased coronary arteries. Arginine administration reversed the effect of nitric oxide inhibitors and improved coronary vasomotor tone in angina patients with angiographically narrowed coronary arteries [127]. The treatment resulted in greater dilation of diseased arteries, including stenoses.

Arginine administered at the end of surgery reduces intraperitoneal adhesion formation, a major cause of infertility and pain following surgery [70]. Administration of arginine after trauma-hemorrhage significantly improves depressed macrophage function and decreasing plasma interleukin-6 levels [128].

Arginine is one agent found to increase sperm count and motility, although the response is quite variable. Of fifteen patients with low sperm counts who received arginine hydrochloride over a 3-month spermatogenesis cycle, three pregnancies (20%) were obtained [71]. Enhancement of penile erection is another effect of supplemental arginine, especially when given along with ornithine to spare the utilization for urea cycle function. The effect on penile erection is mediated by nitric oxide relaxation effects on vascular sphincter muscles [129].

High levels of arginine in plasma or urine may represent a functional block in the urea cycle (Figure 4-11). The arginase enzyme requires manganese, and the failure of this metabolic role may give rise to the early signs of manganese deficiency. If ammonia toxicity symptoms (confusion and disorientation) and signs (elevated urinary citrate) are present, a low protein diet with the avoidance of lysine-containing foods is advisable to reduce the ammonia load. Additional B_6 and α-KG may also be useful to remove excess intracellular ammonia.

Aspartic Acid (Asp)

Aspartic acid is pivotal in amino acid metabolism. Aspartic acid aids in the detoxification of ammonia via the urea cycle (Figure 4-11). It is also a precursor of nucleic acids used in DNA and RNA. Aspartic acid can enter the citric acid cycle by conversion to oxaloacetic acid using B_6 and α-KG. Low levels of aspartic acid may reflect decreased cellular energy generating capacity, experienced as chronic fatigue. Aspartic acid administration has been shown to increase stamina and resistance to fatigue [130]. Equimolar amounts of citric and aspartic acids also tend to specifically drive both the citric acid and urea cycles, when combined with α-KG acid and B_6. Simultaneous administration of aspartic acid and arginine leads to increases in neurostimulatory amino acids in the brain [131]. Further study of this phenomenon revealed multiple organ-specific changes [132]. Increases of GABA in brain, ornithine and proline in liver, and serine, glycine, and alanine in all tissues were found when aspartic acid was given along with arginine.

Asparagine (Asn)

The plasma and urinary levels of asparagine (Asn) can vary over wide ranges with little clinical consequence. Asparagine is needed in all tissues for protein synthesis, and it serves a special function for attachment of carbohydrate residues to membrane proteins by which cells are identified. Asparagine has no physiologic role as a free amino acid, and if plasma asparagine is low, the need for protein synthesis is quickly met by transmination to convert oxaloacetate or aspartate into asparagine as needed. Magnesium insufficiency is the nutritional area most likely associated with failure to maintain adequate production of asparagine.

Monitoring plasma asparagine levels is a useful means of determining dosage schedule of L-asparaginase in the treatment of acute lymphoblastic leukemia [69]. This enzymatic treatment cleaves the asparagine attachment site for polysaccharide chains that provide cell identity. The concurrent cleavage of plasma free asparagine to aspartate results in severe depletion of free asparagine levels. Plasma asparagine does not fall in individuals who are resistant to the enzyme.

Glutamine (Gln)

Alpha-ketoglutaric acid and glutamic acid combine with ammonia, to produce glutamine (Figure 4-11). Supplementation of α-KG supplies additional substrate for ammonia removal and generates additional energy as a component of the citric acid cycle.

Glutamine is one of the most concentrated amino acids in plasma and urine because it is the primary route of nitrogen removal from skeletal muscle activity. This is the primary contributor to plasma glutamine, even in post-prandial plasma specimens because a very high percentage of glutamine entering intestinal enterocytes is metabolized there. Glutamine is the primary energy source for the enterocytes [133]. This is the primary reason that supplementation with L-glutamine can improve gut mucosal integrity [134, 135]. Glutamine acts as both precursor and inhibitor of nitric oxide production [136, 137]. Although glutamine supplementation reduces plasma nitrate in experimental animals, no adverse effects have been reported for humans [138].

Low glutamine levels in plasma may reflect chronic insufficient essential amino acid dietary intake. This may lead to ammonia detoxification problems, particularly if there is impairment in urea cycle function. Starvation, injury, sepsis, and other protein catabolic states result in reduced glutamine in muscle and plasma [14, 72]. Among patients with cancer or HIV infection, increased skeletal muscle catabolism, low plasma glutamine, and high venous glutamate levels are common [139]. These shifts in amino acid metabolism are part of the mechanism for halting the loss of body cell mass in such diseases. Cysteine plays a critical role in this process (see below).

Plasma glutamine is lowered in individuals placed on high protein diets, apparently due to decreased release from muscle and lower net synthesis of glutamine [140]. The average level of 720 mM for healthy men on normal protein intake (1 g/kg/d) fell to 444 mM when protein intake was increased by 2.5 fold.

Citrulline (Cit)

Citrulline must combine with aspartic acid under the action of an enzyme activated by magnesium to complete the urea cycle (Step 3 in Figure 4-11). High plasma citrulline may indicate a functional block in the argininosuccinate synthetase enzyme, leading to a buildup of excess ammonia in the system. Supplementation of magnesium and aspartic acid are required to drive this reaction forward.

Plasma citrulline elevation is used as a marker of postoperative intestinal failure [141]. Nitric oxide (NO) synthesis from arginine results in the production of citrulline. Low production rates of NO may be due to inactivation of nitric oxide synthetase, which may be detected by decreased production of citrulline [142]. Plasma citrulline declined by 30% in adult males receiving an arginine free diet [129], and protein restriction in one infant resulted in lowered plasma citrulline [143]. Arginine loading results in increased citrulline values [144].

Ornithine (Orn)

In the final step in which urea is produced in the urea cycle, ornithine is generated (Figure 4-11). When hepatic failure is induced in pigs, both ornithine and citrulline become greatly elevated [145]. High levels may indicate a blockage of the urea cycle at the critical ammonia incorporation step, leading to a higher body burden of ammonia. α-KG and vitamin B6 are useful in clearing chronic minor excesses of ammonia. When ornithine is elevated, the potential for impaired ammonia clearance is enhanced if signs of intestinal bacterial overgrowth are present because of the large ammonia output of some bacteria. A low protein diet will also decrease bacterial production as a source of ammonia.

Cases of low levels of ornithine are rare, but certain associations can be important. Ornithine is a precursor to the polyamines, putrescine, spermidine, and spermine [146] that regulate cellular metabolism. Low ornithine levels may have effects on these regulatory amines. Ornithine can stimulate release of growth hormone from the pituitary [147] and may serve as an important anabolic substance for the body. Growth hormone responsiveness to ornithine stimulation has been used to assess abnormal growth rates in children [148].

Sulfur-Containing Amino Acids

Humans meet their sulfur amino acid needs largely by intake of protein containing the EAAs cysteine and methionine. Methionine is a major methyl donor in the body and is required for the synthesis of acetylcholine, choline, creatine, and epinephrine (Figure 4-12). Cysteine, a product of methionine, is a component of glutathione, coenzyme A, and taurine. All connective tissue proteins, sulfo-mucopolysaccharides, and antibodies require adequate metabolism of the sulfur-containing amino acids.

Methionine (Met)

Methionine is the dietary essential precursor in sulfur amino acid metabolism. Low methionine levels may adversely affect this crucial pathway and may reflect a poor quality protein diet. Methionine is one of the amino acids most frequently found to be deficient because the methionine content of low quality protein sources is very low. Plasma methionine concentrations reflect total body utilization of cystine for synthetic reactions in tissue maintenance. Dietary intake of tannins found in vegetables increases the rate of methionine utilization [75].

Methionine metabolism disorders have a diverse symptomatology. The pathways are heavily dependent on vitamins B6, B12, and folate [78] (Figure 4-12). Subclinical symptoms include allergic chemical sensitivities, headaches, eyestrain, muscle weakness, brittle hair, hair loss, myopia, mild myopathy, osteoporosis, and cardiovascular symptoms. Cirrhotic patients have significantly subnormal values for cysteine, glutathione, and albumin. In addition, they have elevated methionine during a high protein liquid diet and subnormal taurine during either mixed foods or total parenteral nutrition. Such observations indicate multiple abnormalities in sulfur metabolism in cirrhosis [73]. Urinary levels of methionine may provide the most sensitive measure of sulfur amino acid status [74].

FIGURE 4-12. **Relevant Sulfur-Containing Amino Acid Pathways**

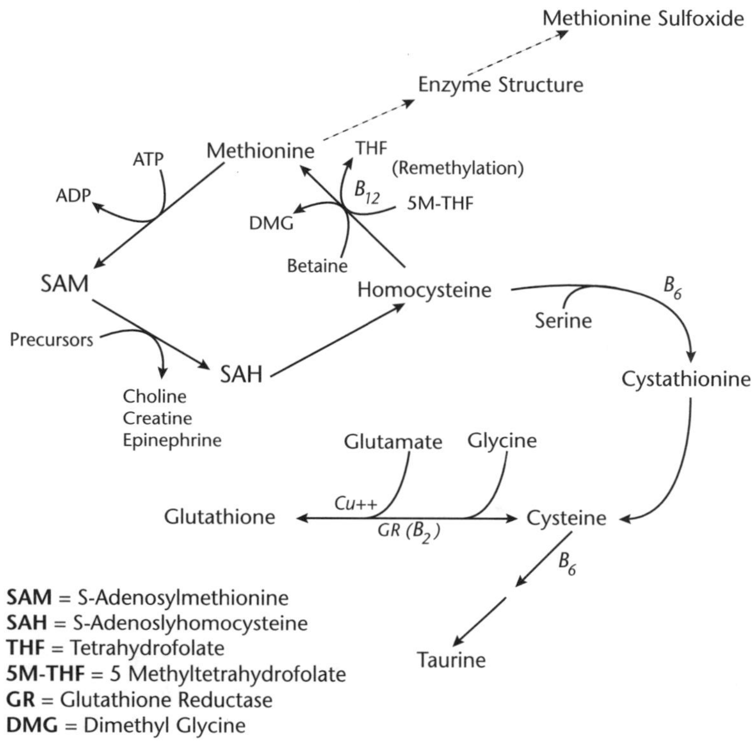

Whenever supplemental methionine is given to a patient who is already replete in the amino acid (normal plasma levels), there is increased production of the heart risk factor homocysteine unless the patient also is replete in the vitamins required to clear homocysteine (folate, B_{12}, and B_6) [78]. This response is the basis of using a methionine load (50 mg per kg body weight) prior to measuring plasma homocysteine to get a reading of the ability to meet the increased vitamin demand. Methionine is elevated in cases of impaired hepatic function such as cirrhosis (65). A reduced excretion of methionine was observed in patients with gastric resection, ileocecal resection, and abnormal gut flora, whereas patients with jejuno-ileal bypass showed an increased excretion of methionine, probably a consequence of the impaired liver function [79].

Methionine may be useful as a supportive therapy in chelation of lead [80] and has a similar effect in cobalt intoxication [81]. The chelating property extends to nutrient elements like copper, which can be depleted by excess methionine.

Cysteine/Cystine (Cys, Cyss, tCys)

The majority of sulfur in food is in the form of protein-bound cysteine/cystine. High-sulfur foods like eggs and legumes, therefore, are high cysteine foods. Cysteine residues are found at the catalytic site of many enzymes because of the unique substrate binding properties offered by the electron-rich sulfur atom. The same electron dense character makes sulfur compounds such as reduced glutathione and DMSA some of the best heavy metal complexing agents (see Chapter 3). The cysteine-rich protein, keratin, which is so abundant in hair, confers extremely high metal binding capacity to hair. Cystine (Cyss) is the oxidized dimer of cysteine (Cys) in which the sulfhydryl group is converted into a disulfide with extraction of two hydrogens.

Under conditions used for specimen preparation for amino acid analysis in most laboratories, cysteine is converted to cystine. In such analyses, the laboratory report may show only cystine. The term tCys indicates the combined presence of the two forms in the original specimen, while cysteine/cystine is generally used to indicate that a response will be elicited by both reduced and oxidized forms. It is important to distinguish between the concentrations of protein-bound and free cysteine. Although virtually all proteins of blood plasma contain cysteine residues, normally there is very little cysteine present in the unbound or free form that is measured in plasma or urine amino acid testing. Both the reduced (sulfhydryl) and oxidized (disulfide) forms occur in proteins. When two cysteine residues in a single polypeptide chain join to form cystine, the structure is greatly stabilized.

Glutathione is a simple tripeptide, γ-glutamyl-cysteinyl-glycine. As much as one half of the total body flux of cysteine may be accounted for by turnover of glutathione [149]. Under oxidizing conditions, two tripeptides join via disulfide formation to create a double molecule (Figures 4-12 and 4-13). The enzyme, glutathione reductase requires copper and FAD from vitamin B_2. The ability to buffer oxidizing and reducing conditions that shift with the flow of NAD+ and NADH allows glutathione to help regulate oxidative cell processes. Hepatic Phase II detoxification also consumes glutathione as it participates in many conjugation reactions (see Chapter 8 – Detoxification). One line of evidence indicating the importance of cysteine in glutathione maintenance is the finding of low plasma tCys in infants receiving an elevated inspired oxygen concentration [150].

Low plasma cystine may reflect a dietary deficiency of methionine and/or cysteine. Patterns of plasma cystine, serine, cystathionine, and taurine can indicate the location of metabolic conversion block. Concurrent low cystine and serine, for example, indicate a restriction of the serine-glycine supply. Low cystine levels may impair cellular synthesis of taurine (see below for low taurine effects), in which case plasma and urinary taurine will be low also. High dietary intake of condensed tannins can lower plasma cystine

FIGURE 4-13. Reduced and Oxidized Glutathione

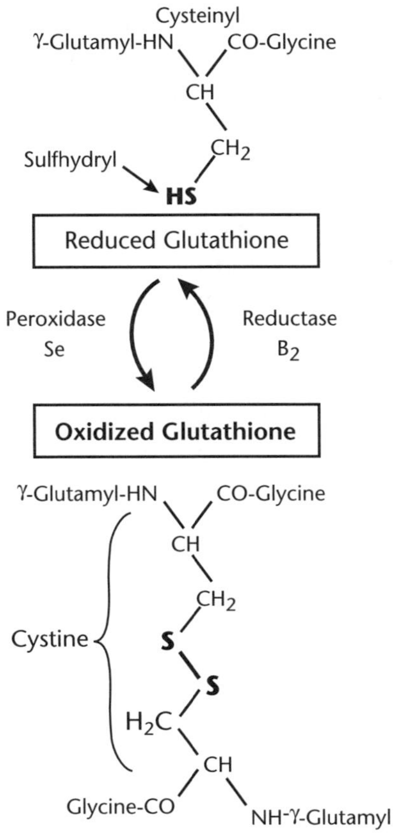

because of increased irreversible loss of cystine and increased utilization of cystine for biosynthesis [75].

Elevated plasma or urinary cystine can reflect either excessive dietary intake of cystine rich foods or impaired metabolism of cystine. Cysteine plays a regulatory role in the physiological control of body cell mass [139]. Plasma tCys is highly correlated with serum creatinine [151]. Plasma tCys rises with age, probably due to declining glomerular filtration rates [152]. This effect will be seen as increasing urinary tCys, and it may explain the elevated tCys found in diabetic patients [153].

Plasma cyst(e)ine both has effects upon and is affected by hepatic biosynthesis. Cysteine has been found to rise in a heterogeneous group of patients receiving intensive care, including those with primary biliary cirrhosis and other forms of liver disease [154]. Hypercholesterolemia can be induced in laboratory animals by feeding excessive cystine [77]. The effect may have little impact on heart disease risk, however, because the increase in cholesterogenesis is accompanied by 65% lower apoprotein fractions, especially lipoprotein(a). In fact, other studies in rats have demonstrated that if, instead of imposing a gross excess of cysteine, the amino acid is added at moderate levels, plasma cholesterol is lowered significantly, and HDL cholesterol is increased [76]. These effects are thought to result from enhanced conversion of cholesterol to bile acids. Dietary supplementation with N-acetylcysteine does not increase the circulating concentrations of tCys [155]. The variation in responses to increased oral cysteine/cystine underscore the benefit of testing for individual amino acid profiles to determine patient status before implementing amino acid therapies.

Plasma cystine undergoes slow degradation even when stored at –20°C. If a specimen takes more than 48 hours to reach the laboratory, the cystine level may be significantly lowered from the level in the freshly drawn blood. This is one reason that it is a good practice to record the specimen collection time on the test request form when ordering laboratory evaluations.

Homocysteine/Homocystine (HCys)

The relation between homocysteine and homocystine is parallel to that described above for cysteine and cystine. Values reported from clinical laboratories are also similarly related so that total homocystine may reflect the sum of both reduced and oxidized forms in the original specimen. Thus, a patient with elevated homocysteine will show a high homocystine on a laboratory report.

The use of HCys as a biochemical marker for vitamin deficiency was discussed in Chapter 2. Age, gender, folate, serum vitamin B12, serum creatinine, and multivitamin usage are determinants of plasma homocysteine concentration. In population studies, total plasma homocysteine concentrations are higher in men than in women, increase markedly with age, and are correlated to concentrations of blood folate, serum vitamin B12, and serum creatine. Users of multivitamins have lower homocysteine levels than nonusers [152]. Moderately elevated levels of homocystine are associated with a significantly increased risk for atherosclerosis. High levels in body fluids of this amino acid have been associated with increased risk of cardiovascular disease as well as ocular, neurological, musculoskelatal, and joint abnormalities. A common cause of this metabolic disorder is impaired function of the enzyme cystathionine β-synthetase for conversion to cystathionine (Figure 4-11). This enzyme is vitamin B6-dependent, and supplementation of the vitamin has been used to treat elevated homocysteine. A second route in metabolism of homocysteine is a re-methylation to methionine using B12, folic acid, and betaine. Supplementation of betaine has proven useful for people with high homocysteine who don't respond to vitamin B6 [83]. Adequate intake of vitamins B6, B12, folic acid, and magnesium is necessary for ensuring proper methionine metabolism to prevent the accumulation of homocysteine.

Elevated homocysteine commonly affects the central nervous system (CNS), primarily as mental retardation, seizures, and stroke

[84]. Elevations indicated by increased plasma or urinary levels are etiologic factors in osteoporosis due to impaired cross-linking of collagen [85]. Smoking has been shown to increase the incidence of homocysteinuria, probably due to lowered levels of the controlling vitamin-derived cofactors.

Cystathionine

The important relations of cystathionine in sulfur amino acid metabolism are discussed above. Cystathionine is the product of the vitamin B6-dependent degradation of homocysteine and high cystathionine levels may indicate a functional vitamin B6 or serine deficiency (Figure 4-12). Elevated cystathionine stimulates superoxide generation in leukocytes [156], placing increased demands on antioxidant capacity.

Taurine (Tau)

Taurine Formula
$H_3N\text{-}CH_2\text{-}CH_2\text{-}SO_3$

Taurine is a two-carbon, beta-amino acid, β-aminoethylsulfonic acid. Since taurine is not found in animal proteins, synthesis by the body is crucial. Taurine is synthesized from cysteine, and the conversion requires vitamin B6 (Figure 4-11).

High plasma taurine is found associated with various stress reactions [86], apparently mediated by release of interleukin [87]. Increased plasma taurine is found in patients suffering from episodic acute psychosis characterized by sensory perceptual distortions. In such patients, oral loading with either serine or glycine can induce psychedelic symptoms [5]. The higher plasma taurine level in serine-responsive patients is caused by an increased synthesis of taurine from homocysteine and serine. Higher plasma taurine values are part of a pattern of amino acids associated with major depression [60]. Lower plasma glycine values and a higher serine/glycine ratio are found in depressed individuals. Depression is accompanied by decreased excitatory amino acids (*e.g.*, glutamate), and increased inhibitory amino acids (*e.g.*, taurine).

A high plasma level can indicate excessive production of taurine due to an inflammatory process mediated by white blood cells. Taurine is found in high concentrations in heart muscle and white blood cells. Taurine is involved in mediation of chemical oxidation by white blood cell phagocytes in response to respiratory burst activity. Low or low normal cystine may indicate rapid conversion to taurine. Inadequate taurine supplies allow the oxidative activity to go unchecked, leading to excess oxidative damage and formation of aldehydes. Individuals with this condition are allergy-prone and often extremely sensitive to environmental chemicals.

Taurine in the form of taurocholic acid is a key component of bile. Low taurine may accompany fat digestion problems [157], fat-soluble vitamin deficiencies, and high serum cholesterol levels [158]. Heart and brain cells require taurine for intracellular retention of calcium, magnesium, and potassium. Taurine has been used successfully in the treatment of congestive heart failure and is implicated in night blindness, arrhythmia, angina, hypercholesterolemia, and atherosclerosis [88].

Taurine deficiencies have been implicated in both neurological (epilepsy) and cardiovascular dysfunction. Beta-agonist drugs cause a reduction in the body pool of taurine [89, 90]. Choline supplementation may stimulate taurine synthesis due to its methyl-donator sparing function on methionine, thus freeing methionine for taurine biosynthesis [91]. Taurine also stabilizes platelets against aggregation. Platelets from taurine-depleted animals are twice as sensitive to aggregation as platelets from those receiving taurine. In addition, human subjects with normal taurine status show increased resistance to platelet aggregation by 30 or 70% when supplemented with taurine at 400 or 1600 mg/d, respectively [92]. Plasma taurine is easily raised by dietary supplementation. Concurrent low cysteine is also relevant in taurine depletion. Cysteine addition to formulas used for home parenteral nutrition normalizes plasma taurine concentrations in children with short gut syndrome [159].

Collagen-related Amino Acids

Collagen, the protein of bone, tendons, ligaments, and basement membranes is the most abundant protein in the human body. Its unique structural properties are due to the recurrence of proline and glycine residues, and its great strength and stability are due to cross-linking afforded by lysine side chains. Cells that are dedicated to collagen synthesis make large demands on proline and glycine supply.

Proline (Pro)

Proline is found in virtually every dietary protein except lactalbumin. Although proline is not classified as an EAA, proline levels in plasma fall by 20-30% when normal individuals are placed on a proline-free diet [93, 94]. Low plasma or urine levels of proline reflect low tissue supply of this amino acid. Proline is a major constituent of collagen and can be metabolized to α-KG, which is important in ammonia detoxification and citric acid cycle function. Low proline levels can indicate a poor quality protein diet and may prevent optimal connective tissue maintenance.

Hyperprolinemia is a recognized genetic disorder characterized by renal and CNS dysfunction [6]. Proline is universally required for protein synthesis and is metabolized to hydroxyproline, an important component in connective tissue. Therefore, high levels may reflect inadequate connective tissue synthesis. Vitamin C is a cofactor in the conversion to hydroxyproline and may be useful if connective tissue symptoms are present. Acute or chronic deficiency of vitamin C produces a significant increase in the proline/hydroxyproline ratio in urine [95]. The relationship between proline and hydroxyproline urinary excretion has been used as an index of collagen catabolism [96]. Proline can also be oxidized to glutamic acid, requiring niacin as a cofactor precursor.

Hydroxyproline (HPro)

Hydroxyproline, a component of collagen, is synthesized from proline using vitamin C and iron as cofactors. High levels in 24-hour urine or plasma correlate with the increased osteocalcin secretion that is characteristic of high bone turnover [97]. This marker may be as sensitive as the measurement of bone collagen-specific peptides [98]. Vitamin C has been used to successfully treat certain types of collagen disorders and to stimulate collagen synthesis [160]. Free plasma hydroxyproline in multiple myeloma is closely related to the extent of skeletal lesions at diagnosis. During the clinical course, plasma hydroxyproline is useful in the follow-up of bone disease [99]. Plasma hydroxyproline also is highly correlated with serum alkaline phosphatase [100], another marker of bone resorption.

Hydroxylysine (HLys)

Hydroxylysine is a derivative of lysine and is used for collagen synthesis in connective tissue. High plasma levels are indicative of connective tissue breakdown. Collagen is synthesized using the cofactors iron, α-KG, and vitamin C. Supplementation of these nutrients, as well as chondroitin sulfate and manganese, may be useful in reversing this condition.

Both hydroxylysine and hydroxyproline are indicators of liver disease; however, hydroxylysine has the strongest correlation, possibly because dietary intake of collagen can influence hydroxyproline levels [101]. Changes in hydroxylysine excretion with cadmium loading in experimental animals indicate that cadmium interferes with collagen synthesis [102].

β-Amino Acids and Their Derivatives

All of the EAAs are alpha amino acids because the amine group is attached to the alpha carbon. Those with amino groups attached to the beta carbon, the β-amino acids, are another source of useful information (Table 4-11). These compounds are not found in proteins. They serve physiological functions ranging from bile acid precursor and antioxidant to neurotransmitter and metabolic control. Some are abundant in certain foods, and others are formed

TABLE 4-11. Beta and Gamma Amino Acids

Taurine (β-Aminoethylsulfonic acid)
β-Alanine
Carnosine (β-Alanylhistidine)
Anserine (β-Alanyl-1-methylhistidine)
γ-Aminobutyrate

in normal human metabolism. The metabolic pathways utilized for the β-amino acids are susceptible to unique genetic influences and vitamin-mineral cofactor demands as outlined below. Taurine is a β-amino acid, but it was discussed under the sulfur amino acids because of the critical relationships with other members of that group.

β-*Alanine*

The pyrimidines cytosine and uracil from DNA and RNA are degraded to β-alanine. β-Alanine can become elevated in plasma or urine due to enzyme deficiency, dietary intake, intestinal microbial overgrowth, or high turnover of muscle tissue (Figure 4-14). Deficient activity of the enzyme, β-alanyl-α-ketoglutarate transaminase, was corrected by oral pyridoxine therapy in one reported case [161]. The patient, a four-year-old girl at diagnosis, was free of the presenting signs of intermittent seizures and lethargy during a two-year period of clinical observation to monitor the effects of the vitamin supplementation. The biochemical pathway involved in this case is the conversion of β-alanine to acetate (Figure 4-14).

FIGURE 4-14. Relevant β-Amino Acid Pathways

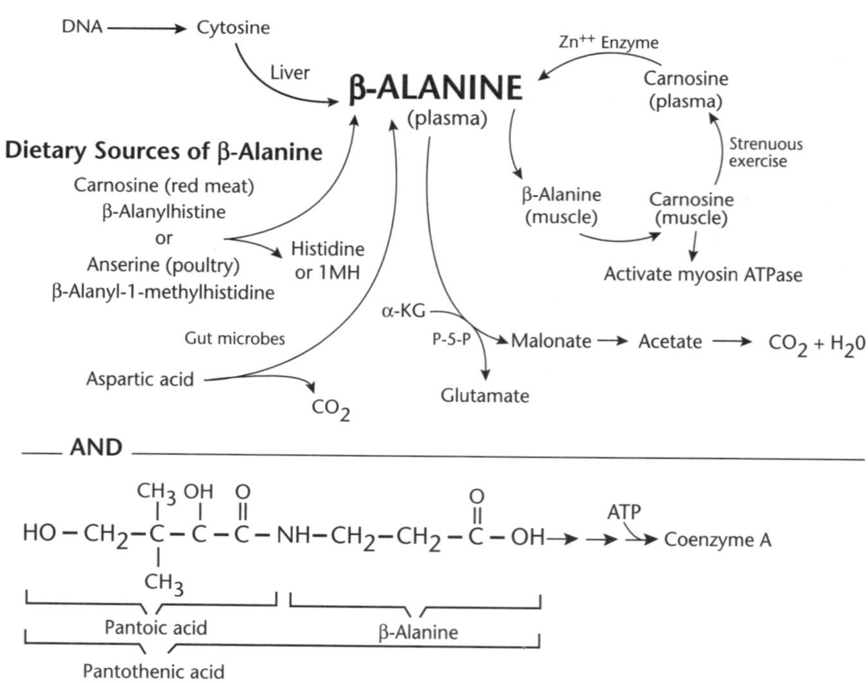

Dysbiosis becomes a larger potential source of β-alanine when aspartate supplements are used since aspartic acid is the microbial precursor. β-Alanine is released from skeletal muscle during strenuous exercise, and it occurs in food mainly as carnosine in red meats or anserine in poultry.

Vitamin B6 deficiency causes lowered levels of the enzymes that degrade β-alanine, resulting in high urinary excretion [162]. High β-alanine is frequently associated with generalized beta-aminoaciduria and concomitant loss of other amino acids (*e.g.*, taurine) due to impairment of renal tubular resorption. Low taurine levels may indicate taurine depletion by this mechanism. This situation, in combination with retarded amine group transfer (due to B6 deficiency) can lead to marked food sensitivity reactions. High levels of β-alanine are frequently accompanied by increases in 1- and 3-methyl-histidine, carnosine, and anserine [103]. Intestinal bacteria and/or *Candida albicans* can also make β-alanine, which can raise plasma levels of β-alanine. A bowel detoxification program may be appropriate with supplementation of a high potency *Lactobacillus acidophilus* and *bifidus* products along with pre-biotics and a high fiber diet to favor growth of the beneficial organisms. Urinary indican and dysbiosis marker organic acids provide confimatory tests for bowel toxicity (see Chapters 6 – Organic Acids and 7 – GI Function).

Excessive taurine may cause muscle weakness due to depletion of carnosine from the competition for β-amino acid entry into cells. Uptake of taurine occurs by a carrier-mediated active transport process specific for β-amino acids [104]. Because of the competition for entry into cells, elevated β-alanine can be a pharmacological effect of taurine administration [105]. The effect was demonstrated with injection of taurine at 0.8 g/kg i.p. This response allows adjustment of taurine supplementation by monitoring β-alanine. Use of taurine should be decreased when β-alanine is elevated.

Excess excretion of taurine may indicate β-aminoaciduria. β-Alanine impairs renal tubular resorption of a variety of amino acids, including taurine, thus contributing to potential taurine and other amino acid deficiencies. If urinary β-alanine is present, a bowel detoxification program to remove a major source (microbial overgrowth) of β-alanine will help to raise the kidney threshold to taurine spill and, therefore, help raise plasma taurine levels.

β-Aminoisobutyric Acid

Thymine, released when RNA and DNA are degraded, enters a catabolic pathway that leads to β-aminoisobutyric acid (β-AIB). High β-AIB in urine is a benign metabolic polymorphism present in some human populations due to deficiency of hepatic β-AIB-pyruvate transaminase. Over 40% of some oriental populations have this condition [163]. Excretion of β-AIB increases in leukemia and severe X-ray radiation exposure due to increased destruction of DNA [164]. Parenteral administration of excessive folic acid also stimulates cell turnover that results in increased β-AIB [165].

Carnosine and Anserine

Carnosine is a dipeptide containing histidine and β-alanine. It is formed in skeletal muscle, where it activates the myosin ATPase enzyme [166]. There is an active turnover of carnosine normally, with exchange between muscle, plasma, and liver. Anserine is structurally like carnosine except that the histidine ring contains a methyl group in the 1-position (see 1-methylhistidine). Both carnosine and anserine are present in the muscle tissue of animals with the proportions varying by species. Their concentrations are lower in humans than in most animals. Anserine predominates in poultry (especially hens), and carnosine is highest in pork. The two peptides are involved in muscle function through the regulation of muscle phosphorylase activity [167].

Both anserine and carnosine increase the respiratory burst and interleukin-1 beta production of human neutrophils. They suppress apoptosis of human neutrophils, showing that carnosine and anserine have the capacity to modulate the immune response in human neutrophils [168]. On the other hand, they potentiate the carcinogenic effects of Nickel(II) by increasing the insertion of oxidized nucleotides into DNA [169]. The pair is also involved in calcium metabolism and cellular control through the activation by anserine and inhibition by carnosine of Ca^{2+} uptake by mammalian mitochondria [170].

Chronic high carnosine in plasma or urine is associated with mental retardation [171]. Inherited traits producing a deficiency of carnosinase, the enzyme that converts carnosine back to histidine plus β-alanine is associated with progressive neurological problems and sensory polyneuropathy [172, 173].

Zinc is the cofactor for carnosinase, and elevations of carnosine or anserine may be one effect of zinc deficiency. A patient with zinc deficiency who consumes a high poultry diet would tend to show increased anserine excretion. There is a special relationship of carnosine and zinc, however that helps to ensure zinc adequacy. Dietary carnosine forms a tight complex with zinc, which greatly assists in the absorption of zinc [174] and acts in an antioxidant role to protect the gastric mucosa from ischemia [175] and ethanol-induced injury [176]. Carnosine can rescue neurons from zinc- and copper-mediated neurotoxicity, and one function of carnosine may be as an endogenous neuroprotective agent [177].

The Methylhistidines

1-Methylhistidine (1-MHis)

One-methylhistidine (1-MHis) is derived mainly from the anserine of dietary flesh sources, especially poultry. The enzyme, carnosinase, splits anserine into β-alanine and 1-MHis (Figure 4-14). High levels of 1-MHis tend to inhibit the enzyme carnosinase and increase anserine levels. Conversely, genetic variants with deficient carnosinase activity in plasma show increased 1-MHis excretions when they consume a high meat diet [172, 178]. Reduced serum carnosinase activity is also found in patients with Parkinson's disease and multiple sclerosis and patients following a cerebrovascular accident. Vitamin E deficiency can lead to 1-methylhistidinuria from increased oxidative effects in skeletal muscle [178]

3-Methylhistidine (3MHis)

The isomer, 3-Methylhistidine (3MHis), a component of muscle, is the result of methylation of histidine bound in the actin-myosin filaments. High urinary excretion of 3MHis indicates active catabolism of muscle and is an abnormal marker for excessive muscle breakdown. It has been used as such a marker in studies of clinical conditions associated with nitrogen loss, including trauma, infection [179], and in uncontrolled diabetes [180]. The ratio of urinary 3-MHis to creatinine is increased in severe injury, thyrotoxicosis, neoplastic disease, prednisolone administration, and sometimes Duchenne muscular dystrophy. In myxoedema, osteomalacia, and hypothermia the ratio of 3MHis to creatinine in urine was decreased [181].

The catabolic effects of cortisol and the opposing anabolic effects of testosterone on muscle catabolism are reflected in the level of 3MHis in urine. Excretion of 3MHis rises with cortisol and falls with administration of testosterone when cortisol is normal [182]. In cases of mild elevation, the contribution of dietary meat intake can be accounted for by checking 1-MHis for elevated excretion [183]. This method has been used to achieve greater accuracy in the assessment of the catabolic effects of infection [184]. Hypothyroidism and reduced dietary protein and energy intake produce decreased output of urinary 3-MHis [185].

Twenty-four-hour urinary 3-MHis excretions can be applied for estimating skeletal muscle mass (SM) in healthy adults on a meat-free diet using the equation:

$$SM = 0.0887 \times 3\text{-}MH + 11.8 \qquad [186]$$

Excessive muscle tissue breakdown can be associated with inadequate antioxidant protection in the muscle. Supplementation of the natural antioxidants -- vitamins A, C, E, B_2, selenium, beta-carotene, and coenzyme Q_{10} -- may be useful in preventing excessive free radical pathology associated with this problem.

Amino Acid Derivatives as Markers of Cell Regulation

Modern instrumental methodologies allow the reporting of amino acid derivatives that do not fall under any of the categories discussed above. They serve as useful biochemical markers of cell regulation and oxidative damage.

3-Nitrotyrosine (3NT) – Tissue Injury Marker

Inadequate nitric oxide synthesis contributes to degeneration of intestinal mucosal immunity and loss of endothelial control of vascular tone. Increased nitric oxide production in individuals under oxidant stress, however, results in the formation of highly reactive peroxynitrite. Tyrosine residues in proteins are nitrated by

FIGURE 4-15. Formation of Nitrotyrosine

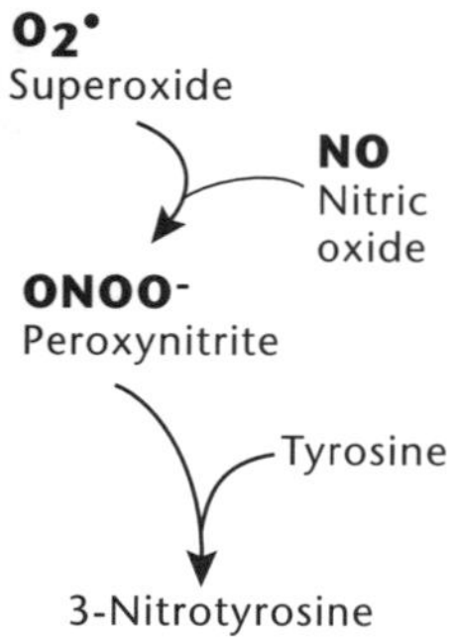

peroxynitrite, forming 3-nitrotyrosine (3NT). Evaluation of 3NT, therefore, can provide insight about which patients need antioxidant supplementation to prevent oxidative damage.

Peroxynitrite ($ONOO^-$) is a 'reactive nitrogen species' that can be formed (among other reactions) by combination of superoxide ($O_2^{\bullet}$) and nitric oxide (NO) radicals. Peroxynitrite is being increasingly proposed as a contributor to tissue injury in several human diseases. The evidence presented for peroxynitrite participation usually includes the demonstration of increased 3-nitrotyrosine (3NT) levels in the injured tissue. Although nitrating species other than peroxynitrite can participate in the formation of 3NT, peroxynitrite is the predominant active compound. Increases in 3NT, therefore, reflect increases in nitric oxide and superoxide production (Figure 4-15). Nitrotyrosine formation takes place in proteins that are present in the nitric oxide-stimulated cells. Nitration of the tyrosine residues in proteins leads to changes of protein structure and function. The proteins are degraded at cell death to release the 3NT residues that then will circulate in blood plasma.

Premature infants are susceptible to bronchopulmonary dysplasia (BPD), a chronic lung disease of infancy. The disease appears to be caused in part by oxidative stress from hyperoxia. Plasma 3NT concentrations are significantly higher in infants with BPD, increasing approximately fourfold during the first month of life and correlating with the fraction of inspired oxygen that the infant receives [187]. Peroxynitrite-mediated oxidant stress may contribute to the development of this disease in premature infants, and 3NT is useful as an early plasma indicator of infants at risk for developing BPD. Through similar mechanisms, the oxidative stress that leads to most degenerative diseases is associated with increased 3NT production.

Nitrotyrosine formation increases with endotoxin-induced kidney injury, and the changes occur in patterns suggesting that protein nitration may play a role in renal regulation and injury in ways that are yet to be defined [188]. The concentration of 3NT found in patients with sporadic amyotrophic lateral sclerosis was approximately seven times that of controls [189].

Helicobacter pylori infection is associated with gastric cancer [190]. *H. pylori* infection may contribute to cause DNA damage because of the sustained production of reactive nitrogen species that results from the upregulation of inducible nitric oxide synthetase (iNOS) expression in the gastric mucosa. Nitrotyrosine in the gastric mucosa is also significantly higher in *H. pylori*-positive groups than in *H. pylori*-negative individuals [191].

In laboratory animals, iNOS expression and NO synthesis in macrophages and cardiomyocytes are elicited in experimental inflammatory heart disease [192]. Similarly, in inflammatory bowel disease, epithelial cells and immune cells producing reactive oxygen species display induction of iNOS and increased 3NT [193].

We may conclude that patients with elevated nitrotyrosine are in a state of oxidative stress. Arginine supplementation is contraindicated, and, since antioxidant status is compromised, general support with antioxidant nutrients is indicated (see Chapter 9 – Oxidative Stess).

Assymetric Dimethylarginine (ADMA) – Regulator of Nitric Oxide Synthesis

Arginine is the amino acid from which nitric oxide (NO) is formed, and assymetric dimethylarginine (ADMA) is a naturally occurring amino acid derivative that inhibits NO synthases (Figure 4-16). ADMA is formed by methylation of arginine residues in proteins. As discussed above, high plasma 3-nitrotyrosine signals excessive nitric oxide and oxygen radical production, whereas high plasma ADMA signals lack of nitric oxide stimulation of cell activities.

FIGURE 4-16. **Arginine and ADMA in Nitric Oxide Synthesis**

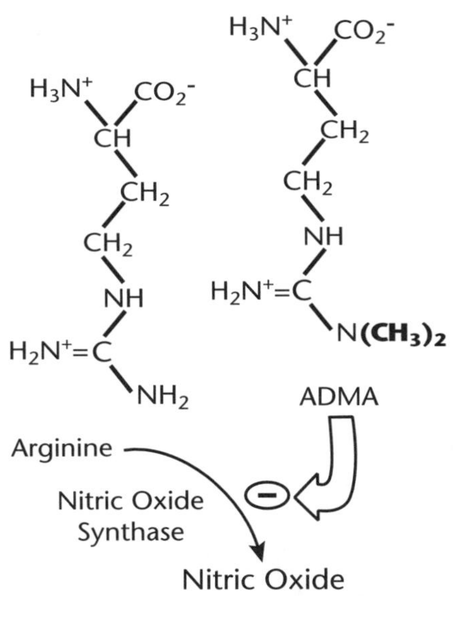

Levels of ADMA high enough to inhibit nitric oxide synthase have been found in renal failure [194, 195], hypertension, hypercholesterolemia, preeclampsia [196], diabetes mellitus [197], tobacco use, and aging [198]. This inhibitory effect may be overcome with the use of supplemental arginine [199]. ADMA is an endogenous inhibitor of angiogenesis [200]. Many angiogenic factors require endothelium-derived NO to exert their effects. Patients with atherosclerosis or with risk factors for atherosclerosis have a defect in endothelium-dependent vasodilator function, which is related to decreased activity of NO. In hypercholesterolemia, elevated plasma levels of ADMA reduce NO synthesis. The lipid-induced impairment of angiogenesis can be reversed by oral administration of arginine.

Atherosclerosis is marked by impaired vasodilation in response to normal physiological stimuli [201]. A number of standard heart disease risk factors (*e.g.*, smoking, hypertension, hyperlipidemias) are related to vasodilatory impairment, but the strongest correlation may be the ratio of arginine to ADMA in plasma [202]. This ratio has been used to assess the balance of stimulatory and inhibitory effects on NO synthesis [203]. Elaboration of endothelium-derived nitric oxide affects the behavior of circulating T lymphocytes and monocytes [204]. Oral administration of 14–21 grams of arginine normalizes plasma arginine/ADMA ratios and attenuates monocyte and T-lymphocyte adhesiveness. These changes lead to lower rates of atherosclerotic plaque formation and lower risk of heart disease.

In patients with type 2 diabetes mellitus, plasma ADMA increases from 1.0 to 2.5 μmol/L five hours after ingestion of a high-fat meal. ADMA may contribute to abnormal blood flow responses and to atherogenesis in type 2 diabetics [197]. In hypertensives, the urinary excretion rate of nitrite plus nitrate, an index of endogenous NO production, was lowered from 77 to 56 micromol/mmol creatinine, while plasma levels of ADMA increased from 1.1 to 2.4 μM [205]. Other findings suggest that ADMA plays an important role in the pathogenesis of hypertension associated with the experimental focal and segmental glomerulosclerosis [206].

Methionine Sulfoxide (MetSO) - Protein Oxidative Damage Marker

Methionine is frequently found near the active site of enzymes, and it is susceptible to oxidation. The major oxidatitive product, methionine sulfoxide, has been shown to reflect tissue antioxidant status. Methionine sulfoxide reveals the earliest stages of serum lipoprotein oxidation [207]. This marker provides a sensitive view of the impact of free radical pathology on cellular proteins.

The subject of oxidative stress and the resulting free radical damage is dealt with in a subsequent chapter. Laboratory testing is available for markers that reveal the extent of damage and provide guidance for the optimal level of antioxidant intervention.

Free-Form Amino Acid Supplement Formulas

Results of individual amino acid testing can be used to guide the formulation of a customized amino acid mixture. Clinical efficacy of this approach has been demonstrated for patients with chronic fatigue syndrome [16], where the formulation was calculated from results of plasma amino acid testing. Preliminary favorable results have been reported for developmental and behavioral disorders and many other conditions.

Only the essential amino acids are included in the supplement. The cofactors α-KG and pyridoxal-5-phosphate are added to assure interconversions leading to dispensable amino acids. The percentages of amino acids in the base formula are set according to studies estimating optimum intake for maintenance and growth [208]. For the indispensable amino acids with reported levels below their low reference limits, additional amounts are added as calculated by formulas like that shown below, where the constant is adjusted to provide safe and effective increments to the base.

Added amount (gm) = (Low limit - Reported level)/Low limit * Constant

The use of such a customized mixture affords several advantages over single amino acid interventions. Adverse interference from excessive amounts of any one essential amino acid is greatly minimized while maintaining a scaled supplementation of deficient components. Wide dosage ranges are possible by varying the daily amount of the mixture. Because of their free (non-peptide linked) form, the amino acids are very efficiently absorbed, even in the absence of stomach acid and pancreatic secretions. Prescribing a customized intake of amino acids by the purchase of bottles of individual amino acids is difficult and time-consuming. Patients also appreciate the ease and savings. The taste of free-form amino acids is unpleasant. Some patients go the trouble of personal encapsulation to avoid contact with taste buds, but in most cases, mixing with flavorful juices or food is sufficient. Effective adult dosage is 10 – 30 grams of the mixture daily.

Optimization of amino acid supply to the tissues of the body has multiple potential benefits. Now that the scientific knowledge base for amino acid testing has been reviewed, we will summarize the clinical benefits of improved amino acid balance. Normal cell regulations can by restored through neurotransmitter and hormone synthesis. The physical and immunologic barriers in the gut can be restored as the rate of secretory and structural protein production is increased. Blood glucose can be stabilized through better gluconeogenesis and utilization of fatty acids for energy is enhanced by better lipoprotein synthesis and mitochondrial performance. Hepatic and gastrointestinal detoxification reactions are improved when amino acids used in conjugation reactions are available, and the oxidative damage from such activity is reduced when sulfur amino acid supply is adequate to maintain glutathione.

Effects at the biochemical level have been emphasized so the reader can comprehend the extreme variety of favorable clinical responses to improved amino acid status. Whether it is seen as a child able to sit and learn or the grandmother able to remember, amino acids play critical roles.

References

[1] Brillon, D.J., *et al.*, Effect of cortisol on energy expenditure and amino acid metabolism in humans. *Am J Physiol*, 268(3 Pt 1). E501-13, 1995.

[2] Rudman, D., *et al.*, Fasting plasma amino acids in elderly men. *Am J Clin Nutr*, 49(3). 559-66, 1989.

[3] Irwin, M.I. and D.M. Hegsted, A conspectus of research on amino acid requirements of man. *J Nutr*, 101(4). 539-66, 1971.

[4] Irwin, M.I. and D.M. Hegsted, A conspectus of research on protein requirements of man. *J Nutr*, 101(3). 387-429, 1971.

[5] Bruinvels, J. and L. Pepplinkhuizen, Impaired glycine-serine conversion and increased plasma taurine levels in episodic psychotic patients with psychedelic symptoms. *J Psychiatr Res*, 18(3). 307-18, 1984.

[6] Scriver, C. and L. Rosenberg, Amino Acid Metabolism and It's Disorders. *Major problems in clinical pediatrics*, ed. A. Schaffer. Vol. X. 1973, Philadelphia: W. B. Saunders. 491.

[7] Scott, C.R., Sarcosinemia, in *The Metabolic and Molecular Bases of Inherited Disease*, C. Scriver, et al., Editors. 1995, McGraw-Hill, Inc.: New York. p. 1329-1335.

[8] Nyhan, W., *Abnormalities in amino acid metabolism in clinical medicine*. 1984, Norwald: Appleton-Century-Crofts. 463.

[9] Friedman, M., *Absorption and utilization of amino acids*. Vol. 1. 1989, Boca Raton: CRC. 257.

[10] Cynober, L.E., *Amino acid metabolism and therapy in health and nutritional disease*. 1995, Boca Raton: CRC Press. 459.

[11] Scriver, C., *et al.*, *The metabolic and molecular bases of inherited disease*. 7th ed. Vol. 1 - 3. 1995, New York: McGraw-Hill.

[12] Filho, J.C., *et al.*, Effect of protein intake on plasma and erythrocyte free amino acids and serum IGF-I and IGFBP-1 levels in rats. *Am J Physiol*, 277(4 Pt 1). E693-701, 1999.

[13] Dai, X.H. and B.G. Sun, [Changes in serum free amino acid level in patients with spleen deficiency syndrome]. *Chung Kuo Chung Hsi I Chieh Ho Tsa Chih*, 14(7). 403-5, 1994.

[14] Askanazi, J., *et al.*, Muscle and plasma amino acids after injury: the role of inactivity. *Ann Surg*, 188(6). 797-803, 1978.

[15] Maddison, J.E., *et al.*, Plasma GABA-like activity in rats with hepatic encephalopathy is due to GABA and taurine. *Hepatology*, 11(1). 105-10, 1990.

[16] Bralley, J. and R. Lord, Treatment of chronic fatigue syndrome with specific amino acid supplementation. J App Nutr, 46(3). 74-9, 1994.

[17] Hammarqvist, F., *et al.*, Stress hormone and amino acid infusion in healthy volunteers: short-term effects on protein synthesis and amino acid metabolism in skeletal muscle. *Metabolism*, 43(9). 1158-63, 1994.

[18] Kirvela, O., *et al.*, Respiratory and sleep patterns during nocturnal infusions of branched chain amino acids. *Acta Anaesthesiol Scand*, 34(8). 645-8, 1990.

[19] Gil, K.M., *et al.*, Parenteral nutrition and oral intake: effect of branched-chain amino acids. *Nutrition*, 6(4). 291-5, 1990.

[20] Nair, K.S., R.G. Schwartz, and S. Welle, Leucine as a regulator of whole body and skeletal muscle protein metabolism in humans. *Am J Physiol*, 263(5 Pt 1). E928-34, 1992.

[21] Gerber, D.A., Low free serum histidine concentration in rheumatoid arthritis. A measure of disease activity. *J Clin Invest*, 55(6). 1164-73, 1975.

[22] Ngo, T.M. and H.S. Winchell, Alterations in histidine catabolism in normal rats given pharmacological doses of folic acid and cyanocobalamin. *Proc Soc Exp Biol Med*, 132(1). 168-70, 1969.

[23] Buehring, K.U., K.K. Batra, and E.L. Stokstad, The effect of methionine on folic acid and histidine metabolism in perfused rat liver. *Biochim Biophys Acta*, 279(3). 498-512, 1972.

[24] Roon-Djordjevic, B.v. and S. Cerfontain-van, Urinary excretion of histidine metabolites as an indication for folic acid and vitamin B_{12} deficiency. *Clin Chim Acta*, 41. 55-65, 1972.

[25] Flodin, N.W., The metabolic roles, pharmacology, and toxicology of lysine. *J Am Coll Nutr*, 16(1). 7-21, 1997.

[26] Ruyechan, W.T. and J.W. Olson, Surface lysine and tyrosine residues are required for interaction of the major herpes simplex virus type 1 DNA-binding protein with single- stranded DNA. *J Virol*, 66(11). 6273-9, 1992.

[27] Thein, D.J. and W.C. Hurt, Lysine as a prophylactic agent in the treatment of recurrent herpes simplex labialis. *Oral Surg Oral Med Oral Pathol*, 58(6). 659-66, 1984.

[28] Lo, J.C., *et al.*, Fanconi's syndrome and tubulointerstitial nephritis in association with L-lysine ingestion. *Am J Kidney Dis*, 28(4). 614-7, 1996.

[29] Takada, G., *et al.*, Carnitine deficiency in lysinuric protein intolerance: lysine-sparing effect of carnitine. *Tohoku J Exp Med*, 153(4). 331-4, 1987.

[30] Fernandez Ortega, M.F., Effect of dietary lysine level and protein restriction on the lipids and carnitine levels in the liver of pregnant rats. *Ann*

Nutr Metab, 33(3). 162-9, 1989.
[31] Berner, Y.N., *et al.*, Low plasma carnitine in patients on prolonged total parenteral nutrition: association with low plasma lysine. *JPEN J Parenter Enteral Nutr*, 14(3). 255-8, 1990.
[32] Olszewski, A.J. and W.B. Szostak, Homocysteine content of plasma proteins in ischemic heart disease [published erratum appears in Atherosclerosis 1991 May;88(1):97-8]. *Atherosclerosis*, 69(2-3). 109-13, 1988.
[33] Manders, A.J., *et al.*, alpha-Aminoadipic aciduria and persistence of fetal haemoglobin in an oligophrenic child. *Eur J Pediatr*, 136(1). 51-5, 1981.
[34] Trijbels, J.M., *et al.*, Biochemical studies in the cerebro-hepato-renal syndrome of Zellweger: a disturbance in the metabolism of pipecolic acid. *J Inherit Metab Dis*, 2(2). 39-42, 1980.
[35] Shaw, S. and C.S. Lieber, Increased hepatic production of alpha-amino-n-butyric acid after chronic alcohol consumption in rats and baboons. *Gastroenterology*, 78(1). 108-13, 1980.
[36] Molina, J.A., *et al.*, Decreased cerebrospinal fluid levels of neutral and basic amino acids in patients with Parkinson's disease. *J Neurol Sci*, 150(2). 123-7, 1997.
[37] Kurosaki, M., *et al.*, Correlation of plasma hepatitis C virus RNA levels with serum alanine aminotransferase in non-A, non-B chronic liver disease. J Med Virol, 39(3). 246-50, 1993.
[38] Reilmann, R., L.H. Rolf, and H.W. Lange, Decreased plasma alanine and isoleucine in Huntington's disease. *Acta Neurol Scand*, 91(3). 222-4, 1995.
[39] Moghaddam, B., Recent basic findings in support of excitatory amino acid hypotheses of schizophrenia. *Prog Neuropsychopharmacol Biol Psychiatry*, 18(5). 859-70, 1994.
[40] Monaco, F., *et al.*, Plasma amino acid alterations in idiopathic generalized epilepsy: an investigation in probands and their first-degree relatives. *Ital J Neurol Sci*, 15(3). 137-44, 1994.
[41] Petty, F., *et al.*, Stability of plasma GABA at four-year follow-up in patients with primary unipolar depression. *Biol Psychiatry*, 37(11). 806-10, 1995.
[42] Gotoh, Y., M. Ishida, and H. Shinozaki, *Excitatory amino acids control the gastric function in the rat, in Amino Acids*, G. Lubec and G. Rosenthal, Editors. 1990, ESCOM: Leiden. p. 1195.
[43] Plaitakis, A., *Altered glutamate metabolism in amyotrophic lateral sclerosis and treatment with branched chain amino acids., in Amino Acids*, G. Lubec and G. Rosenthal, Editors. 1990, ESCOM: Leiden. p. 1195.
[44] Eck, H.P., *et al.*, T4+ cell numbers are correlated with plasma glutamate and cystine levels: association of hyperglutamataemia with immunodeficiency in diseases with different aetiologies. *Int Immunol*, 4(1). 7-13, 1992.
[45] Manjarrez, G., *et al.*, Free tryptophan as an indicator of brain serotonin synthesis in infants. *Pediatr Neurol*, 18(1). 57-62, 1998.
[46] Neumeister, A., *et al.*, Effects of tryptophan depletion in drug-free depressed patients who responded to total sleep deprivation. *Arch Gen Psychiatry*, 55(2). 167-72, 1998.
[47] Kuhn, E., *et al.*, Changes in blood tryptophan level during sleep deprivation. *Experientia*, 32(9). 1117-8, 1976.
[48] Lanoir, J., *et al.*, Long-term effects of a tryptophan-free diet on serotonin metabolism and sleep-waking balance in rats. *Exp Brain Res*, 41(3-4). 346-57, 1981.
[49] Bellodi, L., *et al.*, Plasma tryptophan levels and tryptophan/neutral amino acid ratios in obsessive-compulsive patients with and without depression. *Psychiatry Res*, 69(1). 9-15, 1997.
[50] Anderson, I.M., *et al.*, Decreased plasma tryptophan concentration in major depression: relationship to melancholia and weight loss. *J Affect Disord*, 20(3). 185-91, 1990.
[51] Thomson, J., *et al.*, The treatment of depression in general practice: a comparison of L-tryptophan, amitriptyline, and a combination of L-tryptophan and amitriptyline with placebo. *Psychol Med*, 12(4). 741-51, 1982.
[52] Wurtman, R., *Nutrition and the brain*. Vol. 6. 1983, Philadelphia: Raven Press.
[53] Siersbaek-Nielsen, K., Determination of the plasma tyrosine in thyroid disorders. A new test of thyroid function. *Acta Med Scand*, 179(4). 417-26, 1966.
[54] Belanger, R., *et al.*, Tyrosine and glutamic acid in plasma and urine of patients with altered thyroid function. *Metabolism*, 21(9). 855-65, 1972.
[55] Gelenberg, A.J., *et al.*, Tyrosine for the treatment of depression. *American Journal Psychiatry*, 137:5(May). 622-623, 1980.
[56] Lehmann, W.D. and H.C. Heinrich, Impaired phenylalanine-tyrosine conversion in patients with iron- deficiency anemia studied by a L-(2H5)phenylalanine-loading test. *Am J Clin Nutr*, 44(4). 468-74, 1986.
[57] Jackson, A.A., *et al.*, Urinary excretion of 5-oxo-proline (pyroglutamic aciduria) as an index of glycine insufficiency in normal man. *Br J Nutr*, 58(2). 207-14, 1987.
[58] Wang, P.J., *et al.*, The controversy regarding diagnostic criteria for early myoclonic encephalopathy [In Process Citation]. *Brain Dev*, 20(7). 530-5, 1998.
[59] Dudman, N.P., P.A. Tyrrell, and D.E. Wilcken, Homocysteinemia: depressed plasma serine levels. *Metabolism*, 36(2). 198-201, 1987.
[60] Altamura, C., *et al.*, Plasma concentrations of excitatory amino acids, serine, glycine, taurine

and histidine in major depression. *Eur Neuropsychopharmacol*, 5 Suppl. 71-5, 1995.
[61] Devor, E.J. and R. Waziri, A familial/genetic study of plasma serine and glycine concentrations. *Biol Psychiatry*, 34(4). 221-5, 1993.
[62] Riedel, E., *et al.*, Changes in the concentrations of hydroxyproline, glycine and serine in the plasma of haemodialysis patients undergoing erythropoietin therapy. *J Clin Chem Clin Biochem*, 27(11). 851-6, 1989.
[63] Kitagawa, M., *et al.*, Evaluating exocrine function tests for diagnosing chronic pancreatitis. *Pancreas*, 15(4). 402-8, 1997.
[64] Whyte, M.P., *et al.*, Adult hypophosphatasia. Clinical, laboratory, and genetic investigation of a large kindred with review of the literature. *Medicine (Baltimore)*, 58(5). 329-47, 1979.
[65] Klunk, W.E., *et al.*, Inactivity of phosphoethanolamine, an endogenous GABA analog decreased in Alzheimer's disease, at GABA binding sites. *Life Sci*, 56(26). 2377-83, 1995.
[66] Kano-Sueoka, T., *et al.*, Analysis of cytosolic phosphoethanolamine and ethanolamine and their correlation with prognostic factors in breast cancer. *Jpn J Cancer Res*, 82(7). 829-34, 1991.
[67] Gerashchenko, D., *et al.*, Influence of pyruvate, threonine and phosphoethanolamine on activities of some acetaldehyde-producing enzymes. *Alcohol Alcohol*, 28(4). 437-43, 1993.
[68] Molina, J.A., *et al.*, Cerebrospinal fluid levels of non-neurotransmitter amino acids in patients with Alzheimer's disease. *J Neural Transm*, 105(2-3). 279-86, 1998.
[69] Gentili, D., *et al.*, L-Asparagine depletion in plasma and cerebro-spinal fluid of children with acute lymphoblastic leukemia during subsequent exposures to Erwinia L-asparaginase [see comments]. *Ann Oncol*, 7(7). 725-30, 1996.
[70] Kaleli, B., *et al.*, The effect of L-arginine and pentoxifylline on postoperative adhesion formation. *Acta Obstet Gynecol Scand*, 77(4). 377-80, 1998.
[71] Aydin, S., O. Inci, and B. Alagol, The role of arginine, indomethacin and kallikrein in the treatment of oligoasthenospermia. *Int Urol Nephrol*, 27(2). 199-202, 1995.
[72] Askanazi, J., *et al.*, Muscle and plasma amino acids after injury: hypocaloric glucose vs. amino acid infusion. *Ann Surg*, 191(4). 465-72, 1980.
[73] Chawla, R.K., *et al.*, Plasma cysteine, cystine, and glutathione in cirrhosis. *Gastroenterology*, 87(4). 770-6, 1984.
[74] Lakshmanan, F.L., *et al.*, Plasma and urinary amino acids and selected sulfur metabolites in young men fed a diet devoid of methionine and cystine. *Am J Clin Nutr*, 29(12). 1367-71, 1976.
[75] Wang, Y., *et al.*, The effect of condensed tannins in Lotus corniculatus on plasma metabolism of methionine, cystine and inorganic sulphate by sheep. *Br J Nutr*, 72(6). 923-5, 1994.
[76] Yoshida, K., M. Yahiro, and K. Ahiko, Effects of addition of arginine, cystine, and glycine to the bovine milk-simulated amino acid mixture on the level of plasma and liver cholesterol in rats. *J Nutr Sci Vitaminol (Tokyo)*, 34(6). 567-76, 1988.
[77] Serougne, C., *et al.*, Hypercholesterolemia induced by cholesterol- or cystine-enriched diets is characterized by different plasma lipoprotein and apolipoprotein concentrations in rats. *J Nutr*, 125(1). 35-41, 1995.
[78] Verhoef, P., *et al.*, Homocysteine metabolism and risk of myocardial infarction: relation with vitamins B_6, B_{12}, and folate. *Am J Epidemiol*, 143(9). 845-59, 1996.
[79] Martensson, J., R. Sjodahl, and P. Tobiasson, Effect of gastrointestinal surgery and bacterial overgrowth on the urinary excretion of sulfur amino acids and their main degradation products. *Scand J Gastroenterol*, 19(4). 507-14, 1984.
[80] Kachru, D.N., S. Khandelwal, and S.K. Tandon, Influence of methionine supplementation in chelation of lead in rats. *Biomed Environ Sci*, 2(3). 265-70, 1989.
[81] Domingo, J.L. and J.M. Llobet, Treatment of acute cobalt intoxication in rats with L-methionine. *Rev Esp Fisiol*, 40(4). 443-8, 1984.
[82] Chang, Y.F., Lysine metabolism in the human and the monkey: demonstration of pipecolic acid formation in the brain and other organs. *Neurochem Res*, 7(5). 577-88, 1982.
[83] Finkelstein, J.D., *et al.*, Regulation of hepatic betaine-homocysteine methyltransferase by dietary betaine. *J Nutr*, 113(3). 519-21, 1983.
[84] Abbott, M.H., *et al.*, Psychiatric manifestations of homocystinuria due to cystathionine beta-synthase deficiency: prevalence, natural history, and relationship to neurologic impairment and vitamin B_6-responsiveness. *Am J Med Genet*, 26(4). 959-69, 1987.
[85] Lubec, B., *et al.*, Evidence for McKusick's hypothesis of deficient collagen cross-linking in patients with homocystinuria. *Biochim Biophys Acta*, 1315(3). 159-62, 1996.
[86] Milakofsky, L., N. Harris, and W.H. Vogel, Effect of repeated stress on plasma catecholamines and taurine in young and old rats. *Neurobiol Aging*, 14(4). 359-66, 1993.
[87] Hashiguchi, Y., *et al.*, Interleukin-1 and tumor necrosis factor alter plasma concentration and interorgan fluxes of taurine in dogs. *Shock*, 7(2). 147-53, 1997.
[88] Azuma, J., *et al.*, Taurine for treatment of congestive heart failure. *Int J Cardiol*, 2(2). 303-4, 1982.
[89] Huxtable and Bardeau, *Taurine in Nutrition and*

Neurology. 1982: Plenum.

[90] Waterfield, C.J., F. Carvalho, and J.A. Timbrell, Effect of treatment with beta-agonists on tissue and urinary taurine levels in rats. Mechanism and implications for protection. *Adv Exp Med Biol,* 403. 233-45, 1996.

[91] Inoue, M. and I.M. Arias, Taurine transport across hepatocyte plasma membranes: analysis in isolated rat liver sinusoidal plasma membrane vesicles. *J Biochem (Tokyo),* 104(1). 155-8, 1988.

[92] Hayes, K.C., *et al.*, Taurine modulates platelet aggregation in cats and humans. *Am J Clin Nutr,* 49(6). 1211-6, 1989.

[93] Jaksic, T., D.A. Wagner, and V.R. Young, Plasma proline kinetics and concentrations in young men in response to dietary proline deprivation. *Am J Clin Nutr,* 52(2). 307-12, 1990.

[94] Hiramatsu, T., *et al.*, Plasma proline and leucine kinetics: response to 4 wk with proline-free diets in young adults. *Am J Clin Nutr,* 60(2). 207-15, 1994.

[95] Bates, C.J., Vitamin C deficiency in guinea pigs: changes in urinary excretion of proline, hydroxyproline and total amino nitrogen. *Int J Vitam Nutr Res,* 49(2). 152-9, 1979.

[96] Nusgens, B. and C.M. Lapiere, The relationship between proline and hydroxyproline urinary excretion in human as an index of collagen catabolism. *Clin Chim Acta,* 48(2). 203-11, 1973.

[97] Kaddam, I.M., *et al.*, Comparison of serum osteocalcin with total and bone specific alkaline phosphatase and urinary hydroxyproline: creatinine ratio in patients with Paget's disease of bone. *Ann Clin Biochem,* 31(Pt 4). 327-30, 1994.

[98] Hamdy, N.A., *et al.*, Urinary collagen crosslink excretion: a better index of bone resorption than hydroxyproline in Paget's disease of bone? *Bone Miner,* 22(1). 1-8, 1993.

[99] Bolzonella, S., *et al.*, Urinary hydroxyproline in multiple myeloma: correlation with clinical stages and bone disease. *Tumori,* 70(3). 249-53, 1984.

[100] Gilbertson, T.J., *et al.*, Serum total hydroxyproline assay: effects of age, sex and Paget's bone disease. *J Clin Chem Clin Biochem,* 21(3). 129-32, 1983.

[101] Yamada, S., *et al.*, Urinary hydroxyproline and hydroxylysine excretions in relation to hepatic hydroxyproline content in chronic liver disease. *Clin Biochem,* 22(5). 389-93, 1989.

[102] Nagai, Y., M. Sato, and M. Sasaki, Effect of cadmium administration upon urinary excretion of hydroxylysine and hydroxyproline in the rat. *Toxicol Appl Pharmacol,* 63(2). 188-93, 1982.

[103] Scriver, C. and K. Gibson, Disorders of β- and γ-amino acids in free and peptide-linked forms, in *The Metabolic and Molecular Bases of Inherited Disease,* C. Scriver, *et al.*, Editors. 1995, McGraw-Hill, Inc.: New York. p. 1349-1368.

[104] Korang, K., *et al.*, Levels of taurine, amino acids and related compounds in plasma, vena cava, aorta and heart of rats after taurine administration. *Pharmacology,* 52(4). 263-70, 1996.

[105] Bucuvalas, J.C., A.L. Goodrich, and F.J. Suchy, Hepatic taurine transport: a Na+-dependent carrier on the basolateral plasma membrane. *Am J Physiol,* 253(3 Pt 1). G351-8, 1987.

[106] Schaeffer, M.C., *et al.*, Changes in cerebrospinal fluid and plasma amino acid concentrations with elevated dietary protein concentration in dogs with portacaval shunts. *Life Sci,* 48(23). 2215-23, 1991.

[107] Lepage, N., *et al.*, Age-specific distribution of plasma amino acid concentrations in a healthy pediatric population. *Clin Chem,* 43(12). 2397-402, 1997.

[108] Douglas, A.P. and C.C. Booth, Post-prandial plasma-free amino acids in adult coeliac disease after oral gluten and albumin. *Clin Sci,* 37(3). 643-53, 1969.

[109] Nobels, F., *et al.*, Plasma zinc levels in diabetes mellitus: relation to plasma albumin and amino acids. *Biomed Pharmacother,* 40(2). 57-60, 1986.

[110] Riley, D.R., I. Harrill, and E.D. Gifford, Influence of zinc and vitamin D on plasma amino acids and liver xanthine oxidase in rats. *J Nutr,* 98(3). 351-5, 1969.

[111] Eiduson, S., A. Yuwiler, and E.D. Eberle, The effect of pyridoxine deficiency on L-aromatic amino acid decarboxylase and tyrosine aminotransferase in developing brain. *Adv Biochem Psychopharmacol,* 4. 63-80, 1972.

[112] Rosenlund, B.L., Effects of insulin on free amino acids in plasma and the role of the amino acid metabolism in the etiology of diabetic microangiopathy. *Biochem Med Metab Biol,* 49(3). 375-91, 1993.

[113] Fryburg, D.A., *et al.*, Insulin and insulin-like growth factor-I enhance human skeletal muscle protein anabolism during hyperaminoacidemia by different mechanisms. *J Clin Invest,* 96(4). 1722-9, 1995.

[114] Hauser, S.L., *et al.*, An antispasticity effect of threonine in multiple sclerosis. *Arch Neurol,* 49(9). 923-6, 1992.

[115] Young, V.R. and A.M. Ajami, Glutamate: an amino acid of particular distinction. *J Nutr,* 130(4S Suppl). 892S-900S, 2000.

[116] Meldrum, B.S., Glutamate as a neurotransmitter in the brain: review of physiology and pathology. *J Nutr,* 130(4S Suppl). 1007S-15S, 2000.

[117] Matthews, D.E. and R.G. Campbell, The effect of dietary protein intake on glutamine and glutamate nitrogen metabolism in humans. *Am J Clin Nutr,* 55(5). 963-70, 1992.

[118] Reeds, P.J., *et al.*, Enteral glutamate is the preferential source for mucosal glutathione synthesis in fed piglets. *Am J Physiol*, 273(2 Pt 1). E408-15, 1997.

[119] Patel, D.K., *et al.*, Depletion of plasma glycine and effect of glycine by mouth on salicylate metabolism during aspirin overdose. *Hum Exp Toxicol*, 9(6). 389-95, 1990.

[120] Kiss, Z., *et al.*, Ethanolamine, but not phosphoethanolamine, potentiates the effects of insulin, phosphocholine, and ATP on DNA synthesis in NIH 3T3 cells-- role of mitogen-activated protein-kinase-dependent and protein-kinase- independent mechanisms. *Eur J Biochem*, 250(2). 395-402, 1997.

[121] Peters, H. and N.A. Noble, Dietary L-arginine in renal disease. *Semin Nephrol*, 16(6). 567-75, 1996.

[122] Milner, J.A., Metabolic aberrations associated with arginine deficiency. *J Nutr*, 115(4). 516-23, 1985.

[123] Molderings, G.J., *et al.*, Potential relevance of agmatine as a virulence factor of Helicobacter pylori. *Dig Dis Sci*, 44(12). 2397-404, 1999.

[124] Galea, E., *et al.*, Inhibition of mammalian nitric oxide synthases by agmatine, an endogenous polyamine formed by decarboxylation of arginine. *Biochem J*, 316(Pt 1). 247-9, 1996.

[125] Reis, D.J. and S. Regunathan, Is agmatine a novel neurotransmitter in brain? *Trends Pharmacol Sci*, 21(5). 187-93, 2000.

[126] Algert, S.J., *et al.*, Assessment of dietary intake of lysine and arginine in patients with herpes simplex. *J Am Diet Assoc*, 87(11). 1560-1, 1987.

[127] Tousoulis, D., *et al.*, Effects of changing the availability of the substrate for nitric oxide synthase by L-arginine administration on coronary vasomotor tone in angina patients with angiographically narrowed and in patients with normal coronary arteries. *Am J Cardiol*, 82(9). 1110-3, A6, 1998.

[128] Angele, M.K., *et al.*, L-arginine: a unique amino acid for restoring the depressed macrophage functions after trauma-hemorrhage. *J Trauma*, 46(1). 34-41, 1999.

[129] Castillo, L., *et al.*, Plasma arginine, citrulline, and ornithine kinetics in adults, with observations on nitric oxide synthesis. *Am J Physiol*, 268(2 Pt 1). E360-7, 1995.

[130] Rotondo, G. and A.M. De Angelis, [On the use of acetyl-aspartic acid and of citrulline in the treatment and prevention of flight fatigue]. *Riv Med Aeronaut Spaz*, 29(1). 85-105, 1966.

[131] Campistron, G., *et al.*, Pharmacokinetics of arginine and aspartic acid administered simultaneously in the rat: II. Tissue distribution. *Eur J Drug Metab Pharmacokinet*, 7(4). 315-22, 1982.

[132] Campistron, G., *et al.*, Pharmacokinetics of arginine and aspartic acid administered simultaneously in the rat--III: Changes in the levels of amino acids in the plasma, liver and brain after simultaneous administration of arginine and aspartic acid. *Eur J Drug Metab Pharmacokinet*, 8(3). 281-6, 1983.

[133] Wiren, M., K.E. Magnusson, and J. Larsson, The role of glutamine, serum and energy factors in growth of enterocyte- like cell lines. *Int J Biochem Cell Biol*, 30(12). 1331-6, 1998.

[134] Yoshida, S., *et al.*, Effect of total parenteral nutrition, systemic sepsis, and glutamine on gut mucosa in rats. *Am J Physiol*, 263(2 Pt 1). E368-73, 1992.

[135] Chen, K., *et al.*, Glutamine-supplemented parenteral nutrition improves gut mucosa integrity and function in endotoxemic rats. *JPEN J Parenter Enteral Nutr*, 18(2). 167-71, 1994.

[136] Bellows, C.F. and B.M. Jaffe, Glutamine is essential for nitric oxide synthesis by murine macrophages. *J Surg Res*, 86(2). 213-9, 1999.

[137] Roth, E., L-argine-nitric oxide metabolism. Glutamine: a new player in this metabolic game? [editorial; comment]. *Clin Nutr,* 17(1). 1-2, 1998.

[138] Houdijk, A.P., *et al.*, Dietary glutamine supplementation reduces plasma nitrate levels in rats [see comments]. *Clin Nutr*, 17(1). 11-4, 1998.

[139] Kinscherf, R., *et al.*, Low plasma glutamine in combination with high glutamate levels indicate risk for loss of body cell mass in healthy individuals: the effect of N- acetyl-cysteine. *J Mol Med,* 74(7). 393-400, 1996.

[140] Forslund, A.H., *et al.*, Inverse relationship between protein intake and plasma free amino acids in healthy men at physical exercise. *Am J Physiol Endocrinol Metab*, 278(5). E857-67., 2000.

[141] Crenn, P., *et al.*, Post-absorptive plasma citrulline concentration: a marker of intestinal failure in humans. *Transplant Proc*, 30(6). 2528, 1998.

[142] Lagerwerf, F.M., *et al.*, Assessment of nitric oxide production by measurement of [15N]citrulline enrichment in human plasma using high-performance liquid chromatography- mass spectrometry. *Anal Biochem*, 257(1). 45-52, 1998.

[143] Iafolla, A.K., S.G. Kahler, and Y.T. Chen, Low plasma citrulline concentrations during protein restriction in an unaffected infant at risk for ornithine transcarbamylase deficiency [letter; comment]. *J Pediatr,* 120(3). 496-7, 1992.

[144] Kamoun, P., *et al.*, Citrulline concentrations in human plasma after arginine load. *Clin Chem*, 37(7). 1287, 1991.

[145] Solaini, G., *et al.*, [Variations in the plasma concentration of ornithine, citrulline and arginine in acute experimental liver failure]. *Boll Soc Ital Biol Sper*, 57(7). 705-10, 1981.

[146] Holtta, E. and P. Pohjanpelto, Control of ornithine decarboxylase in Chinese hamster ovary cells by polyamines. Translational inhibition of synthesis and acceleration of degradation of the enzyme by putrescine, spermidine, and spermine. *J Biol Chem*, 261(20). 9502-8, 1986.

[147] Krassowski, J., *et al.*, The effect of ornithine alpha-ketoglutarate on growth hormone (GH) and prolactin (PRL) release in normal subjects. *Endokrynol Pol*, 37(1). 11-5, 1986.

[148] Lecointre, C. and E. Mallet, [Study of tall stature in children. Low paradoxically low answer of growth hormone to stimulation with ornithine (letter)]. *Arch Fr Pediatr*, 47(10). 763, 1990.

[149] Fukagawa, N.K., A.M. Ajami, and V.R. Young, Plasma methionine and cysteine kinetics in response to an intravenous glutathione infusion in adult humans. *Am J Physiol*, 270(2 Pt 1). E209-14, 1996.

[150] White, C.W., *et al.*, Plasma cysteine concentrations in infants with respiratory distress. *J Pediatr*, 125(5 Pt 1). 769-77, 1994.

[151] Jacobson, S.H. and P. Moldeus, Whole blood-, plasma- and red blood cell glutathione and cysteine in patients with kidney disease and during hemodialysis. *Clin Nephrol*, 42(3). 189-92, 1994.

[152] Brattstrom, L., *et al.*, Homocysteine and cysteine: determinants of plasma levels in middle-aged and elderly subjects. *J Intern Med*, 236(6). 633-41, 1994.

[153] Wollesen, F., *et al.*, Plasma total homocysteine and cysteine in relation to glomerular filtration rate in diabetes mellitus [In Process Citation]. *Kidney Int*, 55(3). 1028-35, 1999.

[154] Davies, M.H., *et al.*, Plasma cysteine and sulphate levels in patients with cirrhosis of the liver. *Clin Sci (Colch)*, 87(3). 357-62, 1994.

[155] Bernhard, M.C., *et al.*, Time course of total cysteine, glutathione and homocysteine in plasma of patients with chronic hepatitis C treated with interferon-alpha with and without supplementation with N-acetylcysteine. *J Hepatol*, 28(5). 751-5, 1998.

[156] Zhang, J., *et al.*, Effect of cystathionine ketimine on the stimulus coupled responses of neutrophils and their modulation by various protein kinase inhibitors. *Biochem Biophys Res Commun*, 218(1). 371-6, 1996.

[157] Backus, R.C., *et al.*, Diets causing taurine depletion in cats substantially elevate postprandial plasma cholecystokinin concentration. *J Nutr*, 125(10). 2650-7, 1995.

[158] Park, T. and K. Lee, Dietary taurine supplementation reduces plasma and liver cholesterol and triglyceride levels in rats fed a high-cholesterol or a cholesterol- free diet. *Adv Exp Med Biol*, 442. 319-25, 1998.

[159] Helms, R.A., *et al.*, Cysteine supplementation results in normalization of plasma taurine concentrations in children receiving home parenteral nutrition. *J Pediatr*, 134(3). 358-61, 1999.

[160] Mahmoodian, F., A. Gosiewska, and B. Peterkofsky, Regulation and properties of bone alkaline phosphatase during vitamin C deficiency in guinea pigs. *Arch Biochem Biophys*, 336(1). 86-96, 1996.

[161] Higgins, J., *et al.*, Pyridoxine-resonsive hyper-beta-alaninemia associated with Cohen's syndrome. *Neurology*, 44(9). 1728-32, 1994.

[162] Mizota, C., *et al.*, Effect of pyridoxine deficiency and prednisolone on beta-alanine- oxoglutarate aminotransferase and D-3-aminoisobutyrate aminotransferase in rat liver and kidney. *J Nutr Sci Vitaminol (Tokyo)*, 34(2). 223-36, 1988.

[163] Scriver, C. and K. Gibson, Disorders of β- and γ-amino acids in free and peptide-linked forms, in *The Metabolic and Molecular Bases of Inherited Disease*, C. Scriver, *et al.*, Editors. 1995, McGraw-Hill, Inc.: New York. p. 1349.

[164] Nielsen, H.R. and S.A. Killmann, Urinary excretion of beta-aminoisobutyrate and pseudouridine in acute and chronic myeloid leukemia. *J Natl Cancer Inst*, 71(5). 887-91, 1983.

[165] Zeis, P.M., *et al.*, DNA degradation in the kidney of folic acid-treated guinea pigs [In Process Citation]. *Cytobios*, 102(400). 107-13, 2000.

[166] Avena, R.M. and W.J. Bowen, Effects of carnosine and anserine on muscle adenosine triphosphatases. *J Biol Chem*, 244(6). 1600-4, 1969.

[167] Johnson, P., *et al.*, Regulation of muscle phosphorylase activity by carnosine and anserine. *Biochem Biophys Res Commun*, 109(3). 769-75, 1982.

[168] Tan, K.M. and J.K. Candlish, Carnosine and anserine as modulators of neutrophil function. *Clin Lab Haematol*, 20(4). 239-44, 1998.

[169] Datta, A.K., X. Shi, and K.S. Kasprzak, Effect of carnosine homocarnosine and anserine on hydroxylation of the guanine moiety in 2'-deoxyguanosine, DNA and nucleohistone with hydrogen peroxide in the presence of nickel(II). *Carcinogenesis*, 14(3). 417-22, 1993.

[170] Daniel, R.L., N.J. Osbaldeston, and J.G. McCormack, Activation by anserine and inhibition by carnosine of Ca(2+)-uptake by mammalian mitochondria [published erratum appears in Biochem Soc Trans 1992 Nov; 20(4):911]. *Biochem Soc Trans*, 20(2). 131S, 1992.

[171] Murphey, W.H., *et al.*, Serum carnosinase deficiency concomitant with mental retardation. *Pediatr Res*, 7(7). 601-6, 1973.

[172] Fleisher, L.D., *et al.*, Carnosinase deficiency: a new variant with high residual activity. *Pediatr Res*, 14(4 Pt 1). 269-71, 1980.
[173] Wisniewski, K., *et al.*, Neurological disease in a child with carnosinase deficiency. *Neuropediatrics*, 12(2). 143-51, 1981.
[174] Furuta, S., S. Toyama, and H. Sano, Absorption mechanism of polaprezinc (zinc L-carnosine complex) by an everted sac method. *Xenobiotica*, 24(11). 1085-94, 1994.
[175] Yoshikawa, T., *et al.*, Effect of zinc-carnosine chelate compound (Z-103), a novel antioxidant, on acute gastric mucosal injury induced by ischemia-reperfusion in rats. *Free Radic Res Commun*, 14(4). 289-96, 1991.
[176] Cho, C.H., *et al.*, The cytoprotective effect of zinc L-carnosine on ethanol-induced gastric gland damage in rabbits. *J Pharm Pharmacol*, 44(4). 364-5, 1992.
[177] Horning, M.S., L.J. Blakemore, and P.Q. Trombley, Endogenous mechanisms of neuroprotection: role of zinc, copper, and carnosine. *Brain Res*, 852(1). 56-61, 2000.
[178] Willi, S.M., *et al.*, A deletion in the long arm of chromosome 18 in a child with serum carnosinase deficiency. *Pediatr Res*, 41(2). 210-3, 1997.
[179] Long, C.L., *et al.*, Urinary excretion of 3-methyl-histidine: an assessment of muscle protein catabolism in adult normal subjects and during malnutrition, sepsis, and skeletal trauma. *Metabolism*, 30(8). 765-76, 1981.
[180] Marchesini, G., *et al.*, Muscle protein breakdown in uncontrolled diabetes as assessed by urinary 3-methylhistidine excretion. *Diabetologia*, 23(5). 456-8, 1982.
[181] Elia, M., *et al.*, Clinical usefulness of urinary 3-methylhistidine excretion in indicating muscle protein breakdown. *Br Med J (Clin Res Ed)*, 282(6261). 351-4, 1981.
[182] Santidrian, S., *et al.*, Effect of testosterone on the rate of myofibrillar protein breakdown in castrated and adrenalectomized male rats measured by the urinary excretion of 3-methylhistidine. *Metabolism*, 31(12). 1200-5, 1982.
[183] Sjolin, J., *et al.*, Urinary excretion of 1-methylhistidine: a qualitative indicator of exogenous 3-methylhistidine and intake of meats from various sources. *Metabolism*, 36(12). 1175-84, 1987.
[184] Sjolin, J., *et al.*, Evaluation of urinary 3-methylhistidine excretion in infection by measurements of 1-methylhistidine and the creatinine ratios. *Am J Clin Nutr*, 49(1). 62-70, 1989.
[185] Burini, R., *et al.*, Interaction of thyroid status and diet on muscle protein breakdown in the rat, as measured by N tau-methylhistidine excretion. *Metabolism*, 30(7). 679-87, 1981.
[186] Wang, Z., *et al.*, Urinary 3-methylhistidine excretion: association with total body skeletal muscle mass by computerized axial tomography. *JPEN J Parenter Enteral Nutr*, 22(2). 82-6, 1998.
[187] Banks, B.A., *et al.*, Plasma 3-nitrotyrosine is elevated in premature infants who develop bronchopulmonary dysplasia [see comments]. *Pediatrics*, 101(5). 870-4, 1998.
[188] Bian, K., *et al.*, Nitrotyrosine formation with endotoxin-induced kidney injury detected by immunohistochemistry [In Process Citation]. *Am J Physiol*, 277(1 Pt 2). F33-40, 1999.
[189] Tohgi, H., *et al.*, Remarkable increase in cerebrospinal fluid 3-nitrotyrosine in patients with sporadic amyotrophic lateral sclerosis. *Ann Neurol*, 46(1). 129-31, 1999.
[190] Haruma, K., *et al.*, Helicobacter pylori infection is a major risk factor for gastric carcinoma in young patients. *Scand J Gastroenterol*, 35(3). 255-9, 2000.
[191] Goto, T., *et al.*, Enhanced expression of inducible nitric oxide synthase and nitrotyrosine in gastric mucosa of gastric cancer patients [In Process Citation]. *Clin Cancer Res*, 5(6). 1411-5, 1999.
[192] Bachmaier, K., *et al.*, iNOS expression and nitrotyrosine formation in the myocardium in response to inflammation is controlled by the interferon regulatory transcription factor 1. *Circulation*, 96(2). 585-91, 1997.
[193] Dijkstra, G., *et al.*, Expression of nitric oxide synthases and formation of nitrotyrosine and reactive oxygen species in inflammatory bowel disease. *J Pathol*, 186(4). 416-21, 1998.
[194] Al Banchaabouchi, M., *et al.*, NG, NG-dimethylarginine and NG, NG-dimethylarginine in renal insufficiency. *Pflugers Arch*, 439(5). 524-31, 2000.
[195] Kielstein, J.T., *et al.*, Asymmetric dimethyl-arginine plasma concentrations differ in patients with end-stage renal disease: relationship to treatment method and atherosclerotic disease. *J Am Soc Nephrol*, 10(3). 594-600, 1999.
[196] Boger, R.H., *et al.*, Restoring vascular nitric oxide formation by L-arginine improves the symptoms of intermittent claudication in patients with peripheral arterial occlusive disease. *J Am Coll Cardiol*, 32(5). 1336-44, 1998.
[197] Fard, A., *et al.*, Acute elevations of plasma asymmetric dimethylarginine and impaired endothelial function in response to a high-Fat meal in patients with type 2 diabetes [In Process Citation]. *Arterioscler Thromb Vasc Biol*, 20(9). 2039-44, 2000.
[198] Boger, R.H., *et al.*, Biochemical evidence for impaired nitric oxide synthesis in patients with peripheral arterial occlusive disease [see

comments]. *Circulation*, 95(8). 2068-74, 1997.

[199] Nagase, S., *et al.*, A novel nonenzymatic pathway for the generation of nitric oxide by the reaction of hydrogen peroxide and D- or L-arginine. *Biochem Biophys Res Commun*, 233(1). 150-3, 1997.

[200] Jang, J.J., *et al.*, Angiogenesis is impaired by hypercholesterolemia : role of asymmetric dimethylarginine [In Process Citation]. *Circulation,* 102(12). 1414-9, 2000.

[201] Creager, M.A., *et al.*, Impaired vasodilation of forearm resistance vessels in hypercholesterolemic humans. *J Clin Invest*, 86(1). 228-34, 1990.

[202] Boger, R.H., *et al.*, Plasma concentration of asymmetric dimethylarginine, an endogenous inhibitor of nitric oxide synthase, is elevated in monkeys with hyperhomocyst(e)inemia or hypercholesterolemia. *Arterioscler Thromb Vasc Biol*, 20(6). 1557-64, 2000.

[203] Bode-Boger, S.M., *et al.*, Elevated L-arginine/dimethylarginine ratio contributes to enhanced systemic NO production by dietary L-arginine in hypercholesterolemic rabbits. *Biochem Biophys Res Commun*, 219(2). 598-603, 1996.

[204] Chan, J.R., *et al.*, Asymmetric dimethylarginine increases mononuclear cell adhesiveness in hypercholesterolemic humans. *Arterioscler Thromb Vasc Biol*, 20(4). 1040-6, 2000.

[205] Surdacki, A., *et al.*, Reduced urinary excretion of nitric oxide metabolites and increased plasma levels of asymmetric dimethylarginine in men with essential hypertension. *J Cardiovasc Pharmacol*, 33(4). 652-8, 1999.

[206] Sato, J., *et al.*, Endogenous asymmetrical dimethylarginine and hypertension associated with puromycin nephrosis in the rat. *Br J Pharmacol*, 125(3). 469-76, 1998.

[207] Garner, B., *et al.*, Oxidation of high density lipoproteins. I. Formation of methionine sulfoxide in apolipoproteins AI and AII is an early event that accompanies lipid peroxidation and can be enhanced by alpha-tocopherol. *J Biol Chem*, 273(11). 6080-7, 1998.

[208] Mercer, L., S. Dodds, and D. Smith, *Dispensable, indispensable, and conditionally indispensable amion acid ratios in the diet, in Absorption and utilization of amino acids*, M. Friedman, Editor. 1989, CRC: Boca Raton. p. 1 - 14.k

Chapter 5

Fatty Acids

TABLE 5-1. Summary of Fatty Acid Abnormalities

	Structure	Name		Potential Responses	Metabolic Association
Omega-3 Polyunsaturated	18:3	Alpha linolenic (ALA)	L	Flax and/or fish oils	Essential fatty acid
	20:5	Eicosapentaenoic (EPA)	L		**Eicosanoid substrate**
	22:5	Docosapentaenoic	L	Fish oils	Nerve membrane function
	22:6	Docosahexaenoic (DHA)	L		Neurological development
Omega-6 Polyunsaturated	18:2	Linoleic (LA)	L	Corn or black currant oil	Essential fatty acid
	18:3	Gamma linolenic (GLA)	L	GLA source*	Eicosaoid precursor
	20:2	Eicosadienoic	H	Zinc; 50mg/d	Desaturase inhibition
	20:3	Dihomogammalinolenic	L	Zinc; 50mg/d	**Eicosanoid substrate**
	20:4	Arachidonic (AA)	L	Corn or black currant oil	**Eicosanoid substrate**
			H	Reduce red meats	
	22:2	Docosadienoic	H	Copper Citrate; 6mg	Copper deficiency
	22:4	Docosatetraenoic	H	Glycemic control	Increase in adipose tissue
Omega-9 Polyunsaturated	18:3	Mead		Essential oils	Essential fatty acid deficiency
Monounsaturated	12:1	Vaccenic	H	Biotin; 500 mcg BID	Biotin deficiency
	14:1	Myristoleic	H		
	16:1	Palmitoleic	H	See discussion	Essential fatty acid deficiency
	18:1	Oleic	L	Olive oil	Membrane fluidity
	20:1	11-Eicosenoic	H		
	22:1	Nervonic	L	Fish or canola oils	Neurological development
	24:1	Erucic	L	Peanut oil	Nerve membrane function
Saturated	10:0	Capric	H	Riboflavin; 50 mg TID	MAD^^
	12:0	Lauric	H	Riboflavin; 50 mg TID	MAD^^
	14:0	Myristic	H	Riboflavin; 50 mg TID	MAD^^

TABLE 5-1 Continued. Summary of Fatty Acid Abnormalities

Saturated Cont.	16:0	Palmitic	H	Reduce saturated fats; add Niacin	Cholesterogenic
	18:0	Stearic	H	Reduce saturated fats; add Niacin	Elevated triglycerides
			L		Cancer marker
	20:0	Arachidic	H	Check eicosanoid ratios	D6 desaturase inhibition
	22:0	Behenic	H		Elongation stimulation
	24:0	Lignoceric	H	Cosider rape or mustard oils	Nerve membrane function
	26:0	Hexacosanoic	H		Elongation stimulation
Odd-numbered	15:0	Pentadecanoic	H		
	17:0	Heptadecanoic	H	Biotin, 1000 mcg TID	
	19:0	Nonadecanoic	H	Vitamin B_{12} ; 1000 mcg, TID	Propionate accumulation
	21:0	Heneicosanoic	H	Carnitine, 500 mg TID	
	23:0	Triosanoic	H		
Trans-fatty acids	16:1t	Palmitelaidic	H	Eliminate hydrogenated oils	Eicosanoid interference
	18:1t	Elaidic	H		
Ratios and Indexes		LA/DGLA	H	GLA source*	6 desaturase enzyme
		DGLA/EPA	L	Black currant oils	Eicosanoid imbalance
			H	Fish oils	Eicosanoid imbalance
		AA/EPA	H	Fish oils	Eicosanoid imbalance
		Triene/Tetraene	H	Essential oils	Essential f.a. deficiency
		Vaccenic/Palmitoleic	L	Biotin	Elongase – desaturase inhibition
		Stearic/Oleic (rbc)	L	See comments	Cancer marker
		Total Fatty Acids (pl)	H	Evaluate hyperlipidemias	Serum triglyceride (VLDL)

* Sources of GLA: Evening Primrose oil or Black Currant oil

^^ **M**ultiple **A**cyl-coenzyme A **D**ehydrogenation disorders

Understanding Fatty Acids

Fatty acids are common denominators for all life forms. The same oleic acid that is found in the cell membranes of olive trees is also critical for the cellular structure and functions of bacteria, fungi, and humans. Plant seeds are filled with fatty acids as they are the most efficient form of stored energy. Thin layers of fatty acids form the membranes that separate the inside of living cells from the outside. Powerful hormones are formed from essential fatty acids. Fatty acids flow into multiple metabolic pathways where their constant supply is critical to normal human function. The essential nature of dietary fats and oils in maintaining health is widely appreciated by nutritional scientists, but the public and many health professionals continue to perceive fat as bad and make no distinction between good and bad fats. Rather, the perception is that all fat is bad.

FIGURE 5-1. Metabolic Roles of Fatty Acids

Constant supply of fatty acids from diet or hepatic synthesis is necessary to meet the demands for energy production, cell membrane function, and eicosanoid synthesis. The widths of the arrows in this figure vary according to the magnitude of flow to supply each function.

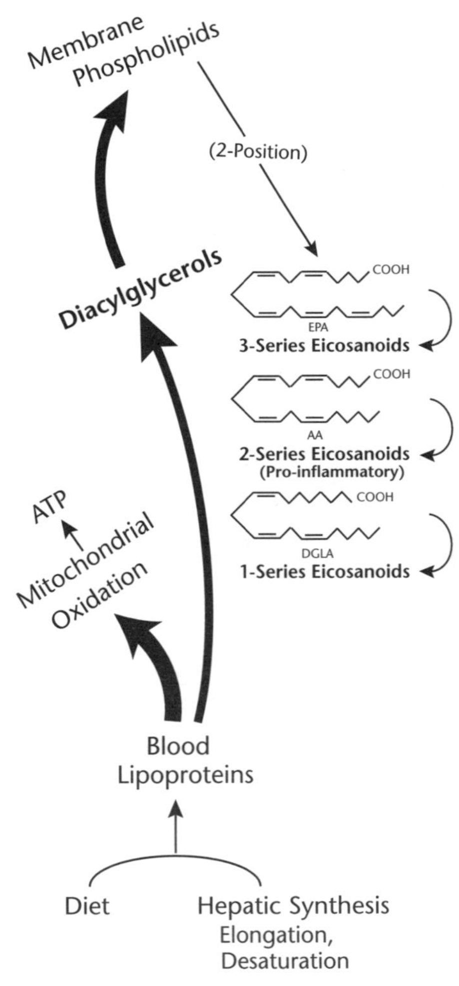

Years of negative associations of dietary fat with calories, body fat, cholesterol, and cancer have resulted in a general attitude that foods containing fats are simply to be avoided. Many food manufacturers have taken advantage of this attitude by modifying fat content and labeling wherever possible to tout "low fat" foods. Advertisements for such foods further instill the notion that dietary fat is bad. Amid the clamor over the largely mistaken problems associated with dietary fats, many very real problems have been created by the large-scale use of modified fats by food suppliers. Individuals do not feel the effects of poor dietary fat intake on the short term because of the presence of many protective mechanisms that cause health threats from fatty acid deficiencies and imbalances to be insidious and unrecognized.

Modern diets of fast foods and packaged dinners tend to be rich in saturated fats and hydrogenated oils and are frequently lacking in the balance of essential fatty acids. We now know that not only the amount but also the type of dietary fat plays a major role in maintaining health. The old concept of three essential fatty acids (linoleic, alpha linolenic, and gamma linolenic) has been replaced by one of critical roles for multiple fatty acids. Dietary fats simultaneously supply the major cellular energy source, influence the function of all biological membranes, determine the integrity of nerve tissue, and serve to form powerful paracrine hormones (Figure 5-1). Polyunsaturated essential fatty acids flow into pathways that supply eicosanoids, which are cell-regulators. The numerous, closely related products that are formed are classified as 1-, 2-, or 3-series eicosanoids, depending on the number of double bonds present. More about these functions will be discussed after general concepts of fatty acid testing and the structural relationships needed for understanding naming conventions are explained.

Control of Fatty Acid Supply and Distribution

Fatty acids are present in the diet as triglycerides in solid fats or liquid oils and as phosphatides in cell membranes of whole foods. They are rarely present in nature as free fatty acids.

Fats are digested first by the action of bile acids and then pass into intestinal epithelial cells along with cholesterol, where they are associated with proteins to form chylomicrons (Figure 5-2). The chylomicrons are passed into the lymphatic system to flow directly through the heart into systemic circulation without action by hepatic enzymes. Enzymes in capillary cell walls cause the transfer of fatty acids from the chylomicrons, which become greatly reduced in size and return for uptake by the liver as chylomicron remnants.

Because of the critical life-supporting functions in forming cell membranes and supplying energy sources and hormone controls, there are mechanisms for assuring that the supply of fatty acids will be continuous, even during short intervals between meals. In the fasting state, fatty acids are produced from either glucose or amino acids or are mobilized from adipose stores. The liver handles the role of supply depot by forming another class of lipid-protein particles called very low-density lipoproteins (VLDL). These particles deliver fatty acids and cholesterol to the tissues by mechanisms similar to those for chylomicrons, except that the entire remnant LDL particles are taken up by binding to receptor sites on extrahepatic cell surfaces. By ensuring a constant supply of fatty acids, the body controls one of the most critical materials for carrying out the various life-sustaining functions of growth and repair.

FIGURE 5-2. Total Body Supply and Distribution of Fatty Acids

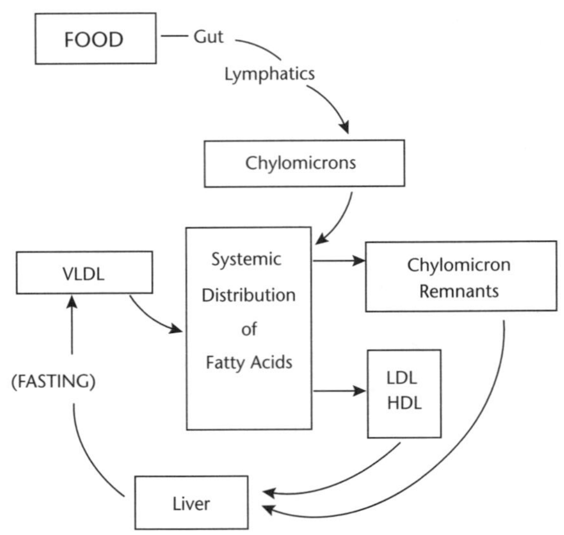

Names for Fatty Acids

There are about forty physiologically significant fatty acids. (We will not deal with derivatives such as hydroxy- and branched fatty acids.) They comprise a set of compounds with similar structure and function. One 20-carbon fatty acid, (stearic) can increase the risk of heart disease, while another (eicosapentanoic acid), that also contains a similar straight chain of 20 carbon atoms can help prevent the disease. The difference lies in the effects of removing a few atoms of hydrogen. Even more subtle differences in bond positions lead to powerful factors that determine the progression of major diseases, define thresholds to pain, and direct responses of the immune system. In order to follow discussions of their clinical relationships it is helpful to acquire familiarity with the basic rules by which fatty acids are named and classified.

Fatty acids are named in various ways, including those with Latin prefixes (penta-, hexa-, hepta-...) for the number of carbon atoms. Hexadecanoic acid is the sixteen-carbon fatty acid known commonly as palmitic acid. Wherever they exist, common names will be used because they are the most widely used among clinicians. Generally, the common names are also shorter and easier to pronounce. Fatty acids also may be designated by a short hand, structural nomenclature in which the number of carbon atoms is followed by a colon and the number of double bonds. Palmitic acid has no double bonds, so it is designated 16:0. Fatty acids that have no double bonds are called "saturated." Short-, medium-, long-, and very long-chain are terms referring to families of fatty acids with less than 10, 10 to 14, 16 to 20, and more than 20 carbon atoms, respectively. Figure 5-3 shows common names (except eicosapentaenoic acid, which has no common name), structural representations, and both delta (Δ) and omega (ω or n) shorthand for six fatty acids.

FIGURE 5-3. Naming Conventions for Unsaturated Fatty Acids

The omega (ω) and delta (Δ) schemes are shown with numbers labeling the double bond positions in the carbon chains, counting from the methyl and carboxyl ends, respectively. Names and designations according to the Δ and ω (or n) shorthand are shown below each fatty acid structure.

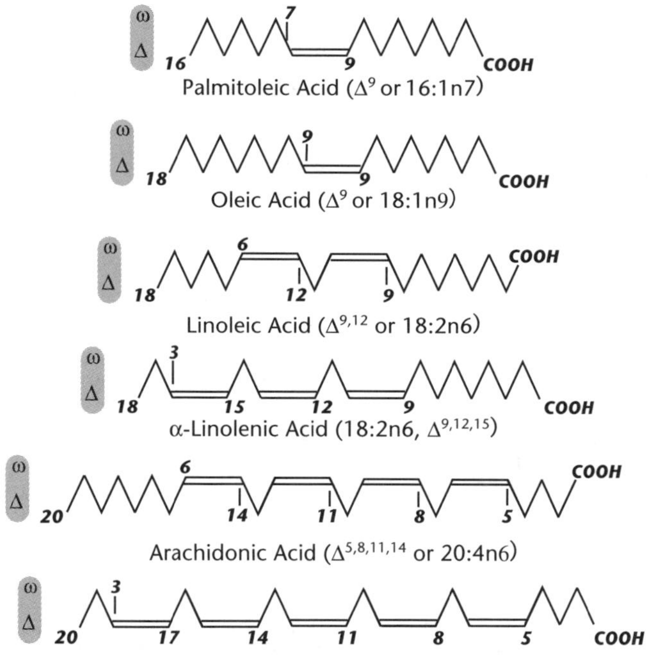

Elcosapentaenoic Acid ($\Delta^{5,8,11,14,17}$ or 20:5n3)

Normal fatty acids in humans may be placed into families with structural similarities. They contain 10 to 26 carbon atoms linked in straight chains. Fatty acids can be synthesized from the acetyl-CoA derived from carbohydrate, protein, and other non-fat sources. Plant and animal enzymes build the chains by linking two-carbon units, producing even-numbered chains. The human hepatic fatty acid synthase enzyme complex produces saturated fatty acids, predominantly palmitic acid (16:0).

Fatty acids containing the maximum number of carbon-hydrogen bonds are called "saturated." They have higher energy or caloric yield than corresponding unsaturated fatty acids. They are present as major components of most foods and are very high in foods such as candy bars. The most abundant saturated fatty acids in human tissues are those that are 14 (myristic), 16 (palmitic), or 18 (stearic) carbons long. The elongation process can be repeated to yield members that are 20, 22, 24, and 26 carbons long. Although the very long chain fatty acids (24 & 26) are minor components of the lipid membranes of the body, they perform valuable functions, apparently helping to stabilize membranes, especially those in peripheral nerve cells.

Double bonds are introduced through a separate process called desaturation. Palmitic acid can be desaturated, forming the n7, mono-unsaturated fatty acid, palmitoleic acid (16:1n7). The Δ^9 desaturase enzyme may operate similarly on other saturated fatty acids (Figure 5-4). In scientific literature, the use of "n" is replacing the older Greek letter, omega (ω), as a position designator. The term omega was chosen because this system counts from the carbon most distant from the carboxyl group. Thus, palmitoleic acid is also written "16:1ω7," and it is a member of the "omega-7" unsaturated fatty acid family. Both "n" and "ω" are shorthand ways to designate that you find the double bond by counting seven carbon atoms from the methyl end of the fatty acid.

FIGURE 5-4. Desaturation of Saturated Fatty Acids

The desaturase enzyme can operate on fatty acids of various chain lengths, placing double bonds up to nine carbons away from the carboxyl group. The first carbon atoms in the double bonds that are produced are at varying positions from the methyl ends of the monounsaturated fatty acid products.

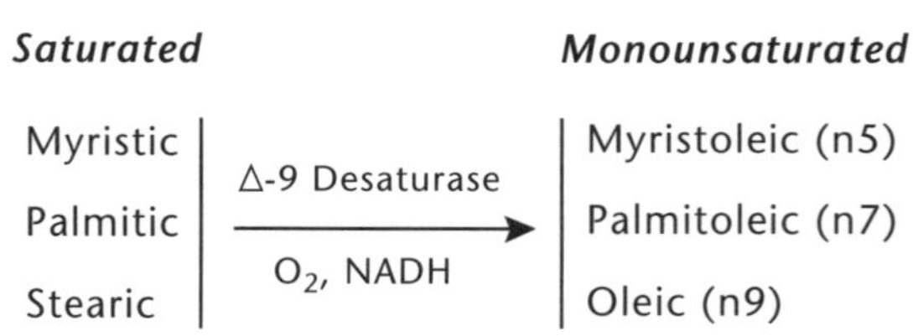

In other literature, yet another naming scheme is used. In this system, the Greek letter, delta (Δ), is superscripted with numbers that specify the positions of all double bonds, counting from the carboxyl or number 1 carbon (Figure 5-3). Linoleic acid is designated as $18{:}2\Delta^{9,12}$ because the double bonds are found at the ninth and twelfth positions from the carboxylic acid carbon.

Since three carbon atoms always separate the double bonds, knowing the position of the first gives the position of all others. The shorthand most frequently used in recent years takes advantage of this relationship to simply state the total number of carbon atoms, the number of double bonds, and the position of the first double bond. Linoleic acid is 18:2n6. Counting from the methyl end of the fatty acid, the first of two double bonds is at the sixth carbon so the second double bond must be at the 6+3 or ninth position.

FIGURE 5-5. Conversion of Linoleic to Arachidonic Acid

After attachment to coenzyme A, a double bond is inserted at the Δ^6 position of linoleic acid to form γ-linoleic acid. The elongase system adds a two-carbon unit at the carboxyl end, and another desaturation step inserts the fourth double bond at the Δ^5 position to form arachidonic acid.

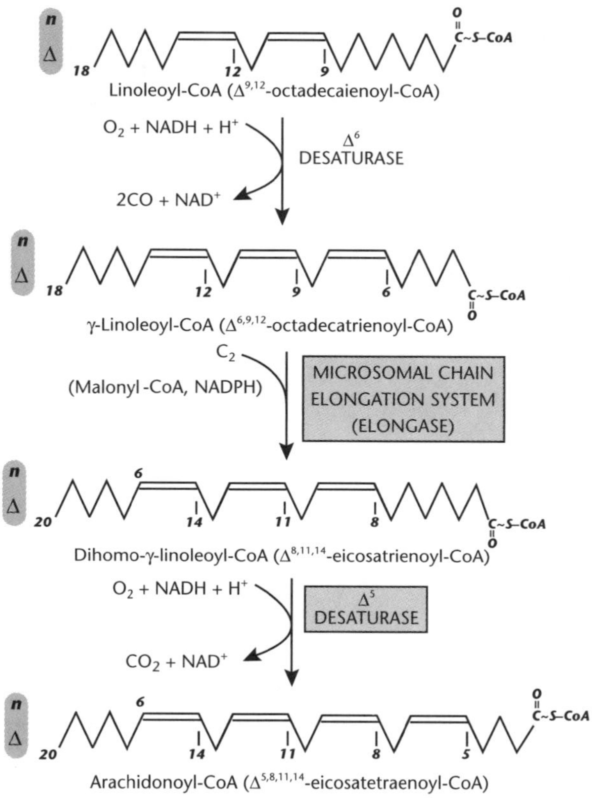

Fatty acids can also be lengthened by elongation reactions that add two-carbon units to the carboxyl end. Linoleic acid (18:2n6) is thus desaturated to form gamma-linolenic acid (18:3n6), which may be lengthened by two carbons to form dihomogamma-linolenic acid (20:3n6) and, finally, desaturated to form arachidonic acid (20:4n6) (Figure 5-5). The n6 and n3 classes of unsaturated fatty acids must be provided from dietary polyunsaturated fats as precursors for desaturation and elongation sequences or pre-formed long-chain polyunsaturated fatty acids.

Members of the n6 and n3 families of fatty acids are listed in Table 5-2. These families are defined by the position of the double bond closest to the terminal methyl group of the fatty acid molecules. In the n6 family, the first double bond occurs between the sixth and seventh carbons, whereas in the n3 family the first double bond occurs between the third and fourth carbons.

Dietary n3 or n6 fatty acids can be lengthened and desaturated but members of the two classes cannot be interchanged because of the positions of the double bonds. The desaturation and elongation reactions can be repeated in various combinations, giving an array of

TABLE 5-2. Polyunsaturated Fatty Acid Families

n6 or omega-6	
18:2	Linoleic (LA)
18:3	Gamma Linolenic (GLA)
20:3	Dihomogamma-linolenic (DGLA)
20:4	Arachidonic (AA)
n3 or omega-3	
18:3	Alpha linolenic (ALA)
20:5	Eicosapentaenoic (EPA)
22:5	Docosapentaenoic (DPA)
22:6	Docosahexaenoic (DHA)

FIGURE 5-6. Elongation and Desaturation Sequences

For fatty acids containing 18 or more carbon atoms, human enzymes cannot insert double bonds in the n6 or n3 positions, but they can elongate or desaturate at Δ^4, Δ^5, and Δ^6. The sequences of reactions produce families of fatty acids of chain lengths of 18 to 22 carbons containing one to six double bonds. The position of the first double bond from the methyl end defines the family of each product as n9, n6, or n3. Linolenic, linoleic, and oleic acids compete for the active site of desaturase enzymes in decreasing order of binding affinity.

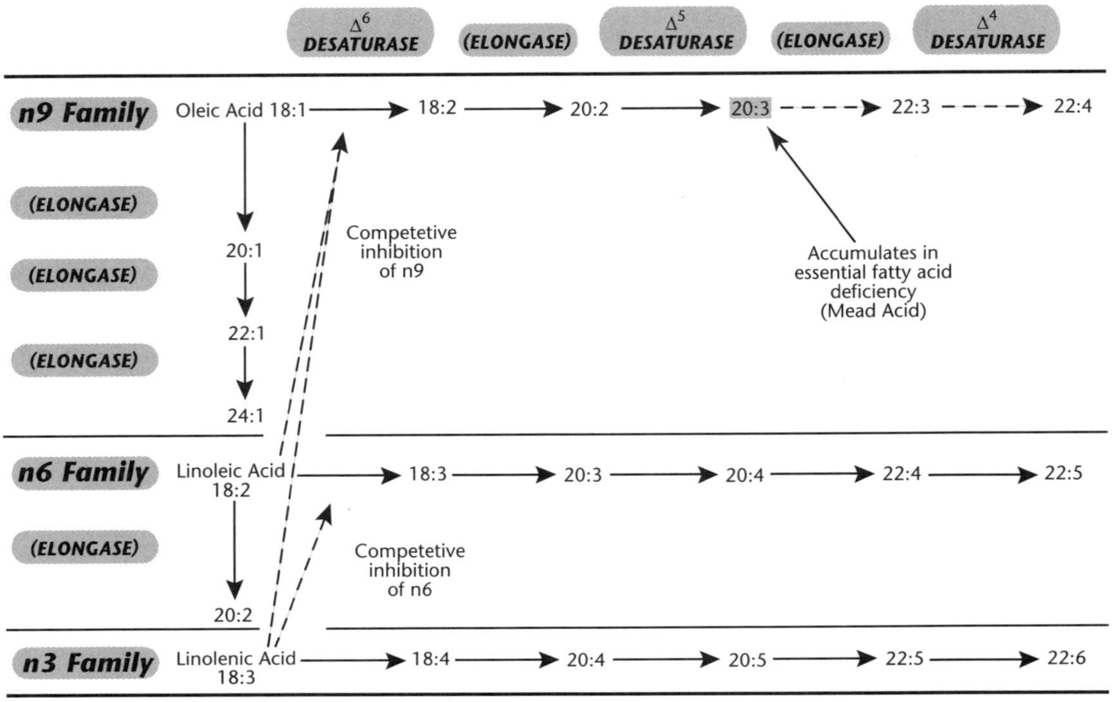

saturated and unsaturated fatty acids for use in the essential functions of tissue maintenance (Fiure 5-6).

The desaturase enzymes function to place double bonds at positions up to nine carbons from the carboxyl end of the molecules. When you count from the other end, the position varies, depending on the length of the fatty acid. Thus, for stearic acid with 18 carbons, a desaturase can form a double bond nine carbon atoms from the carboxyl end, which is also nine carbons from the methyl end (18 - 9 = 9). These differences become important because the type of eicosanoid hormones that can be formed later depend on the position of the first double bond from the methyl end.

The geometry of the desaturase enzyme will not allow insertion further than nine carbons from the carboxyl group (Δ^9). This is the reason that linoleic acid ($\Delta^{9,\,12}$) cannot be synthesized in humans. The human enzyme cannot reach the position 12 atoms from the carboxyl group. Saturated or unsaturated fatty acids of various chain lengths can be acted upon by this enzyme. By utilizing Δ^6 and Δ^5 desaturase enzymes in sequence with an intervening elongation step, human hepatic cells can convert linoleic acid into arachidonic acid. Note that coenzyme A (derived from pantothenic acid) is used as a carrier for all of these reactions. The activities of the desaturase enzymes are critical for maintaining the ratio of saturated and unsaturated components in cell membranes.

Polyunsaturated Fatty Acid Metabolism

Dietary polyunsaturated fatty acids (PUFAs) are largely composed of two groups of fatty acids: the n6 family (*e.g.*, linoleic), which is abundant in vegetable seed oils, and the n3 family (*e.g.*, alpha-linolenic), which is high in vegetable leaves and modest in soybean oil [1]. When the diet supplies adequate linoleic and alpha linolenic acids, they may be used to form the longer chain-length eicosanoid precursors (Figure 5-6).

The enzyme, Δ-6-desaturase, acts as a gateway for the flow of fatty acids through the desaturation and elongation pathway. Although it can act on any long chain fatty acid, the substrate binding affinity increases greatly with the number of double bonds. Thus, linolenic acid (18:3) binds stronger than linoleic acid (18:2), which binds stronger than oleic acid (18:1). The result is that, in the presence of appreciable amounts of the more unsaturated precursors, linolenic and linoleic are bound and quickly converted to their 20:4 (AA) and 20:5 (EPA) products, while oleic acid, although more abundant, is slowly converted into 20:3n9 (Mead acid). When the essential fatty acids are deficient, however, Mead acid accumulates because the desaturase is free to act on oleic acid. The implication of Mead acid presence is further discussed below under polyunsaturated fatty acids. The several markers of essential fatty acid deficiency are:

1. Low levels of ALA, LA, and EPA
2. Elevated Mead acid.
3. Elevated palmitoleic acid
4. Elevated triene/tetraene ratio

A patient who stops eating dietary sources of all essential fatty acids goes through phases of fatty acid metabolism changes that cause markers to appear in sequence (Figure 5-7). The changes are reflected in plasma fatty acid profiles. The initial event is depletion of adipose tissue levels of polyunsaturated fatty acids. When the concentrations reach levels low enough to free up binding sites on the desaturase enzyme, oleic acid will be the first non-essential fatty acid to take their place because of its higher binding affinity relative to the more abundant palmitic acid. Formation of Mead acid is the result of this phase. The ratio, Mead/Arachidonate (T/T ratio discussed later), provides a sensitive marker of the combined effect of non-essential fatty acid desaturation and essential fatty acid deficiency. In cases where Mead acid is high with normal arachidonate, there is evidence of recent intake of arachidonate sources with long-term deficiency of other essential fatty acids. As the deficiency continues, palmitic desaturation to palmitoleic increases, yielding abnormally high palmitoleic acid. The exact course of appearance in each case will depend on which dietary sources are most severely restricted.

FIGURE 5-7. Relative Time Courses for Appearance of Essential Fatty Acid Deficiency Markers

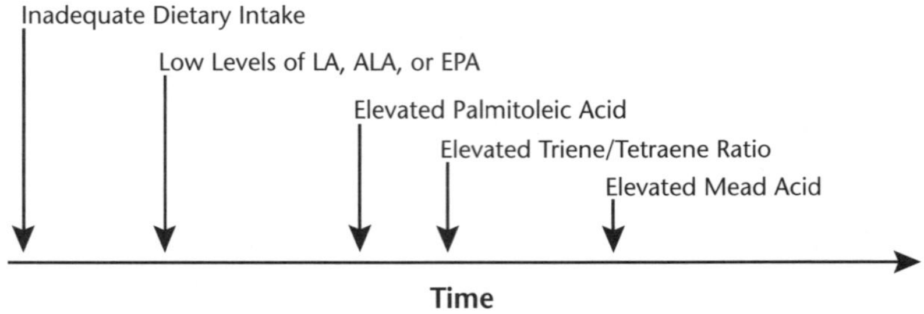

The clinical significance of these interactions may be seen in the reported increase in infant's plasma DHA when the LA:ALA ratio in infant formulas was lowered from 10.1 to 5.1 [2]. The higher ratio might be found in people who use high corn oil diets. In this study, when the ratio was lowered so that LA (18:2) was no longer so abundant, the ALA (18:3) easily passed into the pathway for conversion to DHA. This phenomenon is especially important for pregnant women. See DHA below for discussion of the effects on childhood development.

Δ-6-desaturase enzymes require zinc. For a patient with low EPA, adding ALA-rich flax oil will do little good if zinc deficiency is present. The changes in body growth, organ weights and lipid concentrations of plasma and liver produced by zinc deficiency are reversed by addition of γ-linolenic acid (18:3n6) as primrose oil, but not by linoleic acid (18:2n6) as safflower oil [3]. This data shows the role of zinc in the Δ-6-desaturase enzyme that inserts the extra double bond between carbons 6 and 7 of linoleic acid. The enzyme is also inhibited by magnesium deficiency, elevated insulin, and high concentrations of saturated, monounsaturated, and *trans* fatty acids. *Trans* fatty acids compete for the binding site on the enzyme and yield products that are of less significance to the cell.

The metabolism of PUFAs is also affected by vitamin B_6 deficiency. Mice made B_6 deficient and fed a diet high in ALA show impaired activities of delta 6-dehydrogenase and acyl-CoA oxidase enzymes. The oxidase enzyme is a gateway to peroxisomal fatty acid oxidation. The most pronounced effect of B_6 deficiency is reduction in the production of DHA [4].

Fatty Acid Laboratory Reports

Quantitative profiles of plasma fatty acids are done by the method of gas chromatography with sensitive detection techniques. Fatty acids are extracted from the specimen and converted into volatile derivatives. The analysis of fatty acids in blood provides useful information about fatty acid deficiencies and imbalances and can reveal functional need for specific vitamins and minerals. The laboratory report shows the impact of diet and disease on circulating fatty acid patterns. Clinical interventions involve changes in diet or supplementation with specific dietary oils and needed micronutrients.

In order to aid interpretation, the fatty acids are usually grouped on the report according to structural class. By grouping similar members of a class, relationships that reveal a patient's ability to convert the more simple to the complex members are more easily examined. Classes that include the dietary essential polyunsaturated classes are normally listed first, followed by the mono-unsaturated and saturated classes.

Several ratios may be calculated to draw attention to significant imbalances even when individual concentrations are normal. Summing individual components and expressing the total in mg/dl may be done to calculate the total fatty acid concentration in plasma. The value obtained is roughly equivalent to that for serum triglycerides from an automated serum chemistry report. The profiling of plasma fatty acids may be accurately thought of as a very detailed analysis showing which fatty acids make up the structures called lipoproteins. The option of analyzing fatty acids in erythrocyte membranes is becoming widely available, raising the question of which laboratory profile to order for each case.

Plasma vs. Erythrocyte Fatty Acid Profiles

Since almost all of the fatty acids in plasma are present as glycerol or cholesterol esters contained in lipoproteins, hypertriglyceridemic patients are also hyperlipoproteinemic. These patients will also show elevated plasma fatty acid patterns. An erythrocyte membrane fatty acid profile does not similarly reflect hypertriglyceridemia.

Fatty acids are present in serum or plasma primarily in the form of lipoprotein particles, of which the LDL species is the most abundant in normal fasting plasma. The small fraction of free fatty acids in blood are bound to transport proteins, principally serum albumin. Profiles report individual concentrations of each fatty acid instead of measuring only the total of all fatty acids present as in a serum triglyceride assay.

Patients with high serum triglycerides have higher levels of some fatty acids in plasma. Conditions, such as insulin insensitivity, that lead to elevated serum triglycerides do so by stimulating endogenous synthesis and transport. When glucose or amino acids are used to build fatty acids in liver or adipose tissue, the principal product is the sixteen-carbon long palmitic acid. Plasma palmitic acid has a linear relationship with serum triglyceride concentration (Figure 5-8). Thus an elevated palmitic acid on a plasma profile of fatty acids is evidence of elevated serum triglycerides and the degree of palmitate elevation is directly proportional to the degree of triglyceride elevation.
The relationship may be expressed quantitatively as:

$$\text{Serum Triglycerides} = \frac{(\text{Palmitic Acid}_{\text{serum}} - 142)}{0.83}$$

where Palmitic Acid$_{\text{serum}}$ is the concentration in µmol/L and serum triglycerides are in units of mg/dl. This relationship does not apply at palmitic acid levels below 200/μmol/L.

Other fatty acids do not have such linear relationships with serum triglycerides. Most importantly, total triglyceride levels have little effect on concentrations of the essential fatty acids ALA, GLA, and EPA in plasma. Concentrations of these fatty acids in fasting plasma reflect their composition in adipose tissue. Adipose tissue fatty acid composition is affected by dietary intake and rates of endogenous synthesis.

An individual's pattern of high and low levels will depend on the total effects of endogenous synthesis and recent dietary intake. Because plasma fatty acid levels reveal stored fatty acids, plasma profiling has been frequently used in studies of fatty acid status and dietary intake [5-8]. Some researchers have reported simultaneous plasma and erythrocyte data from which additional insight is gained regarding fatty acid metabolism. For example, the higher levels of long-chain linoleic acid metabolites in erythrocyte phospholipids versus plasma lipoproteins was given as evidence of the importance of assaying the very long chain members [9]. In addition, a finding of simultaneous decreased plasma and erythrocyte levels of DHA helped support the conclusion that premature infants may have higher n3 requirements than term infants [10].

FIGURE 5-8. Linear Relationship Between Plasma Palmitic Acid and Serum Triglyceride Concentrations

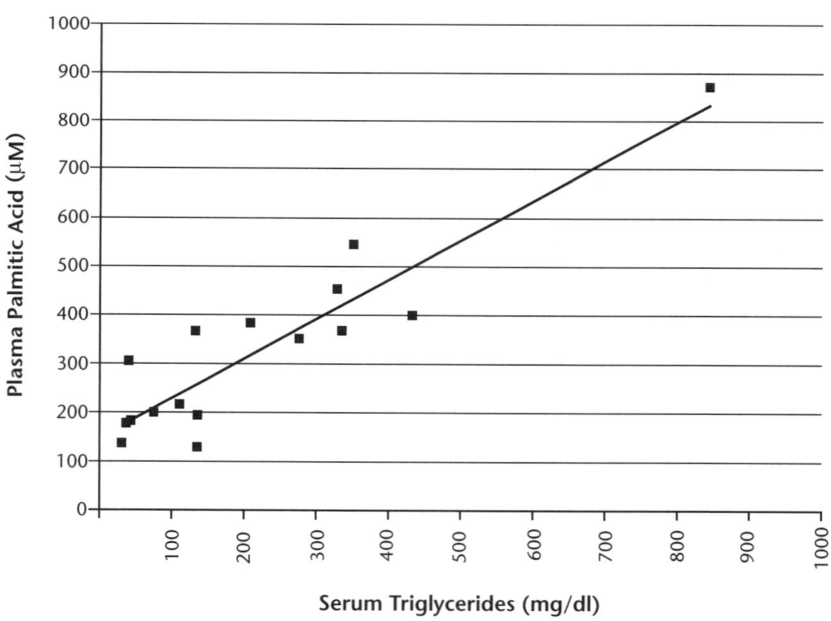

The erythrocyte membrane is 45% fatty acids in the form of various phosphatides and glycolipids. The most common procedure for erythrocyte fatty acid profiling measures concentrations of individual fatty acids in packed erythrocytes. Fatty acid levels in erythrocytes differ from plasma profiles in that they are not as representative of recent dietary intake and adipose composition but rather provide information about the patient's ability to carry out metabolic transformations within the cell. Dietary intake factors are seen on a longer time scale as they influence newly formed red cell populations. Because of the very high binding affinities of elongation and desaturation enzymes for LA, ALA, and GLA, the concentrations of these compounds are lower in membranes relative to their levels in plasma lipoproteins. Drugs that alter fatty acid metabolism will enhance the differences between plasma and erythrocyte fatty acids levels. The triglyceride-lowering drug, gemfibrozil, causes a profound suppression of erythrocyte unsaturated fatty acids even in the presence of high concentrations of the same compounds in plasma. The effect is apparently due to stimulation of the peroxisomal beta-oxidation system [11]. Erythrocyte levels of the long-chain metabolic products of elongation and desaturation are representative of fatty acid content of other tissues. For example, the DHA composition of brain tissue is directly correlated with that in red blood cells in rat and primate studies [12].

Concentration vs. Percentage Reporting of Fatty Acids

Prior to 1996, virtually every clinical laboratory reporting fatty acids had little choice but to express each as a percentage of chromatographic peak areas for all fatty acids present. Actual concentrations could not be reported due to the lack of pure standards necessary to build calibration tables. With the availability of pure standards, it is now possible to measure the concentrations of each fatty acid. In many research studies, it is preferable to express results as percentages of total extracted lipid because the research is not concerned with the question of total body fatty acid composition.

For the routine application of assessing nutritional needs for essential fatty acids, there is little data to support arguments for maintaining the old system of calculating ratios and expressing results as percentages of total. While it is tempting to think that one should normalize for the effect of serum lipoprotein variations, valuable information can be lost by ratio calculations. This is especially true for profiling of erythrocyte or other cell membrane composition.

The reporting of "relative area" or percent data increases the chance of false negatives and false positives. The most abundant fatty acids may be present at concentrations 50 times higher than those of the smaller components [9]. In percent area calculations, variations in the most abundant components have disproportionate impact. If linolenic acid is the same in two cases, for example, while palmitic acid is low in the first and high in the second, the percent linolenic can appear normal in the first and low in the second (Table 5-3). This is because palmitic acid makes up so much of the total and the ratios of the minor components, like linolenic acid, are calculated by dividing by the total. The clinically relevant issue - whether there is a need to increase linolenic acid - is revealed only by reporting individual concentrations, not percentages.

TABLE 5-3. Erythrocyte Fatty Acid Percentage vs. Concentration
When reported as percentages of total fatty acids, fluctuations in a major component can cause large variation in the percentages of all minor components. The doubling of palmitic acid from "Case A" to "Case B" has no effect on the absolute concentration of α-linolenic acid, while it causes the percent value to shift by a factor of two.

	Case A	*Case B*
	Concentration	
Total	2000	4000
Palmitic	1000	2000
α-Linolenic	1 Low	1 Low
	Percentage	
Palmitic	50	50
α-Linolenic	0.05 Normal	0.025 Low

Interpreting Fatty Acid Data

Quantitative reports of all fatty acids that are present at significant concentration in human tissues are now routinely available. The clinical importance of fatty acids lies in both deficiency effects and metabolic problems due to high levels. Childhood development and adult disease resistance are dependent on fatty acid status. Fatty acid status, in turn, is dependent on specific micronutrient supply as well as the quality of dietary fat. A summary of signs and symptoms associated with specific fatty acid abnormalities is presented in Table 5-4. Some clinical associations of fatty acid families will be discussed, followed by detailed associations of each of the individual fatty acids.

Fatty Acids and Human Diseases

Prenatal and neonatal fatty acid supply has lasting effects on childhood development. This should be no surprise when we recall that every tissue growth process depends on the construction of new biological membranes. Once in place, the membranes of every cell become a source of supply for fatty acids that regulate cell function. Thus, the fatty acid composition of tissues laid down in childhood determines health status into adult years. Pre-term infants deprived of vital fatty acids during late pregnancy are likely to have failures of normal development (especially the visual system) unless n3 fatty acids are supplemented at birth [10, 12]. Children with attention deficit hyperactivity disorder have lower concentrations of key fatty

TABLE 5-4. Signs and Symptoms Associated with Fatty Acid Abnormalities

Signs & Symptoms	Fatty Acid Association	Intervention
Emaciation, weakness, disorientation	Caloric deprivation	Balanced dietary fat, protein, and carbohydrate
Reduced growth, renal dysplasia, reproductive deficiency, scaly skin	Classic essential fatty acid deficiency	Good quality fats and oils
Eczema-like skin eruptions, loss of hair, liver degeneration, behavioral disturbances, kidney degeneration, increased thirst, frequent infections, poor wound healing, sterility (m) or miscarriage (f), arthralgia, cardiovascular disease, growth retardation	Linoleic acid insufficiency	Corn or safflower oils
Growth retardation, weakness, impairment of vision, learning disability, poor coordination, tingling in arms/legs, behavioral changes, mental disturbances, low metabolic rate, high blood pressure, immune dysfunction	Alpha or gamma linolenic acid insufficiency	Flax, primrose, or black current oils
Depression, anxiety, learning behavioral and visual development, or cardiovascular disease risk	Long chain PUFA-dependent neuromembrane function, prostanoid balance	Fish oils Avoid hydrogenated oils
Cancer	Low stearic to oleic ratio, prostanoid imbalance	n3 PUFAs (use n6 PUFAs with caution)
Rheumatoid arthritis	Low GLA & DGLA	Primrose oil
Myelinated nerve degeneration	Increased very long chain fatty acids	High-erucate rape or mustard oils
Fatty liver	Saturated and ω9 accumulation in liver	Restrict alcohol Add lecithin Increase methionine

acids (AA and DHA) in plasma and in red blood cells [13, 14]. Of the many adult manifestations of fatty acid abnormalities, heart disease and cancer top the list for mortality and intensity of investigation relative to physiology of fats.

Higher palmitic and lower n3 fatty acids in serum are correlated with higher incidence of coronary heart disease in middle-aged men at high risk for cardiovascular disease [15]. Improvements in plasma fatty acids and vitamins E and C were the only factors found related to improvements in life expectancy and 70% lowering of heart disease in one study population [1].

Because of reports showing various degrees of increases in serum triglycerides with higher saturated fat intake, lowering intake of fat in the American diet has long been a goal of the American Heart Association and other consumer information groups. Although it may seem enigmatic, increased intake of good fats reduces risk factors for cardiovascular disease. Serum triglyceride and cholesterol may be controlled with garlic (900 mg/day) and fish oils (12 gm/day) [16]. Almonds and walnuts were shown to lower cholesterol and triglycerides [17]. Individuals on a high fish oil intake are protected from arrhythmias arising at the onset of ischemia and reperfusion [18]. All of these effects are mediated by the beneficial actions of polyunsaturated fatty acids.

Drugs designed to lower blood triglyceride levels may cause long-term undesirable effects because their actions lead to worsening of essential fatty acid status. Gemfibrozil lowers total plasma triglycerides and cholesterol by selectively increasing the peroxisomal oxidation of unsaturated fatty acids. In addition to the effects noted above regarding changes induced in plasma and erythrocyte fatty acids, gemfibrozil increases plasma palmitoleic acid while lowering linoleic acid [19]. Valuable answers to the question of how much a given patient will benefit from increasing good dietary fats versus using drugs to control serum triglycerides may be found in laboratory profiles of fatty acids.

Tumor tissue and virus-transformed cells have a higher content of unsaturated fatty acids. Oleic acid increases relative to the amount of stearic acid as a part of the process of tumor development. The effect in tumor cells is due to increased activities of enzymes that convert stearic acid into oleic acid, rather than a preferential oxidation of stearic acid. The resulting shifts in the proportion of saturated to unsaturated fatty acids increase the metabolic rates of many lipid-dependent enzymes and are associated with a higher capacity for cell division [20]. The mean value for the stearic/oleic ratio in healthy people is 1.5 while patients with malignancies consistantly have values below 1.0 [21].

If tumor growth causes changes in membrane fatty acids, then one might guess that changes in dietary fatty acid composition would have effects on tumor growth. Antitumor activity is modulated by dietary fat, which affects macrophage function of the immune system. The macrophage plays an important role in anti-tumor activity. For example, menhaden oil was shown to suppress macrophage activity more than safflower oil in mice [22].

Nutritional deficiencies of vitamins and minerals have direct impact on fatty acid metabolism (see Table 5-1). As described above, zinc is required for critical steps in the conversion of dietary fatty acids into powerful cell control molecules (prostanoids). For this reason, it may be said that the clinical effects of zinc deficiency are due to induced essential fatty acid deficiency. Copper deficiency causes an increase in stearic acid and a decrease in oleic acid, resulting in measurable loss of membrane fluidity [23]. This relationship takes on extra significance when coupled with the finding that copper is depleted following myocardial infarction [24]. Testing for depletion of oleic acid and dietary changes involving the inclusion of olive oil are indicated for patients with signs of heart disease.

Adrenoleukodystrophy and related conditions are associated with the accumulation in nerve sheath membranes of saturated fatty acids more than 20 carbons long [25]. The sum of concentrations of the 22 through 26-carbon length, even-numbered fatty acids in erythrocytes may be used as an indicator for such accumulation. This situation may be made worse by the lack of sufficient long-chain polyunsaturated fatty acids.

The Saturated Family of Fatty Acids

Capric acid (10:0), Lauric acid (12:0), and Myristic acid (14:0)

These medium-chain fatty acids accumulate in the fatty acid catabolic disorders known as multiple acyl-coenzyme A dehydrogenation (MAD) disorders [26]. Some patients with these enzyme defects are responsive to riboflavin at levels far above normal intakes [27, 28]. All patients with the genetic defect called long-chain acyl-CoA dehydrogenase (LCAD) deficiency have elevated myristic acid due to impairment in the oxidation of oleic acid. The blockage causes a partial oxidation with release of myristic acid. When the electron-accepting role of riboflavin nucleotides is increased, the oxidation can proceed to completion.

A pattern of low levels of the shorter chain fatty acids and high levels of longer chain fatty acids may indicate a fatty acid restricted diet. In fatty acid restriction, there is a stimulation of hepatic palmitic acid synthesis and elongation enzymes.

Palmitic acid (16:0)

The liver can convert fatty acids into cholesterol. Any fatty acids can enter this pathway and stimulate LDL cholesterol synthesis, increasing the risk of atherosclerosis, cardiovascular disease, and stroke. Fatty acids also modulate the production of LDL receptors, leading to variations in the removal of LDL particles from blood.

The effect of high palmitic acid intake on serum lipoprotein levels depends on the total composition of the diet. On a low cholesterol diet, palmitic acid has the greatest effect of increasing LDL cholesterol when compared with myristic or stearic acids [29]. When linoleic acid is adequate, however, there is no effect of dietary palmitic acid on total cholesterol [30]. Substituting palmitic acid for lauric and myristic acid-rich coconut oil can actually raise high-density lipoprotein cholesterol [31], which is beneficial for cardiovascular health. These data emphasize the importance of profiling to assess the relationships between fatty acids when proposing dietary and supplemental fat changes. Most other fatty acids are either cholesterol neutral or have reverse effects, as in the case of EPA.

Palm kernel and coconut oils are rich sources of palmitic acid (Table 5-5). Palmitic is also the principle fatty acid product of human fatty acid synthesis that is stimulated by dysinsulinemia and by diets high in simple carbohydrates. Thus, both poor quality dietary fat and high simple sugar intake contribute to elevated palmitic acid in plasma or erythrocyte membranes. These factors generally result in plasma fatty acid profiles with multiple elevations of saturated fatty acids, especially palmitic and stearic. Patients with chronic hypercortisolemia have shown unusual patterns of greatly elevated palmitic and palmitoleic acids with normal levels of other members of these classes, indicating stimulation of hepatic fatty acid synthase activity.

The overall effect of higher levels of dietary polyunsaturated fatty acids is suppression of hepatic lipogenesis [32], which can help to lower serum triglyceride levels. This finding confirms the desirability of lower palmitic acid levels. Normal plasma and erythrocyte palmitate levels are associated with diets that have good fatty acid balance.

TABLE 5-5. Dietary Sources of Even-numbered and Long-chain Saturated Fatty Acids (See Table 5-7 for greater detail)

	Name	Source
10:0	Capric Acid	Most vegetables and butter
12:0	Lauric	Palm kernel, coconut, laurels, butter
14:0	Myristic	Nutmeg, palm kernel, coconut, myrtle
16:0	Palmitic	Palm and coconut
18:0	Stearic	Animal fats
20:0	Arachidic	Peanut (arachis)
22:0	Behenic	Nuts & seeds
24:0	Lignoceric	Brazil, peanuts

Stearic acid (18:0)

Stearic acid is one of the major saturated fatty acids in the diet and in the body. It is a saturated fatty acid that is two carbon atoms longer than palmitic and is similarly cholesterogenic. High levels in plasma are a risk factor in atherosclerotic vascular disease because they signal elevated serum triglycerides. (See palmitic acid above.)

Stearic and palmitic acids are preferentially utilized as energy sources compared to unsaturated fatty acids during strenuous exercise. During two thirty-minute sessions, male handball players had an average increase of the ratio of unsaturated to saturated fatty acids from 1.1 to 1.6 [33]. This finding gives direct evidence of the beneficial effect of exercise on saturated fatty acids as well as the general triglyceride lowering effect. Lower dietary saturated fat and exercise are effective ways to manage patients with elevated plasma stearic acid.

Abnormally low levels of stearic acid cause increased membrane fluidity that is associated with active tumor proliferation. The best way to evaluate the significance of low levels of stearic acid is to examine its relationship with oleic acid. Erythrocyte membrane profiling is used to determine the stearic/oleic ratio, a prognostic tool in cancer therapy (see Red Cell Stearic/Oleic Index below). A low ratio in plasma signals potential cell distress from loss of membrane fluidity, but does not carry the significance for cancer therapy as the erythrocyte ratio.

Arachidic acid (20:0)

Arachidic acid can be utilized as an energy source or to build membrane phospholipids, but it serves no other special function. Its accumulation is most likely to competitively inhibit essential fatty acid metabolism as it binds to the Δ-6 desaturase enzyme and inhibits the insertion of double bonds needed to produce DGLA, EPA, and AA. For this reason, elevated arachidic acid exacerbates essential fatty acid deficiencies.

Very Long Chain Fatty Acids: Behenic acid (22:0), Lignoceric acid (24:0), and Hexacosanoic acid (26:0)

These three fatty acids are the largest, most slowly oxidized members of the saturated fatty acid family. They are the saturated members in the family of very long chain fatty acids (VLCFAs).

Adrenoleukodystrophy (ALD) and its adult variant, adrenomyeloneuropathy (AMN), are X-linked diseases caused by a deficiency of lignoceroyl-CoA ligase, a peroxisomal membrane enzyme needed for the degradation of VLCFAs. Elevations in VLCFAs, particularly C24:0 and C26:0, have proven useful in the diagnosis of the childhood, adolescent, and adult cerebral forms of ALD and AMN [25, 34]. VLCFA levels are increased at day of birth, providing the potential for neonatal mass screening. The detection of a positive peroxisomal disorder is increased by examining for elevation of the C26:0/C24:0 and C26:0/C22:0 ratios [35]

Dietary treatment with a 4:1 mixture of triglycerides made from oleic and erucic acids ("Lorenzo's oil") normalizes plasma VLCFA concentrations, but may not arrest the rapid progression of neurological symptoms [36]. Early detection and use of the mixture of oils may prevent further demyelination in some mildly affected boys [37]. (See erucic acid below for further discussion of fatty acids in ALD.)

The Monounsaturated Family of Fatty Acids

Monounsaturated or monoenoic fatty acids contain a single double bond positioned near the center of the chain. The double bond causes a bending of the chain of carbon atoms so that they don't stack together in the same way as saturated fatty acids. When they line up in a sheet to form a biological membrane, the manner in which fatty acids stack together determines physical properties, including fluidity, and interactions with other molecules such as proteins and lipoproteins. The fatty acids discussed in this section contain only bonds in the *cis* configuration where the chains of carbon atoms are on the same side of the double bond. (*Trans* fatty acids are discussed below.) The extra bond causes a rigid bend in the chain (Figure 5-9). The geometry of the *cis* double bond is defined as having the two hydrogen atoms on one side and the chain carbons on the other. The desaturase enzymes discussed above always produce *cis* double bonds. Oleic acid is the most significant member, exerting its effects in a myriad of cell responses and cell-cell interactions due to cell membrane and organelle membrane fluidity modulation. In the analysis of fatty acids from normal human plasma, the third largest peak on the chromatogram (exceeded only by palmitic and linoleic) is oleic acid.

Myristoleic acid (14:1n5)

Myristoleic is a fatty acid of medium-chain length with a single double bond. It accumulates in adipose tissue with the consumption of milk products, which are rich sources of myristoleic acid. Its levels correlate with those of another special fatty acid in milk products, conjugated linoleic acid (CLA) [38]. CLA has strong anti-carcinogenic properties. Being unsaturated and having short chain length makes myristoleic acid particularly good at increasing membrane fluidity. This characteristic lends generally favorable health consequences but high levels may raise concern when found in cancer patients because of the tumor-promoting effect of high membrane fluidity (see Red Cell Stearic/Oleic Index below).

Palmitoleic acid (16:1n7)

Palmitoleic acid is the first of several markers for general essential fatty acid deficiency that will be discussed. For a summary of how the markers may reflect various phases of deficiency, see above under Polyunsaturated Fatty Acid Metabolism. Palmitoleic acid is the desaturation product of palmitic acid. Since the latter is a predominant fatty acid in human tissues where desaturase enzyme activity is present, one might expect relatively high levels of palmitoleic acid. Such levels are not found in healthy humans, however, because the enzyme is normally occupied by the essential fatty acids that have much stronger binding affinity. When intake of essential fatty acids is low, palmitoleic acid formation is increased [39]. Thus, high palmitoleic acid is a marker of essential fatty acid deficiency.

Vaccenic acid (18:1n11)

Vaccenic acid is a positional isomer of oleic acid. The double bond is in the n11 rather than the n9 position. The ratio of vaccenic acid to palmitoleic acid has been reported to be a valuable indicator of biotin deficiency [40]. Vaccenic acid also seems to have large effects on membrane fluidity, possibly because the n11 double bond does not align with the much more abundant n9 positions of the majority of fatty acids present in the membrane. Inhibition of tumor growth in cell culture has been reported for vaccenic acid [41].

Oleic acid (18:1n9)

Oleic acid is present in all foods that contain fat and is easily produced by the fatty acid synthetic pathways present in normal human hepatic cells. It constitutes 15% of the fatty acids in erythrocyte membranes. Because of the presence of the double bond in the center of the molecule (Figure 5-9), it helps govern the critical nature of fluidity of the membrane matrix.

The pathogenesis of atherosclerosis involves accumulation of foam cells along the arterial wall. Oxidized low-density lipoproteins (LDLs) transform macrophages into foam cells. Oleic acid is the principal lipid of a class that makes LDLs resistant to oxidation. A diet rich in this fatty acid reduces foam cell accumulation rates and thereby lowers the chances of atherosclerosis [42].

FIGURE 5-9. The Double Bond of Oleic Acid

Since the double bond occupies the two central atoms, it is 9 atoms distant from both the carboxyl and methyl ends. The Δ numbering scheme is shown here.

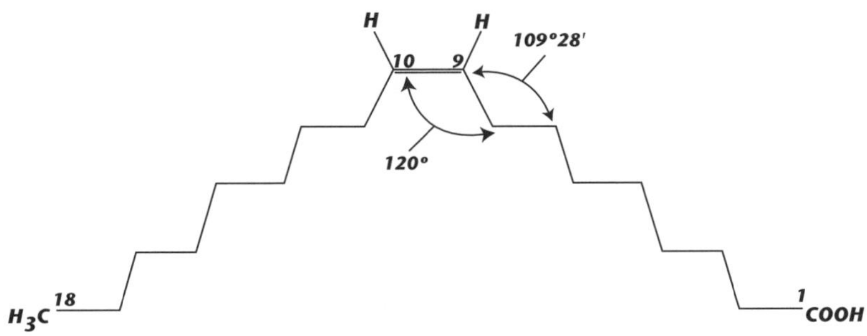

High oleic acid intake (usually as olive oil) can raise total oleic acid content of plasma above the reference range. This finding alone is not highly significant since oleic acid is a "cholesterol neutral" fatty acid and is produced in humans from stearic acid, so it is not an essential fatty acid. Concurrent increases in oleic acid in erythrocytes may not be seen because of the active metabolism of these cells. The action of the triglyceride-lowering drug, gemfibrozil, in potentiating oxidation of oleic and other fatty acids is discussed above under the topic, Plasma vs. Erythrocyte Fatty Acid Profiles. The use of erythrocyte stearic/oleic ratio in cancer therapy is discussed below under The Red Cell Stearic/Oleic Index.

Erucic acid (22:1n9)

Erucic acid is a very long chain, monoenoic fatty acid. It is used in the treatments for peroxisomal oxidation disorders discussed above under Very Long Chain Fatty Acids. Levels in plasma and erythrocytes may rise during treatment. Although there are no reported neurological consequences, side effects have been noted, namely mild increases in liver enzymes, thrombocytopenia, gastrointestinal complaints, and gingivitis [36].

In Zellweger syndrome, where peroxisomes are absent, erucic and adrenic (docosatetraenoic) acids accumulate. The peroxisomes of normal fibroblasts actively shorten erucic to oleic acid. The lack of peroxisomes makes Zellweger fibroblasts deficient in this activity [43].

Nervonic acid (24:1n9)

Nervonic acid contains the longest chain of carbon atoms of all monoenoic fatty acids present at appreciable levels in plasma and erythrocytes. The unique structure of nervonic acid confers special properties to nerve membranes where it is found in highest concentrations. The myelin sheath is especially rich in nervonic acid. Nervonic acid accumulation in red blood cells yields information on cerebrum maturation in premature infants [44]. Demyelination in adrenoleukodystrophy (ALD) is associated with an accumulation of very long chain saturated fatty acids, such as 26:0, stemming from a genetic defect in the peroxisomal beta oxidation system responsible for the chain shortening of these fatty acids. Long chain monoenoic acids, such as erucic acid, 22:1(n9), can normalize elevated serum levels of 26:0 in ALD by depressing their biosynthesis from shorter chain saturated fatty acids [45]. Impairment in the provision of nervonic acid in demyelinating diseases is indicated, suggesting that dietary therapy with oils rich in very long chain, monoenoic fatty acids may be beneficial in such conditions.

Polyunsaturated Fatty Acids

The primary function of the pathway described above in which LA and ALA are converted into DGLA, AA, and EPA is to supply the parent compounds for the 1-, 2-, and 3- series prostanoid and leukotriene pathways. We will discuss these relationships in detail before proceeding with interpretations of abnormalities for individual

FIGURE 5-10. Eicosanoids From Cell Membrane Fatty Acids
Polyunsaturated fatty acids normally occupy the second glycerol atom of membrane phospholipids. The fatty acid abundances vary with dietary supply. When phospholipase A2 is stimulated, it cleaves at the two-positions, releasing the fatty acids to be acted upon by lipoxygenase and cyclooxygenase enzymes. The ultimate products control responses of immune, smooth muscle, epithelial, and other tissues.

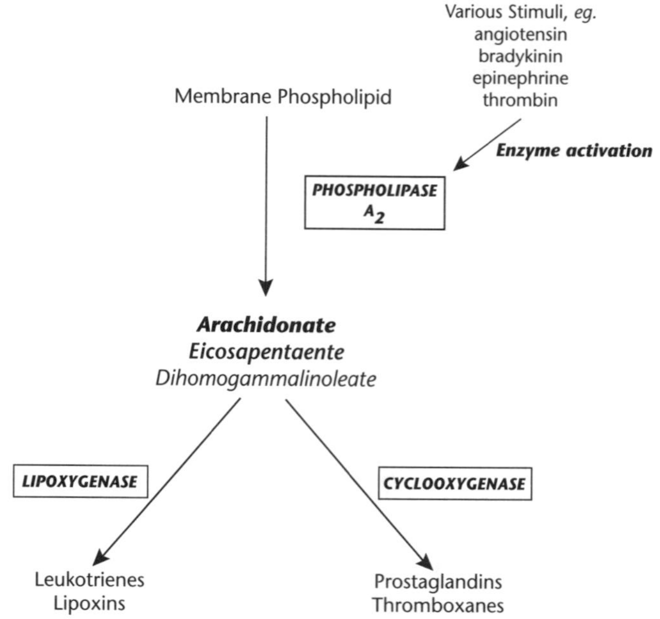

polyunsaturated fatty acids (PUFAs). PUFAs have great impact on health due to their conversion to the compounds collectively called "eicosanoids" (from Greek eikosa, which means twenty). Eicosanoids possess extremely potent biological activities, and their homeostatic functions in regulating blood vessel leaking, lipid accumulation, and immune cell behavior are relevant to the initiation and progress of heart and blood vessel disease [46]. These compounds, in turn, are used to amplify and balance signals to the organs, the blood clotting system, and the immune system.

When a tissue receives distress signals in the form of cell death products or their amplification through the cytokines of the immune system, the local cell response is elicited through a process starting in the membrane of affected cells. Phospholipase A2 within the cell translates stimuli into the release of arachidonate and other fatty acids from the 2-position of membrane phosphodiglycerides (Figure 5-10). Lipoxygenase and cyclooxygenase enzymes within the cell then convert the fatty acids into various classes of compounds containing 20 carbon atoms. Leukotrienes, lipoxins, and

FIGURE 5-11. Lipoxygenase and Cycloxygenase Enzyme Products

Eicosanoid products of lipoxygenase and cyclooxygenase enzymes help maintain the balance of tissue responses to stress signals. The series-1, -2, and -3 products are defined by the number of double bonds they contain. Prostaglandin formation causes loss of two double bonds, so the products of DGLA, AA, and EPA contain 1, 2, or 3 double bonds, respectively. These numbers become the subscripts of the product names (*e.g.*, PGE1, 2, or 3).

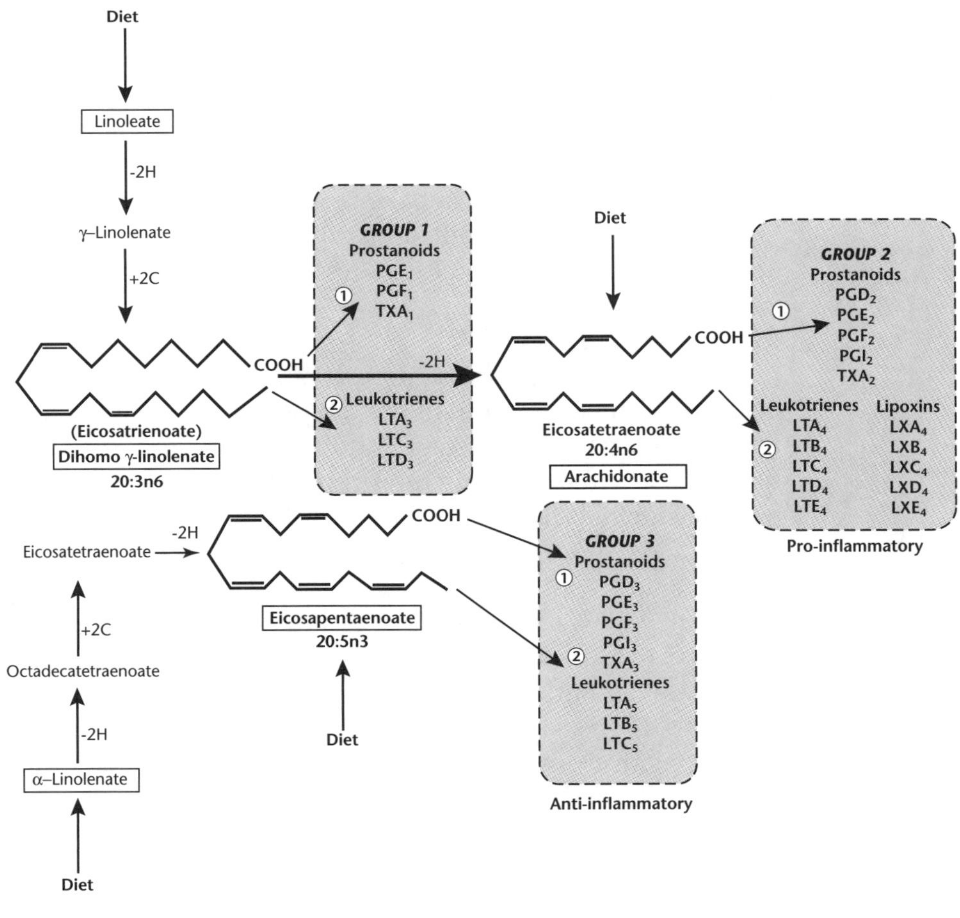

prostaglandins are classes of eicosanoids within which the Series 1-, 2-, and 3- subclasses are defined by the number of double bonds they possess (Figure 5-11). Groups 1, 2, and 3 in Figure 5-11 are synonymous with Series 1-, 2-, and 3-, and in both cases the subscript indicates the number of double bonds present. Thus, PGE_1, PGE_2, and PGE_3 are similar compounds that differ by possessing 1, 2, or 3 double bonds because they are derived from fatty acids containing 2, 3, or 4 double bonds. One of the fatty acid double bonds is lost when the enzyme forms the new class of compound.

The parent compounds are the polyunsaturated fatty acids DGLA, AA, and EPA, and the pathway utilizes the same enzyme for all three series of products. The concentrations of the three fatty acids present in the phospholipid of the cell membranes determine which product will predominate.

The lipoxygenase enzyme initiates the sequence of reactions leading to leukotriene formation (Figure 5-11). The 2-series that is derived from AA is by far the most pro-inflammatory. Antagonism of AA by DGLA and EPA for lipoxygenase active sites moderates inflammatory responses because the 1-series and 3-series leukotrienes are much less inflammatory than those of the 2-series. The anti-inflammatory effects of DGLA and EPA on rheumatoid arthritis and other diseases are mediated via this mechanism. They bind to the cyclooxygenase in place of AA and stimulate the production of the series 1 (DGLA) and series 3 (EPA) eicosanoids. Dietary intake is a primary determinant in the flow of fatty acids through this pathway. This mechanism explains the direct link between the balance of specific fats in the diet and long-term health maintenance. Eating a diet rich in fish oils shifts the dynamic equilibrium towards less inflammatory responses. The fatty acid profile allows specific intervention where the imbalance is most prevalent so that pro-inflammatory and anti-inflammatory responses remain in balance, and other local control processes respond properly to cellular stress.

Mead Acid (20:3n9)

Mead acid is a polyunsaturated fatty acid, but it is not essential because it can be produced in human tissues under certain conditions as discussed above (see Polyunsaturated Fatty Acid Metabolism). It serves as a membrane structural component in place of the normal polyunsaturated fatty acids derived from essential precursors. It cannot serve, however, in the critical role of precursor to eicosanoid cell regulators.

There is direct substrate competition for the desaturase enzymes (Figure 5-6). Linoleic acid competes with oleic acid for the Δ-6 desaturase, and linolenic acid competes with both linoleic and oleic acids for the enzyme. When the essential fatty acids are available, there is relatively little conversion of oleic to Mead acid. The precursor fatty acids have the preference of ALA > LA > OA, and this preference explains the relatively low production of Mead acid in the presence of adequate ALA and LA.

The appearance of Mead acid, then, is a marker for chronic essential fatty acid deficiency, similar to the indication of an elevated triene/tetraene (T/T) ratio (discussed below). Elevated Mead acid can be used in conjunction with the actual concentrations of the essential fatty acids and the T/T ratio in plasma and erythrocyte membranes to confirm degrees of essential fatty acid deficiencies [47, 48].

The n3 Family of Fatty Acids

This family includes the essential fatty acids alpha linolenic (ALA) and eicosapentaenoic (EPA). Dietary deficiency of these fatty acids may be epidemic due to poor intake associated with the popular diets of low fat, commercially-prepared foods. Oils used for fast food preparation do not contain significant polyunsaturated fatty acids. As explained above, human tissues cannot produce these from any other compounds. From the initial observations of skin and reproductive disorders, deficiency symptoms have expanded to a long list that includes most degenerative diseases. The more recent evidence of requirement of the 22 carbon-long, six double bond member docosahexaenoic acid for neonatal development makes the n3 family candidate for close inspection in clinical evaluations from pre-conception to old age.

Alpha-linolenic acid (ALA) (18:3n3)

ALA cannot be formed from any other fatty acids in humans and, therefore, is an essential fatty acid. Some symptoms of ALA deficiency are listed in Table 5-4. Many plants produce ALA, but because of the small amounts of fresh vegetables consumed by most people, it is one of the least abundant of the essential fatty acids in most diets. It is found in flax, hemp, rape (canola) seed, soybean, and walnut oils, and in dark green leaves and must be supplied by eating such foods. Because of the central importance of this fatty acid and its counterpart, GLA, a table showing the relative values of their clinical use has been constructed (Table 5-6). The wide range of symptoms is due to the functions of ALA in critical cell processes of membrane integrity and eicosanoid hormone production. Production of eicosanoids utilizes the fatty acid EPA (see below) that can be produced from ALA by elongation and desaturation. This conversion is very efficient in most individuals, so the content of ALA in cell membranes is very low. Low EPA and DHA in the presence of normal ALA indicate inadequate conversion, most often due to zinc deficiency and blocking of Δ-6 desaturase.

Eicosapentaenoic acid (EPA) (20:5n3)

EPA is the parent of the 3-series prostanoids and leukotrienes that moderate the proinflammatory effects of the 2-series derived from arachidonic acid. Although EPA can be produced from the essential fatty acid, ALA, dietary intake of ALA is generally poor. The conversion also requires the action of the Δ-6 desaturase enzyme that may be low due to inadequate zinc, magnesium, or vitamins B3, B6, and C. Such enzyme impairment would be indicated if EPA is low and ALA is normal or high. High levels of saturated, monounsaturated, and *trans* fatty acids, and cholesterol slow the conversion of ALA to EPA (as well as GLA to DGLA).

Insufficiency of EPA is likely the most prevalent fatty acid abnormality affecting the health of individuals in western societies. Low levels in plasma or in erythrocytes are indicative of insufficiency. Arthritis, heart disease, and aging result from direct or indirect effects of unchecked inflammatory response. Supplementation with EPA-rich fish oils aids in the prevention of cardiac arrhythmias [18]. Significant reduction in total cholesterol and triglyceride has been achieved with a combination of garlic concentrate (900 mg/d) and fish oil supplementation [16]. Reductions of inflammatory cell responses have been shown to occur within ten days of implementing a diet rich in EPA and DHA [49].

Docosapentaenoic acid (DPA) (22:5n3) and Docosahexaenoic acid (DHA) (22:6n3)

Although the common name of cervonic acid (CA) is found in some discussions, the abbreviation DHA is more widely used for the six-double bond docosahexaenoic acid. The growth and development of the central nervous system is particularly dependent upon the presence of an adequate amount of the very long chain, highly unsaturated fatty acids DPA and DHA [10, 12]. Attention deficit hyperactivity disorder [14] and failures in development of the visual system in essential fatty acid deficiencies are two examples of this dependency. DHA is an important member of the very long chain fatty acids (C22 to C26) that characteristically occur in glycosphingolipids, particularly those in brain. Since this fatty acid is so important in early development, it is worth noting that the levels in breast milk are correlated with the mother's intake of fish oils [50], which are rich sources of DHA and DPA.

The n6 Family of Fatty Acids

Linoleic acid (LA) (18:2n6)

LA is by far the most abundant polyunsaturated fatty acid in most human tissues. It is one of the essential fatty acids because it contains a double bond at the n6 position that is beyond the reach of the human desaturase enzyme. Low levels indicate dietary insufficiency that leads to a variety of symptoms (Table 5-4). Symptoms as diverse as hair loss and high blood pressure result from combined effects of LA deficiency on membrane integrity and failure to produce the 1-series eicosanoids. LA is the starting point for this pathway. Normal neonatal status of LA is marginal, if not insufficient. Fetal linoleic and cervonic acid (DHA) are correlated with maternal RBC levels [51]. Since dietary sources (especially corn oil) are abundant, however, linoleic acid may be found above normal in some adults. Because of the need for balanced prostanoid and leukotriene synthesis, excessive linoleic acid can contribute to an overproduction of the proinflammatory 2-series local hormones. When it was substituted for oleic acid, linoleic acid caused growth inhibition of *Helicobacter pylori*, the bacterium associated with gastric ulcer [52].

Gamma linolenic acid (GLA) (18:3n6)

GLA is the precursor of DGLA, the parent of the 1-series prostanoids, as well as the precursor of AA, the parent of the 2-series prostanoids. It is found in hemp, black currant, and evening primrose oils. It can be produced in human tissues by the action of desaturase enzymes on linoleic acid. Use of 1.4 g/d in the form of borage seed oil resulted in clinically important reductions in the signs and symptoms of rheumatoid arthritis [53]. See Table 5-6 for other clinical associations. In cases of cancer, it is especially important that low levels of GLA not be supplemented without added n3 fatty acids because n6s enhance tumor formation and growth, while n3s inhibit tumors [54]. Gamma linolenic acid corrects most of the biological effects of zinc deficiency [3], indicating that the requirement of the Δ-6 desaturase enzyme for zinc is a first-order essential function of zinc.

TABLE 5-6. **Relative Clinical Responses to ALA and GLA**

Application	GLA	ALA
Lower blood pressure, lower blood cholesterol, and lower risk of stroke and heart attack	X	XX
Normalize fat metabolism in diabetes and decrease the amount of insulin required by diabetics	X	X
Prevent liver damage due to alcoholism and decrease withdrawal symptoms after discontinuing the habitual use of alcohol	X	X
Provide adjunctive treatment for schizophrenics	X	XX
Cause weight loss by increasing metabolic rate and fat burn-off	X	XX
Relieve premenstrual breast pain (masalgia) and premenstrual syndrome of bloating, irritability, depression, and aggressive behavior	XX	X
Prevent drying and atrophy of lacrimal and salivary glands (Sjogren's syndrome)	X	–
Prevent arthritis in animals	X	X
Improve the condition of hair, nails, and skin	X	XX
Slow down or stop deterioration in multiple sclerosis	X	X
Help treat diabetic neuropathy in Type II diabetes	X	X
Kill cancer cells in tissue culture without harming normal cells	X	XX

(X = moderate, XX = strong clinical response)

Dihomogammalinolenic acid (DGLA) (18:3n6)

Diets low in the essential fatty acid, LA, are almost universally also low in DGLA. The 1-series prostanoids and leukotrienes are derived from DGLA, so an insufficiency of this fatty acid impairs a wide range of cellular functions and tissue responses. The 1-series compounds act like the 3-series derived from ALA to moderate the proinflammatory 2-series. In tumor response, however, they are unique in being promoters of growth, whereas the 3-series are inhibitors. When testing reveals low levels of DGLA, supplementation with black currant or evening primrose oils should be considered, but if a history of tumor formation is known, always consider ALA sources (*i.e.*, black currant) as well. See EPA above for other factors in common regarding the desaturase enzyme required for conversion of GLA into DGLA.

Arachidonic acid (AA) (18:4n6)

AA is the essential fatty acid least likely to be found insufficient. Because of the prevalence of corn and corn oil products in feed for cattle and hogs, diets high in these red meats are rich in AA. AA is a 20-carbon or eicosanoate fatty acid that serves as substrate for the cylcooxygenase and lipoxygenase enzymes, leading to the production of the 2-series prostanoids and leukotrienes. Several of these products have potent proinflammatory and thrombogenic activity. High AA also promotes gall stone formation by stimulating mucin production in the gall bladder mucosa [55].

Docosadienoic acid (22:2n6) and Docosatetraenoic acid (22:4n6)

Docosadienoic acid is a very long-chain fatty acid (VLCFA) produced as an elongation product of DGLA. Docosatetraenoic acid (DTA) is also known by the common name of adrenic acid. When n6 dietary fatty acids are constantly supplied in overabundance, the intermediate products of further desaturation and elongation are not used as fast as they are produced. Under these conditions, the process of modification can continue through docosadienoic acid to the 22-carbon, four double bonded DTA. These fatty acids can then accumulate in adipose tissue. Data from unpublished studies by the authors indicate an association of DTA acid in plasma with obesity. Apparently, this fatty acid accumulates in adipose tissue from which it is mobilized in the fasting state. No such association has been seen in erythrocyte fatty acid profiles, indicating rapid utilization of these fatty acids once they are mobilized. Insulin insensitivity is a common condition that stimulates elongation enzymes associated with high levels of VLCFA.

The level of DTA in plasma may have usefulness in monitoring the metabolic aspects of weight control, especially in avoiding the risk of essential fatty acid deficiencies while maintaining a very low fat diet. A low level of DTA in an individual who is on a fat-restricted weight loss program may indicate a state of essential fatty acid deprivation. DTA accumulates to high levels in Zellweger syndrome where peroxisomal oxidation is impaired (see erucic acid above).

The Odd-numbered Family of Fatty Acids

Pentadecanoic (15:0), Heptadecanoic (17:0), Nonadecanoic (19:0), Heneicosanoic (21:0) and Tricosanoic (23:0) acids

Since the normal synthetic pathway for fatty acids leads to only even-numbers of carbon atoms in the products, it may seem curious to find any odd-numbered chains present in human tissues. The three-carbon compound, propionic acid, initiates the synthetic pathway producing fatty acids with odd numbers of carbon atoms. Propionic acid is a normal metabolic intermediate, and it can take the place of the two-carbon acetyl group in the first step of fatty acid synthesis. Anything that causes propionate accumulation will lead to increased levels of odd chain fatty acids. Vitamin B_{12} is required for the conversion of propionate into succinate for oxidation in the central energy pathways. Deficiency of vitamin B_{12} results in accumulation of propionate and subsequent buildup of the odd-numbered fatty acids [56, 57].

The reason that odd-numbered fatty acids are normally produced in such minor amounts is that the supply of the two-carbon precursor is so much greater. Acetyl-CoA production from carbohydrate, amino acid, and fat metabolism easily overwhelms the small amount of propionate present. The initiation of fatty acid synthesis from acetyl-CoA, however, is a biotin-dependent step. In biotin deficiency, the rate of this step in which acetyl-CoA is converted into malonyl-CoA is reduced, allowing propionate to enter in greater amounts (Figure 5-12). Elevated levels of odd chain fatty acids have been shown to occur in human plasma phospholipids [58, 59] and in plasma, heart, and liver in experimental animals with biotin deficiency [58, 59].

The bacteria in the gut of ruminants (cattle, sheep, etc.) produce large amounts of propionate, which is absorbed and enters the metabolism. Consequently, intake of animal and dairy products

FIGURE 5-12. Formation of Odd-Numbered Fatty Acids in Vitamin Deficiencies

Biotin deficiency impairs even-numbered fatty acid synthesis while vitamin B_{12} deficiency contributes to propionate accumulation, which enhances odd-numbered fatty acid synthesis.

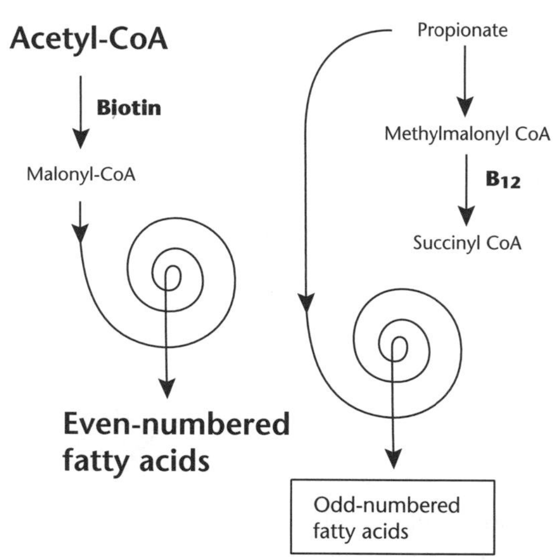

results in higher levels of odd-numbered fatty acids. Alternatively, the bacteria in the human gut can produce sufficient amounts of propionate to lead to elevation in the odd-carbon fatty acids. This would only occur under conditions of significant gut dysbiosis.

The association between vitamin B12 and abnormal fatty acid synthesis provides a rationale for the neuropathy of vitamin B12 deficiency. Odd-chain fatty acids would build up in membrane lipids of nervous tissue with inadequate B12, resulting in altered myelin integrity and demyelination, leading eventually to impaired nervous system functioning [60].

Trans Fatty Acids

In the tissues of plants and higher animals, the insertion of double bonds always results in the *cis*-geometry discussed above. The words "partially hydrogenated vegetable oil" on food labels indicate that a natural oil has been chemically modified in a process that converts some of the *cis* unsaturated fatty acids to the *trans* form. Oleic acid is transformed into its *trans* isomer, elaidic acid, which is the most abundant *trans* fatty acid in most hydrogenated oils. Comparing the structure of elaidic acid (Figure 5-13) with that for oleic acid (Figure 5-9), the placement of the two carbon chains on opposite sides of the double bond causes a straightening of the chain so that the packing in cell membranes is more like that of saturated fatty acids. Of equal importance is the shift in oxidative metabolism. The *trans* configuration is processed at higher rates, possibly even faster than those of saturated fatty acids.

FIGURE 5-13. The *Trans* Double Bond of Elaidic Acid

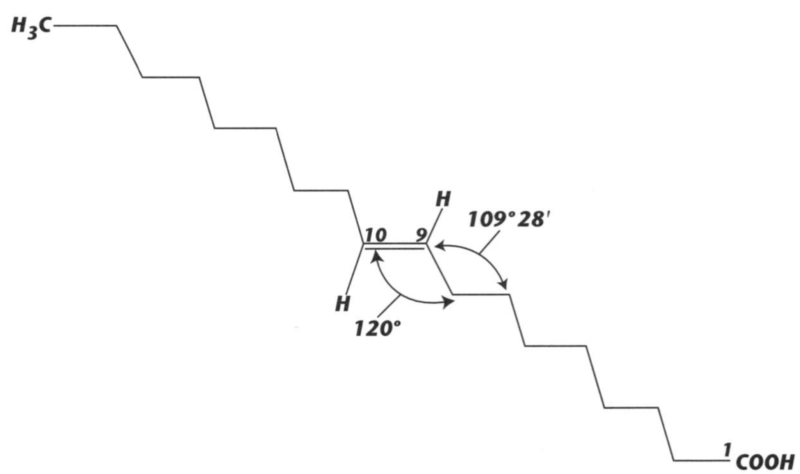

In human tissues, *trans* unsaturated fat behaves as if it were saturated, meaning more risky for heart disease, cancer, diabetes, and all other degenerative diseases related to fat metabolism. Trans fatty acids behave metabolically like saturated fats on the one hand, leading to higher cholesterol levels [61]. On the other hand they mimic unsaturated fats in binding to desaturase enzymes and interfering with the normal conversions of unsaturated fatty acids. The net effect is to raise plasma LDL and lower HDL. Adverse impact on HDL or LDL cholesterol is widely reviewed in the medical research literature [62-64]. Interference with eicosanoid production is also suspected, and research in the area of health risks of *trans* fatty acids is accelerating. *Trans* fatty acid levels in cell membranes are determined by dietary intake of hydrogenated oils [65]. As an example, a person eating a doughnut for breakfast (3.19 g of *trans* fatty acids), a small order of French fries with lunch (3.43 g), two teaspoons of margarine on bread with dinner (1.24 g), and two cookies for a snack (1.72 g) would ingest 9.58 g of *trans* fatty acids. This amount is enough to negate the serum cholesterol-lowering effects of decreasing dietary saturated fat to 10 percent of total energy intake [66]. Up to 15% of tissue fatty acids have been found at autopsy to contain *trans* double bonds.

Plasma fatty acid profiling reveals *trans* fatty acid body content in the concentrations of the major *trans* fatty acids in hydrogenated oils, elaidic and palmitelaidic acids. When high levels are found, the simple dietary change of avoiding hydrogenated oils is effective control, although the change must be a long-term lifestyle modification because the replacement in the cell membranes of the major organs can take several months.

Elaidic acid is generally the most abundant of the *trans* fatty acids because it is produced by the hydrogenation of the most common oils used in the food industry, including corn, soybean, and safflower. These oils contain relatively high amounts of oleic and linoleic acids, both of which may convert to elaidic acid during the hydrogenation process. Hard margarine may contain as much as 60% of the total unsaturated fatty acids as elaidic acid. Palmitelaidic acid is a *trans* fat that can be formed from hydrogenation of natural fatty acids containing palmitoleic acid or partial saturation of PUFAs with isomerization of the remaining *cis* bond. Some addition to total body *trans* fatty acid content may be from linolelaidic acid that has the same length and double bond positions as linoleic acid, but both of the double bonds are in the *trans* configuration.

Palmitelaidic acid (16:1n9t), Elaidic acid (18:1n9t) and Other Trans-Isomers

Although elaidic acid is the most common *trans* fatty acid, others may be found in foods because of unusual oils that can be used for the hydrogenation process. Other isomers that may be present are the 16-carbon-long palmitelaidic and two 18-carbon-long isomers, petroselaidic (18:1n12t) and *trans*-vaccenic (18:1n6t) acids. The latter may be reported as a part of a total 18:1 *trans* fatty acid value in order to simplify the analysis and reporting. All of these fatty acids are minor components, and there is no clinical significance beyond the total *trans* fatty acid content of a specimen. The clinically significant information is the total level of accumulation signaling risk of membrane function impairment.

The growing consensus among experts in lipid nutrition is that foods containing hydrogenated oils are to be avoided. Increased intake of *trans* fatty acids has been implicated as a factor in the continuing high incidence of coronary heart disease [63]. Women consuming 5.7 g/day were 50% more likely to develop coronary heart disease than those who consumed only 2.4 g/day [67]. In another study of serum cholesterol response, men and women had higher LDL and lower HDL while consuming diets with 11% *trans* fatty acids compared to the same subjects put on a diet containing 0% *trans* fatty acids. The blood lipid profile was worse when using *trans* fatty acids than when the diet was shifted to high saturated fat content [62]. The bacteria in the gut of ruminant animals also produce these fatty acids. That is the reason that beef and milk contain small amounts (1-3%) of elaidic acid. Moderate use of these foods does not provide *trans* fatty acids at levels that are of concern.

TABLE 5-7. Trans-Fatty Acid Content of Common Foods

Food	Total *Trans*-Fatty Acids per serving
Cake	1.04 g
Cookies	0.86
Corn Chips	1.42
Crackers	0.12
Danish Pastries	3.03
Doughnuts	3.19
French fries	2.92
Pie	1.00
Pizza	0.13
Potato chips	0.11
Soft margarine	0.27
Stick margarine	0.62
Vegetable shortening	0.63

Because of the high *trans* fatty acid content of many common foods (Table 5-7), individuals who regularly eat such foods can easily ingest 10 g/day of the *trans* isomers [66]. If plasma or RBC *trans* fatty acids are elevated, foods containing hydrogenated oils need to be avoided. Foods that generally contribute the greatest amounts of *trans* fatty acids include stick margarine, peanut butters, and bakery products such as breads, rolls, cookies, crackers, pies, and cakes. Products with labels that simply say "vegetable oil" should be substituted for those containing hydrogenated oils.

Calculated Values

A calculated ratio of two values allows a comparison of the relative magnitudes of the two individual measurements. The added value lies in the use of the ratio as a quick measure of the balance between two parameters. Sometimes valuable insights into imbalances are found in ratio values even though the individual components are within normal limits. Precursor–product relationships among fatty acids lend themselves to the use of ratios for evaluation.

The LA/DGLA Ratio

Inhibition of Δ-6 desaturase by zinc deficiency or excess saturated, monoenoic, or *trans* fatty acids leads to an elevated ratio of linoleic to dihomogammalinoleic acids (LA /DGLA). Under these conditions, the enzyme cannot convert the substrate (LA) to the product (DGLA) fast enough. Supplementation to improve zinc status should be used along with specific oils to enhance EPA, DHA, and AA as indicated by the levels of these products requiring desaturase activity. When the LA/DGLA ratio is elevated, the typical pattern of individual fatty acids is: low DGLA with normal or elevated LA.

The EPA/DGLA Ratio

The EPA/DGLA ratio reflects the balance of n3 and n6 polyunsaturated fatty acids. The balance of these two 20-carbon or eicosenoic fatty acids is critical for proper supply of the eicosanoid 3- and 1-series local hormones that control a host of cellular functions and responses. It will be low when DGLA is elevated relative to EPA, indicating a need for EPA sources like fish oils. When the ratio is high, sources of DGLA (black currant or evening primrose oil) are indicated.

The AA/EPA Ratio

This ratio allows a quick comparison of the relative amounts of these two eicosanoid precursors. An abundance of AA is quite common in Western high meat and corn oil-rich diets and can result in elevation of the AA/EPA ratio. This is one of the indicators that extra n3 fatty acids, including EPA of fish oils, would be beneficial. Several weeks of supplementation with fish oils may be necessary to see the full effects. Cells in all tissues must have time to build the EPA into their membranes. Once the levels are raised, EPA offsets the production of pro-inflammatory series-2 eicosanoids from AA. One of the most common conditions associated with high AA/EPA ratios is dry, itching skin that worsens with cold weather [68].

The Triene/Tetraene (T/T) Ratio

The triene/tetraene (TT) ratio in plasma is used as an index of essential fatty acid deficiency [69]. It is calculated as the ratio of Mead acid (20:3n9) to arachidonic acid (20:4n6), the major essential fatty acid in plasma. The ratio is diagnostic of essential fatty acid deficiency in experimental animals [70] and humans [47]. The formation of triene derivatives of non-essential fatty acids has a protective effect in maintaining the overall double bond density of membrane phospholipids [71]. This phenomenon is sometimes referred to as molecular mimicry because the body is producing a substitute or mimic for the normal molecules that maintain cellular function in the well-fed state. The mimic, Mead (20:3n9) acid, acts in some ways like the usual membrane component, AA (20:4n6). Mead acid, however, cannot enter into the eicosanoid pathways in place of AA, so the effects due to lack of cell control molecules produce fatty acid deficiency symptoms. The T/T ratio adds insight beyond a finding of elevated Mead acid because it reflects the degree of inhibition of arachidonate conversion to eicosanoids in the presence of overall polyunsaturated fatty acid deficiency.

Vaccenic/Palmitoleic Ratio

Biotin deficiency is indicated by a low Vaccenic/Palmitoleic ratio [40]. This relationship has been demonstrated for fatty acid patterns in lymphocytes. Biotin deficiency causes elongation and desaturation of unsaturated fatty acids to be impaired. The extension of the relationship to plasma or erythrocyte patterns is presumptive at this time.

Red Cell Stearic/Oleic Index

In tumors, changes in desaturase enzymes cause a profound decrease in the ratio of stearic to oleic acids in erythrocyte membranes [20]. One of the most abundant fatty acids in cell membranes, stearic, is more actively converted into oleic acid. One likely outcome of this shift is increased fluidity of the tumor cell membrane, perhaps allowing for faster metabolic rates. The stearic/oleic ratio is used as a monitor of the effectiveness of cancer therapy [72]. The ratio rises with tumor regression.

The ratio changes in parallel to changes in malignant tissue cell membranes [73]. Values below 1.1 are associated with the presence of malignancy, although they are not diagnostic of cancer. The ratio responds to hormonal therapy for prostatic cancer, and, in studies following surgical cures, individuals who maintained this ratio above 1.1 had no tumor recurrences. When individuals with recurrent tumors show low stearic/oleic ratios, there is too much oleic acid relative to stearic acid and encouraging a diet high in olive oil may be contraindicated. Use of other polyunsaturated fatty acids should be guided by test profile results.

Total Fatty Acid Concentration

When the concentrations of fatty acids present in plasma are summed and expressed in units of mg/100 dL, the result is closely equivalent to that of the routine serum triglyceride assay. (See Plasma vs. Erythrocyte Fatty Acid Profiles above for a discussion of the use of palmitic acid as a means for calculating triglyceride values.) The treatment of elevated saturated fatty acids or triglycerides by various actions involving dietary restriction of fats, the use of pharmacological doses of niacin, and lipid-lowering drugs is common practice.

When total fatty acids are low, there is possibility of stimulation of peroxisomal oxidation of fatty acids. A number of structurally diverse chemicals have been found to cause proliferation of peroxisomes in liver and other tissues [74]. Paradoxically, high fat diets cause such proliferation also. The increased oxidative activity leads to higher levels of hydrogen peroxide and development of liver tumors in laboratory animals [75]. In humans, the related disorder, hypobetalipoproteinemia is associated with increased risk for a variety of cancers, pulmonary, and gastrointestinal diseases [76, 77].

Conclusion

Balanced intake of mono- and polunsaturated fatty acids, not just a high PUFA/Saturated ratio, is required for optimal control of almost every tissue in the body [78, 79]. The long standing emphasis on low fat diets has led to fatty acid deficiencies in large numbers of people in developed countries where, ironically, the availability of high quality fats is abundant. Many people do not eat the fresh nuts, seeds, whole grain breads, and seafood products that are rich sources of essential fatty acids. The same individuals are likely to have high intakes of saturated and *trans* fatty acids. The recognition of multiple health effects of fatty acid inadequacies at all stages of the human life cycle has led to a surge of interest in fatty acids. It is no longer adequate (or even correct) simply to recommend the replacement of saturated fats with those that contain unsaturated fatty acids.

The laboratory evaluation of fatty acid profiles in blood plasma or erythrocytes is a valuable tool in the clinical assessment of fatty acids as factors in the maintenance of health and the amelioration of disease. Highly accurate routine analysis is made possible by the availability of analytical capabilities such as gas chromatography with mass spectral detection.

Recommendations of dietary oil use to overcome imbalances revealed by laboratory profiles of fatty acids involve choosing appropriate products as neat liquids or gel caps. Food oils are generally not highly purified in any one fatty acid, so that familiarity with their relative fatty acid compositions is helpful (Table 5-8). The dosage of supplementation with a specific fatty acid may be calculated from data on composition.

TABLE 5-8. Typical Fatty Acid Composition of Dietary Fats

			% of Total Oil				% Fat* in
Source	ALA✦	GLA	LA	Oleic	Stearic	Palmitic	***Source***
Almond			17	78	5		54
Avocado			10	70	20		12
Black currant	**15**	**17**	**46**	**12**	**1**	**7**	**100**
Borage		**24**	**39**	**17**	**1**	**11**	**100**
Brazil			24	48	24		67
Cashew			6	70	18		42
Chia	30		40				30
Coconut			3	5		91	35
Corn			59	24	17		4
Cottonseed			50	21	25		40
Primrose		**9**	**72**	**2**	**3**	**17**	**100**
Flax (linseed)	**58**		**14**	**19**	**4**	**5**	**100**
Grape			71	17	12		20
Hemp	35	2	60	12	2	6	35
Macadamia			10	71	12		
Olive			8	76	16		20
Palm kernel			2	13		85	35
Peanut			29	49	18		48
Pumpkin	12		45	34	0	9	47
Rape (canola)		7	30	54‡	7		30
Safflower			75	13	12		60
Soybean	7		50	26	6	9	18
Sunflower			65	23	12		47
Walnut	12		58	18	4	9	70
Wheat germ	5		50	25	18		11

✦ALA: alpha linolenic acid, GLA: gamma linolenic acid, LA: linoleic acid
** % of whole food*
‡ Includes up to 5% erucic acid
***Note: Fish oil** is usually supplied in soft gel capsules that state the mass of EPA (360 mg/g) and DHA (240 mg/g).*

How to Calculate Supplemental Oil Intake

- To find the amount of oil to use for a given amount of fatty acid, divide the desired amount of fatty acid by the decimal equivalent of the percentage in the oil. For a patient that should take 6 g of ALA and 6 g of GLA,
    1. To achieve 6 g of ALA (omega 3 fatty acids): 6/0.5 = 12 g of flax oil.
    2. To achieve 6 g of GLA (omega 6 fatty acids): 6/0.72 = 8.3 g of evening primrose (EPO).
- If the product is a liquid oil, then 1 tsp. is about 4 grams, so the dosage would be 3 tsp. of flax and 2 tsp. of primrose.
- If the product is packaged as one-gram soft gel capsules, then 12 flax and 8 EPO per day may be used.
- To prevent toxic oxidation product formation from the polyunsaturated fatty acids, 10 mg of vitamin E should be added per gram of oil.

References

[1] Siguel, E.N., Dietary sources of long-chain n-3 polyunsaturated fatty acids [letter; comment]. *Jama*, 275(11). 836-7, 1996.

[2] Makrides, M., et al., A randomized trial of different ratios of linoleic to alpha-linolenic acid in the diet of term infants: effects on visual function and growth. *Am J Clin Nutr*, 71(1). 120-9, 2000.

[3] Huang, Y.S., *et al.*, Most biological effects of zinc deficiency corrected by gamma-linolenic acid (18: 3 omega 6) but not by linoleic acid (18: 2 omega 6). *Atherosclerosis*, 41(2-3). 193-207, 1982.

[4] Tsuge, H., N. Hotta, and T. Hayakawa, Effects of vitamin B-6 on (n-3) polyunsaturated fatty acid metabolism. *J Nutr*, 130(2S Suppl). 333S-334S., 2000.

[5] Lorentzen, B., *et al.*, Fatty acid pattern of esterified and free fatty acids in sera of women with normal and pre-eclamptic pregnancy. *Br J Obstet Gynaecol*, 102(7). 530-7, 1995.

[6] Innis, S.M., *et al.*, Marine and freshwater fish oil varying in arachidonic, eicosapentaenoic and docosahexaenoic acids differ in their effects on organ lipids and fatty acids in growing rats. *J Nutr*, 125(9). 2286-93, 1995.

[7] Parkinson, A.J., *et al.*, Elevated concentrations of plasma omega-3 polyunsaturated fatty acids among Alaskan Eskimos. *Am J Clin Nutr*, 59(2). 384-8, 1994.

[8] Seigneur, M., *et al.*, Serum fatty acid profiles in type I and type II diabetes: metabolic alterations of fatty acids of the main serum lipids. *Diabetes Res Clin Pract*, 23(3). 169-77, 1994.

[9] Manku, M.S., *et al.*, Fatty acids in plasma and red cell membranes in normal humans. *Lipids*, 18(12). 906-8, 1983.

[10] Innis, S.M., n-3 fatty acid requirements of the newborn. *Lipids*, 27(11). 879-85, 1992.

[11] Hashimoto, F., *et al.*, Effect of gemfibrozil on lipid biosynthesis from acetyl-CoA derived from peroxisomal beta-oxidation. *Biochem Pharmacol*, 49(9). 1213-21, 1995.

[12] Hoffman, D.R. and R. Uauy, Essentiality of dietary omega 3 fatty acids for premature infants: plasma and red blood cell fatty acid composition. *Lipids*, 27(11). 886-95, 1992.

[13] Burgess, J.R., *et al.*, Long-chain polyunsaturated fatty acids in children with attention- deficit hyperactivity disorder. *Am J Clin Nutr*, 71(1 Suppl). 327S-30S, 2000.

[14] Stevens, L.J., *et al.*, Essential fatty acid metabolism in boys with attention-deficit hyperactivity disorder. *Am J Clin Nutr*, 62(4). 761-8, 1995.

[15] Simon, J.A., *et al.*, Serum fatty acids and the risk of coronary heart disease. *Am J Epidemiol*, 142(5). 469-76, 1995.

[16] Adler, A.J. and B.J. Holub, Effect of garlic and fish-oil supplementation on serum lipid and lipoprotein concentrations in hypercholesterolemic men. *Am J Clin Nutr*, 65(2). 445-50, 1997.

[17] Abbey, M., *et al.*, Partial replacement of saturated fatty acids with almonds or walnuts lowers total plasma cholesterol and low-density-lipoprotein cholesterol. *Am J Clin Nutr*, 59(5). 995-9, 1994.

[18] Nair, S.S., *et al.*, Prevention of cardiac arrhythmia by dietary (n-3) polyunsaturated fatty acids and their mechanism of action. *J Nutr*, 127(3). 383-93, 1997.

[19] Tavella, M., C.N. Corder, and W. McConathy, Effect of gemfibrozil on fatty acids in lipid fractions of plasma from patients with hypertriglyceridemia. *J Clin Pharmacol*, 33(1). 35-9, 1993.

[20] Wood, C.B., *et al.*, Reduction in the stearic to oleic acid ratio in human malignant liver neoplasms. *Eur J Surg Oncol*, 11(4). 347-8, 1985.

[21] Wood, C.B., *et al.*, Increase of oleic acid in erythrocytes associated with malignancies. Br *Med J* (Clin Res Ed), 291(6489). 163-5, 1985.

[22] Erickson, K.L., N.E. Hubbard, and R. Chakrabarti, Modulation of signal transduction in macrophages by dietary fatty acids. *J Nutr*, 125(6 Suppl). 1683S-1686S, 1995.

[23] Abu-Salah, K.M., A.A. al-Othman, and K.Y. Lei, Lipid composition and fluidity of the erythrocyte membrane in copper-deficient rats. *Br J Nutr*, 68(2). 435-43, 1992.

[24] Bakos, S.N., H.K. Ahmed, and T.A. Nasser, Serum copper, magnesium, zinc, calcium, and potassium changes following acute myocardial infarction. *Angiology*, 39(5). 413-6, 1988.

[25] Powers, J.M., *et al.*, A correlative study of the adrenal cortex in adreno-leukodystrophy--evidence for a fatal intoxication with very long chain saturated fatty acids. *Invest Cell Pathol*, 3(4). 353-76, 1980.

[26] Saudubray, J.M., *et al.*, Approach to the patient with a fatty acid oxidation disorder. *Prog Clin Biol Res*, 375. 271-88, 1992.

[27] Rhead, W.J., *et al.*, Clinical and biochemical variation and family studies in the multiple acyl-CoA dehydrogenation disorders. *Pediatr Res*, 21(4). 371-6, 1987.

[28] Roettger, V., *et al.*, Multiple acyl-coenzyme A dehydrogenation disorders (MAD) responsive to riboflavin: biochemical studies in fibroblasts. *Prog Clin Biol Res*, 375. 317-26, 1992.

[29] Salter, A.M., *et al.*, The effect of different dietary fatty acids on lipoprotein metabolism: concentration-dependent effects of diets enriched in oleic, myristic, palmitic and stearic acids. *Br J Nutr*, 79(2). 195-202, 1998.

[30] Clandinin, M.T., *et al.*, The effect of palmitic acid on lipoprotein cholesterol levels and endogenous cholesterol synthesis in hyperlipidemic subjects. *Lipids*, 34(Suppl). S121-4, 1999.

[31] Lindsey, S., *et al.*, Dietary palmitic acid (16:0) enhances high density lipoprotein cholesterol and low density lipoprotein receptor mRNA abundance in hamsters. *Proc Soc Exp Biol Med*, 195(2). 261-9, 1990.

[32] Blake, W.L. and S.D. Clarke, Suppression of rat hepatic fatty acid synthase and S14 gene transcription by dietary polyunsaturated fat.

J Nutr, 120(12). 1727-9, 1990.
[33] Mougios, V., *et al.*, Exercise-induced changes in the concentration of individual fatty acids and triacylglycerols of human plasma. *Metabolism*, 44(5). 681-8, 1995.
[34] Moser, H.W., Clinical and therapeutic aspects of adrenoleukodystrophy and adrenomyeloneuropathy. *J Neuropathol Exp Neurol*, 54(5). 740-5, 1995.
[35] Moser, A.B., *et al.*, Plasma very long chain fatty acids in 3,000 peroxisome disease patients and 29,000 controls. *Ann Neurol*, 45(1). 100-10, 1999.
[36] Van Geel, B.M., *et al.*, Progression of abnormalities in adrenomyeloneuropathy and neurologically asymptomatic X-linked adrenoleukodystrophy despite treatment with "Lorenzo's oil". *J Neurol Neurosurg Psychiatry*, 67(3). 290-9, 1999.
[37] Rizzo, W.B., *et al.*, Dietary erucic acid therapy for X-linked adrenoleukodystrophy. *Neurology*, 39(11). 1415-22, 1989.
[38] Jiang, J., A. Wolk, and B. Vessby, Relation between the intake of milk fat and the occurrence of conjugated linoleic acid in human adipose tissue. *Am J Clin Nutr*, 70(1). 21-7, 1999.
[39] Siguel, E.N., *Essential fatty acids in health and disease*. 1994, Brookline: Nutrek, Inc. 396.
[40] Liu, Y.Y., *et al.*, Abnormal fatty acid composition of lymphocytes of biotin-deficient rats. *J Nutr Sci Vitaminol (Tokyo)*, 40(3). 283-8, 1994.
[41] Awad, A.B., *et al.*, 18:1 n7 fatty acids inhibit growth and decrease inositol phosphate release in HT-29 cells compared to n9 fatty acids. *Cancer Lett*, 91(1). 55-61, 1995.
[42] Parthasarathy, S., *et al.*, Low density lipoprotein rich in oleic acid is protected against oxidative modification: implications for dietary prevention of atherosclerosis. *Proc Natl Acad Sci USA*, 87(10). 3894-8, 1990.
[43] Christensen, E., T.A. Hagve, and B.O. Christophersen, The Zellweger syndrome: deficient chain-shortening of erucic acid (22:1 (n-9)) and adrenic acid (22:4 (n-6)) in cultured skin fibroblasts. *Biochim Biophys Acta*, 959(2). 134-42, 1988.
[44] Babin, F., *et al.*, Nervonic acid in red blood cell sphingomyelin in premature infants: an index of myelin maturation? *Lipids*, 28(7). 627-30, 1993.
[45] Sargent, J.R., K. Coupland, and R. Wilson, Nervonic acid and demyelinating disease. *Med Hypotheses*, 42(4). 237-42, 1994.
[46] Stamler, J., Nutrition-related risk factors for the atherosclerotic diseases-- present status. *Prog Biochem Pharmacol*, 19. 245-308, 1983.
[47] Lee, E.J., K. Simmer, and R.A. Gibson, Essential fatty acid deficiency in parenterally fed preterm infants. *J Paediatr Child Health*, 29(1). 51-5, 1993.
[48] Siguel, E.N. and R.H. Lerman, Altered fatty acid metabolism in patients with angiographically documented coronary artery disease. *Metabolism*, 43(8). 982-93, 1994.
[49] Jolly, C.A., *et al.*, Dietary (n-3) polyunsaturated fatty acids suppress murine lymphoproliferation, interleukin-2 secretion, and the formation of diacylglycerol and ceramide. *J Nutr*, 127(1). 37-43, 1997.
[50] Henderson, R.A., *et al.*, Effect of fish oil on the fatty acid composition of human milk and maternal and infant erythrocytes. *Lipids*, 27(11). 863-9, 1992.
[51] Al, M.D., *et al.*, Maternal essential fatty acid patterns during normal pregnancy and their relationship to the neonatal essential fatty acid status. *Br J Nutr*, 74(1). 55-68, 1995.
[52] Khulusi, S., *et al.*, The effects of unsaturated fatty acids on Helicobacter pylori in vitro. *J Med Microbiol*, 42(4). 276-82, 1995.
[53] Leventhal, L.J., E.G. Boyce, and R.B. Zurier, Treatment of rheumatoid arthritis with gamma-linolenic acid [see comments]. *Ann Intern Med*, 119(9). 867-73, 1993.
[54] Noguchi, M., *et al.*, The role of fatty acids and eicosanoid synthesis inhibitors in breast carcinoma. *Oncology*, 52(4). 265-71, 1995.
[55] Hayes, K.C., A. Livingston, and E.A. Trautwein, Dietary impact on biliary lipids and gallstones. *Annu Rev Nutr*, 12. 299-326, 1992.
[56] Metz, J., Cobalamin deficiency and the pathogenesis of nervous system disease. *Annu Rev Nutr*, 12. 59-79, 1992.
[57] Kishimoto, Y., *et al.*, Branched-chain and odd-numbered fatty acids and aldehydes in the nervous system of a patient with deranged vitamin B 12 metabolism. *J Lipid Res*, 14(1). 69-77, 1973.
[58] Mock, D.M., S.B. Johnson, and R.T. Holman, Effects of biotin deficiency on serum fatty acid composition: evidence for abnormalities in humans. *J Nutr*, 118(3). 342-8, 1988.
[59] Mock, D.M., *et al.*, Effects of biotin deficiency on plasma and tissue fatty acid composition: evidence for abnormalities in rats. *Pediatr Res*, 24(3). 396-403, 1988.
[60] Frenkel, E.P. and J.D. White, Characterization of an animal model of vitamin B_{12} deprivation. *Lab Invest*, 29(6). 614-9, 1973.
[61] Manteca, X., E. Fernandez, and E. Gonalons, Effects of oleic and elaidic acids on in vitro intestinal uptake of cholesterol in the rat. Arch *Int Physiol Biochim Biophys*, 102(3). 231-2, 1994.
[62] Mensink, R.P., *et al.*, Effect of dietary cis and trans fatty acids on serum lipoprotein[a] levels in humans. *J Lipid Res*, 33(10). 1493-501, 1992.
[63] Longnecker, M., Do trans fatty acids in margarine and other foods increase the risk of coronary heart disease? *Epidemiology*, 4(6). 492-494, 1993.
[64] Anonymous, Trans fatty acids, blood lipids, and cardiovascular risk: Where do we stand? *Nutr Rev*, 51(11). 340-343, 1993.
[65] Pettersen, J. and J. Opstvedt, trans fatty acids. 5. Fatty acid composition of lipids of the brain and other organs in suckling piglets. *Lipids*, 27(10). 761-9, 1992.
[66] Litin, L. and F. Sacks, Trans fatty acid content of common foods. *NEJM*, 329(26). 1969-1970, 1993.
[67] Willett, W.C., *et al.*, Intake of trans fatty acids and risk of coronary heart disease among women [see comments]. *Lancet*, 341(8845). 581-5, 1993.
[68] Wander, R.C., *et al.*, The ratio of dietary (n-6) to (n-3) fatty acids influences immune system function, eicosanoid metabolism, lipid peroxidation and vitamin E status in aged dogs. *J Nutr*, 127(6). 1198-205, 1997.
[69] Sardesai, V.M., The essential fatty acids. *Nutr Clin*

Pract, 7(4). 179-86, 1992.

[70] Mascioli, E.A., *et al.*, Essential fatty acid deficiency and home total parenteral nutrition patients. *Nutrition*, 12(4). 245-9, 1996.

[71] Garda, H.A., A.M. Bernasconi, and R.R. Brenner, Possible compensation of structural and viscotropic properties in hepatic microsomes and erythrocyte membranes of rats with essential fatty acid deficiency. *J Lipid Res*, 35(8). 1367-77, 1994.

[72] Apostolov, K., *et al.*, Reduction in the stearic to oleic acid ratio in leukaemic cells--a possible chemical marker of malignancy. *Blut*, 50(6). 349-54, 1985.

[73] Persad, R.A., *et al.*, Erythrocyte stearic to oleic acid ratio in prostatic carcinoma. *Br J Urol*, 65(3). 268-70, 1990.

[74] Reddy, B.S., *et al.*, Metabolic epidemiology of large bowel cancer: fecal bulk and constituents of high-risk North American and low-risk Finnish population. *Cancer*, 42(6). 2832-8, 1978.

[75] Reddy, J. and N. Lalwani, Peroxisome proliferation and hepatocarcinogenesis, in *Mechanism of Carcinogenesis in Risk Identification*, H. Vainio, *et al.*, Editors. 1992, Int. Agency Res. Cancer: Lyon. p. 225-235.

[76] Averna, M., et al., Familial hypobetalipoproteinemia is not associated with low levels of lipoprotein(a). *Arterioscler Thromb Vasc Biol*, 15(12). 2165-75, 1995.

[77] Schonfeld, G., The hypobetalipoproteinemias. *Annu Rev Nutr*, 15. 23-34, 1995.

[78] Horrobin, D.F., Loss of delta-6-desaturase activity as a key factor in aging. *Med Hypotheses*, 7(9). 1211-20, 1981.

[79] Horrobin, D., ed. Clinical uses of essential fatty acids. 1981, Eden Press: Montreal.

Chapter 6

Organic Acids

TABLE 6-1. Summary of Organic Acids in Urine

	Name		Potential Intervention	Metabolic Pathway
Fatty Acid Oxidation	Adipate	H	L-carnitine, 250 mg TID; B2, 100 mg BID; B5, 500 mg BID; Choline, 100 mg TID; C, 1000 mg TID; CoQ10, 150 mg	Fatty acid oxidation
	Suberate	H		
	Ethylmalonate	H		
Carbohydrate Metabolism	Pyruvate	H	B-complex, 1 capsule TID; B_1, 100 mg TID; Lipoic acid, 100 mg TID	Aerobic/anaerobic energy production
	Lactate	H	Coenzyme Q_{10}, 50 mg TID	
	β-Hydroxybutyrate	H	Chromium picolinate, 200 mcg BID	Glucose utilization
	α-Hydroxybutyrate	H		
Energy Production (Citric Acid Cycle)	Citrate	H	Arginine, 1-3 gm/day	Citric Acid Cycle Intermediates
		L	Aspartic acid, 500 mg	
	cis-Aconitate	H	Lipoate, 25 mg/kg/day	
	Isocitrate	H	Magnesium, 400 mg; Manganese, 20 mg	
	α-Ketoglutarate	L	α-KG, 300 mg; Arginine, 1000 mg; Glutamine, 1-5 g	
		H	B-complex, 1 TID; Lipoic acid 100 mg	
	Succinate	L	Isoleucine, 1000 mg TID; Valine, 1000 mg TID	
		H	Magnesium, 500 mg, CoQ10, 50 mg TID	
	Fumarate	L	Tyrosine, 1000 mg BID; Phenylalanine, 500 mg BID	
	Malate	H	B_3, 100 mg TID	
	Hydroxymethylglutarate	H/L	CoQ_{10}, 50 mg TID	CoQ_{10} synthesis
B-Complex Vitamin Markers	α-Ketoisovalerate	H	B-complex, 1 capsule TID + 100 mg lipoic acid	Amino acid catabolism
	α-Ketoisocaproate	H		
	α-Keto-β-methylvalerate	H		
	Methylmalonate	H	B_{12}, 1000 mcg TID	Valine catabolism
	β-Hydroxyisovalerate	H	Biotin, 5 mg BID; Magnesium, 100 mg BID	Isoleucine catabolism

TABLE 6-1. Cont.

Detoxication Indicators	p-Hydroxyphenyllactate	**H**	Vitamin C up to 100 mg/kg	Prooxidant and carcinogen
	Orotate	**H**	Arginine, 1-3 gm/day; α-KG, 300 mg TID Aspartic acid, 500 mg BID; Magnesuim, 300 mg	Ammonia clearance, Pyrimidine synthesis
	2-Methylhippurate	**H**	Avoidance of xylene; Glycine, 2-5 g,/day	Hepatic conjugation
	Glucarate	**H**	Glycine, GSH, NAC, 500-5000 mg/day	Detox. liver enzyme induction
	Pyroglutamate	**H**	NAC, 1000 mg, Glutathione, 300 mg	Renal amino acid recovery-
	Sulfate	**L**	Taurine, 500 mg BID; Glutathione, 300 mg	Detox and anti-oxidant functions
Neurotransmitter Metabolism	Vanilmandelate	**L/H**	Tyrosine, 1000 mg BID-TID, between meals Contraindicated for patients taking MAO inhibitors (Vanilmandelate, Homovanillate)	Epinephrine, Norepinephrine catabolism
	Homovanillate	**L/H**		DOPA catabolism
	5-Hydroxindoleacetate	**L/H**	5-Hydroxytryptophan, 100 mg TID	Serotonin catabolism
	Quinolinate	**H**	Magnesuim, 300 mg	Serotonin catabolism
Dysbiosis Markers (Products of Abnormal Gut Microflora)	Benzoate	**H**	(Benzoate through β-Ketoglutarate:) (Glycine, 500 gm + Pantothenate, 100 mg) - TID These compounds reflect intestinal overgrowth, usually accompanied by microbial hyperpermeability. Glutamine, 10-20 gm daily and free-form amino acids to normalize gut permeability. Digestive aids (betaine, enzymes, bile) Take appropriate steps to ensure favorable gut microflora poplulation.	Hepatic Phase II conjugation Bacterial (Benzoate through Dihydroxyphenylpropionate) Numerous interferences in energy pathways and cellular control mechanisms
	Hippurate	**H**		
	Phenylacetate	**H**		
	Phenylpropionate	**H**		
	p-Cresol	**H**		
	p-Hydroxybenzoate	**H**		
	p-Hydroxyphenylacetate	**H**		
	Tricarballylate	**H**		
	Dihydroxyphenylpropionate	**H**		
	Tartarate	**H**		Yeast (Tartarate through β-Ketoglutarate)
	Citramalate	**H**		
	β-Ketoglutarate	**H**		

FIGURE 6-1. Urinary Markers of Nutrients Involved in Central Energy Pathways

Note: Vitamin and mineral requirements for cofactors are shown in black boxes. Elevations of metabolites before these steps indicate functional deficit of the nutrients.

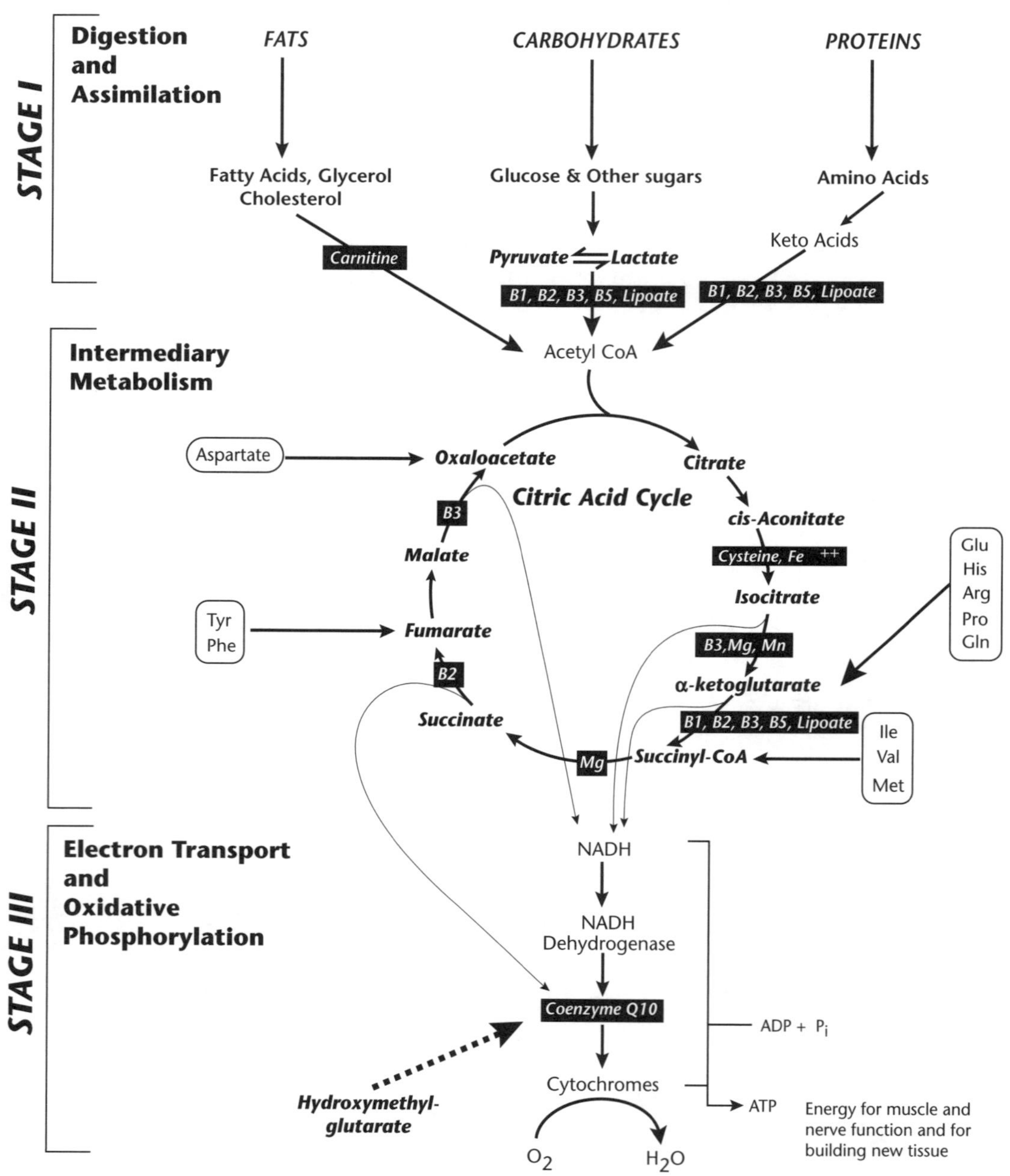

Introduction

The logical sequence of nutrient categories followed throughout the previous four chapters is broken at this point in the text. Unlike amino acids and fatty acids, the category of compounds called organic acids contains no essential nutrients. Instead of tests that measure nutrient concentrations, abnormal concentrations of organic acids in urine provide functional markers for the metabolic effects of vitamin inadequacies, toxic exposure, neuroendocrine activity, and intestinal bacterial overgrowth. The ultimate tool for laboratory evaluations in nutritional medicine is a simple, sensitive test that can reveal evidence of functional inadequacy of specific nutrients. The promise of such a tool is found in profiling of organic acids in urine.

All bodily functions are powered by the release of chemical energy. Each day, the energy content of the food for an average person could raise the temperature of about seven gallons of water to the boiling point. The energy is released through a process of controlled oxidation where chemical bonds are broken and energy is released. Fats, carbohydrates, and amino acids are converted into carboxylic acids before they flow on to the final conversion to carbon dioxide (Figure 6-1). The organic acids that are formed as intermediates in this process are normally absent from urine or present at very low concentrations. When specific reactions are blocked due to the absence of sufficient enzyme or cofactor, the intermediates that precede the blocked step accumulate and spill into urine. Overt nutrient deficiencies are one reason for metabolic inefficiency. Variations in enzyme structure that lead to decreased cofactor binding are another major cause of nutrient deficiencies. Individuals with faulty enzyme binding can have increased nutrient needs that will not be revealed by measures of vitamin concentrations in blood. Various nutrient-related abnormalities that might appear on a typical quantitative report of organic acids in urine are summarized in Table 6-1. The supplementary nutrient amounts are given as guides for starting points to improve clinical outcomes for adults. Maple syrup urine disease provides an example of this concept.

The oxidation of carbon skeletons from amino acids is a major energy production pathway in which an early step is catalyzed by the enzyme, branched chain keto acid dehydrogenase. Variants of maple syrup urine disease are due to degrees of impairment of this enzyme (Table 6-2). Genetic variants in which the enzyme is not expressed result in severe clinical consequences in infancy. Other conditions may develop in adults with mild levels of functional impairment due to a high requirement for thiamin [1].

Urinary organic acid analysis for metabolic profiling has traditionally been used for detection of neonatal inborn errors of metabolism. Since the reporting of isovaleric acidemia in 1966, there has been a rapidly growing list of disorders resulting in elevated

TABLE 6-2. Maple Syrup Urine Disease

Variant	Organic Acid	Clinical Presentation
Severe	α-ketoisocaproate α-ketoisovalerate α-keto-β-methylvalerate	Infant death or prominent neurologic abnormalities and profound mental retardation
Mild	Same as "Severe"	Irritability, ataxia, lethargy, developmental retardation
Thiamin-responsive (10-4000 mg/day)	Same as "Severe" (lower levels in urine)	Same signs and symptoms as above that appear only with the stress of infection, surgery, etc. Bio-chemical abnormalities clear and return as thiamin is administered and withdrawn

excretion of metabolic intermediates [2, 3]. The application of the testing to assess special nutrient requirements of individuals is discussed in a variety of sources [3-7]. Organic acid profiling has also been useful in identification of the source of toxicants from the environment [8] and the gut [9].

The term "organic acid" refers to a broad class of compounds used in fundamental metabolic processes of the body. Chemically, organic acids share the common features of water solubility, acidity, and ninhydrin-negativity (no primary or secondary amines). The term is generally considered to include all carboxylic acids, with or without keto, hydroxyl, or other non-amino functional groups, but does not include most amino acids. Some nitrogen-containing compounds are included, such as pyroglutamate, or amino conjugates like hippurate (benzoylglycine). Short chain fatty acids are also contained in this group. With adjustments of the method for extraction of the compounds from urine, the analysis may include neutral compounds that possess no acid group, in which case the naming of the entire set as "organic acids" is somewhat inaccurate.

Morning urine has become widely accepted for primary study of metabolic disorders. Urination is generally the first post-sleep activity because the cortisol-stimulated, anabolic sleep period requires the metabolic, energy-yielding pathways that produce water. Prior to waking, the cortisol peaks and then drops, initiating a shift to greater utilization of catabolic pathways. For these reasons, the first morning urine usually contains a greater array of metabolic by-products than at any other time of the day and is relatively free of the metabolic effects of strenuous muscle activity. Organic compounds are extracted from the overnight urine sample and analyzed by gas chromatography with mass spectrometric detection. The basic methodologies are well known [10], but improvements are regularly reported [11].

The information content of the profile is high, but the interpretation is simplified by keeping in mind that the data supplies answers to four basic questions of clinical relevance:

(1) Is mitochondrial energy production adversely affected?

(2) Are functional nutrient deficiencies present?
(3) Are symptoms related to excessive growth of bacteria and fungi in the gut?
(4) Is there an undue toxic load and is this adversely affecting detoxification capacity?

Each of several compounds reported in the typical profiling of organic acids in urine will be discussed briefly to indicate why they are related to these clinical questions with key references cited.

Fatty Acid Metabolism Markers

Adipate, Suberate, and Ethylmalonate

Need for carnitine is indicated when any of the dicarboxylic acids (*i.e.*, adipate, suberate, and ethylmalonate) are elevated [12]. Carnitine is required as a carrier for the transport of fatty acids from the cytosol into the mitochondria for beta-oxidation (Figure 6-2). This is the dominant pathway used to derive energy from fat in most cells. When carnitine is inadequate to keep up with demand, the degradation of fatty acids takes place through an alternate, less efficient pathway known as "omega-oxidation." Adipate, suberate, and ethylmalonate are products of incomplete oxidation in the omega-oxidation pathway. All three of the compounds will be elevated in overt enzyme failure, but the patterns of elevations in milder carnitine deficiencies will vary. Symptoms include periodic mild weakness, nausea, fatigue, hypoglycemia, "sweaty feet" odor, and recurrent infections. Abnormalities are frequently found in children with attention deficit disorders. Patients may also exhibit a Reye-like syndrome in dicarboxylic aciduria, which has been associated with various metabolic toxins from viral infections that affect mitochondrial function. Patients with low carnitine respond well to therapeutic riboflavin [13]. The biochemical relationship between these two nutrients can explain this effect. Carnitine is

FIGURE 6-2. Fatty Acid Metabolism Markers

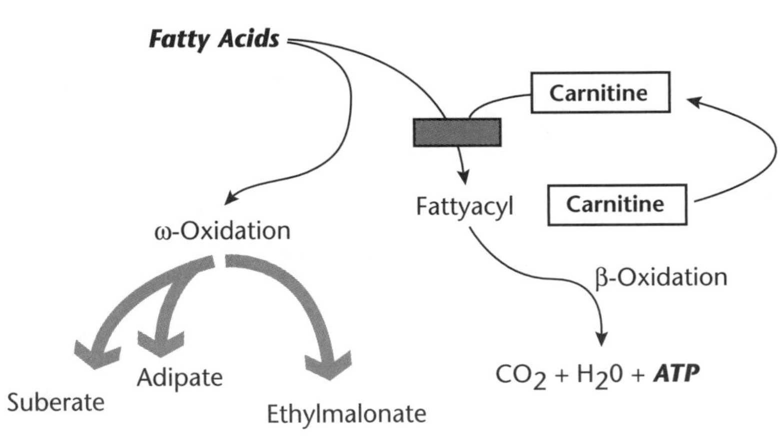

required to allow fatty acids to enter into mitochondria, but fatty acids cannot undergo oxidative metabolism without riboflavin coenzymes. The use of aspirin can change the interpretation of results for the three marker compounds because salicylic acid is an inhibitor of fatty acid β-oxidation and may lead to elevated markers [14]. The β-oxidation enzymes involved also respond to environmental toxin exposure with altered lipid metabolism that can lead to impaired immune responsiveness and mitochondrial DNA damage [15]. Supplementation of carnitine and riboflavin is indicated when adipate, suberate, or ethylmalonate are elevated [16].

Carbohydrate Metabolism Markers

α-Hydroxybutyrate and β-Hydroxybutyrate

These two compounds are sometimes called "ketone bodies" because they are formed along with acetone and acetoacetate. Increased β-hydroxybutyrate is a textbook feature of metabolic acidosis due to failure of glucose utilization as with diabetes. Ketone body production increases in diabetes because the oxidation of free fatty acids is stimulated, and excess acetyl-CoA is converted to the four-carbon acids. This acetyl-CoA spillover phenomenon occurs because the control of ATP production from fatty acids cannot be regulated as well as from carbohydrate oxidation. Individuals with normal blood glucose response to insulin do not produce high concentrations of ketone bodies because their production of energy from glucose is well controlled. Elevations of the hydroxybutyrates in an overnight urine collection may indicate inefficient utilization or mobilization of glucose [17]. In these cases, chromium and vanadium supplementation may support carbohydrate utilization by improving the action of insulin [18-20]. Defects in cytochrome oxidase enzymes of the electron transport system are another reason for elevated hydroxybutyrate [21]. Excessive fatigue on exertion is the most common symptom associated with ketosis.

α-Hydroxybutyrate

$H_3C\text{-}CH_2\text{-}CH(OH)\text{-}CO_2$

β-Hydroxybutyrate

$H_3C\text{-}CH(OH)\text{-}CH_2\text{-}CO_2$

It is relatively rare to find elevated α-hydroxybutyrate in urine because many tissues have high activities of the enzyme alpha hydroxybutyrate dehydrogenase (aHBD) that catalyzes oxidation of the organic acid. The activity of aHBD on the second day after a myocardial infarction is a marker for estimates of infarct size and a measure of reperfusion effectiveness [22]. Smoking, poor diet, and lack of exercise significantly inhibit the activity of aHBD, suggesting that urinary elevation of α-hydroxybutyrate may be related to these factors [23]. High α-hydroxybutyrate is also found during phases of increased lymphocyte destruction in infectious diseases such as measles [24].

Lactate and Pyruvate

Abnormalities of urinary excretion of lactate and pyruvate provide useful insight to basic metabolic factors due to their position in the energy production process. Pyruvate is the anaerobic breakdown product of glucose. Its further conversion to acetyl-CoA requires the pyruvate dehydrogenase enzyme complex. Pyruvate dehydrogenase requires cofactors derived from thiamin, riboflavin, niacin, lipoic acid, and pantothenic acid. Elevated levels of pyruvate may reflect failure of the enzyme due to a functional need for increased B vitamins, particularly thiamin [25] and pantothenic acid [26]. Pyruvate elevation in the brain accompanies the encephalopathy associated with alcohol use in the thiamin-responsive Wernicke-Korsakoff syndrome [27]. Dietary supplementation with pyruvate has been found to prevent fatty liver caused by chronic alcohol ingestion in laboratory animals [28]. The beneficial effect of pyruvate is explained as a stimulation of glucogenesis, which results in inhibition of fatty acid synthesis.

Metabolic Acidosis:

- Increased levels of organic acids due to their increased production or failure to be removed
- Narrowly defined as either diabetic or lactic acidosis but more broadly includes greatly elevated levels of any organic acids in body fluids

Lactate, also known as lactic acid, is a principal product of glucose oxidation in skeletal muscle. Lactate accumulates when there is a block in the final oxidative phosphorylation stage of energy production. Such a block results in inactivation of the citric acid cycle. Because lactate is transported to the liver and converted to pyruvate, they have similar responses to nutrient deficiencies. Coenzyme Q_{10} supplementation is effective when lactate is elevated and energy production is interrupted [29-31]. Elevated lactate, therefore, is a sign of coenzyme Q_{10} deficiency. Increased lactate is a common acidotic condition that can be caused by a variety of metabolic problems. A case of lactic acidosis and marked hepatic steatosis has been normalized with administration of 50 mg of riboflavin per day [32]. Both lactate and pyruvate may be elevated by alcohol consumption prior to specimen collection [33]. This effect may be due to the niacin-depleting effect of alcohol because the conversion of lactate to pyruvate (Figure 6-3) and the further oxidation of pyruvate are NAD-dependent.

FIGURE 6-3. Lactate and Pyruvate Interconversion With Transfer of Electrons to NAD

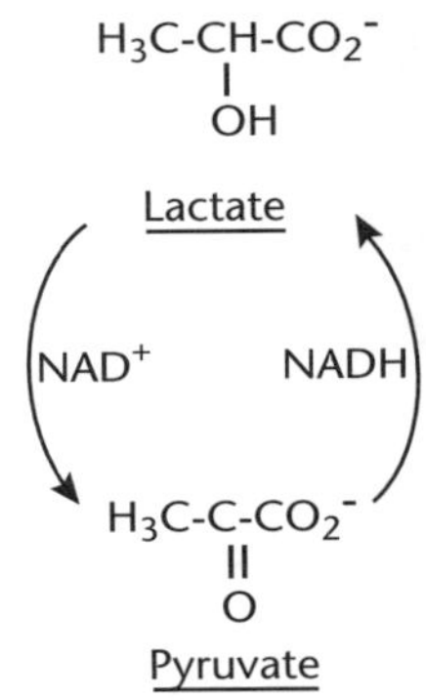

Lactic acidosis with increased excretion of hydroxybutyrate and fumarate were biochemical signs of the metabolic disturbance found in a 42-year-old woman who presented with a 10-year history of ophthalmoplegia, malabsorption resulting in chronic malnutrition, muscle atrophy, and polyneuropathy [21, 34]. The underlying biochemical lesion in this mitochondrial, multi-system disorder was found to be genetically defective cytochrome C oxidase (complex IV of the respiratory chain). This case illustrates the potential for graded metabolic effects from genetic polymorphic expressions. Interestingly, this defect did not manifest until adulthood. Compensatory mechanisms maintained nominal function until the genetic weakness became manifest, which was most likely due to accumulation of environmental insult and dietary inadequacy. As we age, metabolic systems function less efficiently. How many of our "degenerative diseases" are genetic weaknesses manifesting as our systems become less efficient?

Most individuals carry mild enzyme defects that manifest as symptoms only when metabolic stressors exceed the ability to maintain normal function. The age of onset and the degree and duration of functional loss depend on an individual's ability to maintain organ reserve. As we proceed to describe the scientific reports of specific organic acidurias, it is good to consider this central axiom: any report of a case or small group of patients with a serious disorder showing severe organic acidurias is expected in less severe forms in larger numbers of individuals. Those with the milder forms may maintain normal function until adulthood.

Central Energy Pathway Markers

Citrate, Isocitrate and cis-Aconitate

The Krebs or citric acid cycle (CAC) not only is the final common pathway of energy release from food components but is also the source of basic structural or anabolic molecules that are drawn away from the cycle to support organ maintenance and neurological function. Therefore, the CAC serves both anabolic and catabolic functions of the body, representing the "crossroads" of food conversion and utilization (Figure 6-1).

Conversions of the CAC intermediates are under the control of enzymes that often require vitamin-derived cofactors and minerals for their function. Abnormal spilling of CAC intermediates in urine indicates mitochondrial inefficiencies in energy production. This fundamental pathway of energy flow is critical for all organ systems. Elevated excretion of the compounds involved in the cycle can explain the biochemical basis of excessive fatigue and weakness and guide ways to improve energy production by supplying B complex vitamins. Since the compounds are related in a cyclic manner, a block at any step can cause accumulation of multiple compounds that precede that step. In cytochrome C oxidase deficiency, there is inefficient utilization of NADH, the primary product of the CAC, resulting in citrate, malate, fumarate, and α-ketoglutarate elevations [35]. Coenzyme Q_{10} deficiency can also result in elevated CAC intermediates (see HMG below). Toxic effects of drugs like gentamicin are, in part, due to loss of mitochondrial integrity. Isocitrate dehydrogenase, the enzyme that converts isocitrate to cis-aconitate, is inhibited by such drugs. Lipoic acid, used at 25 mg/kg/day, protects against such cytotoxic effects. Thus, elevated isocitrate and citrate are signs of increased need for lipoic acid. Isocitrate dehydrogenase activity also requires the presence of magnesium and manganese ions.

Citrate is frequently the most abundant organic acid in urine. The range of normal values is broad because citrate excretion varies with changes in renal metabolism and acid-base disturbances [36]. The anion properties of citrate are used to balance increased positive ammonia ion excretion. Citrate elevation is a marker of ammonia accumulation due to arginine deficiency [37, 38]. Orotate elevation is an additional, sensitive marker for this condition (see below).

Low levels of these three compounds may occur due to the multiple reactions that utilize CAC intermediates for biosynthesis. Aspartic acid is a preferred substrate for the refilling or "anaplerotic" pathway that restores CAC levels. By a single transamination step, aspartate is converted into oxaloacetate, the immediate precursor of citrate. The reaction yields a net gain of cycle intermediates, a result that cannot be achieved from the flow of acetyl-CoA (Figure 6-1).

α-Ketoglutarate

Alpha ketoglutarate (α-KG), also known as 2-ketoglutarate or 2-oxoglutarate, is formed by oxidation of isocitrate, and it is further oxidized in an energy-releasing step that requires an enzyme system similar to the one described for pyruvate oxidation above. The same five B-vitamins are necessary for the action of this enzyme, so elevations of α-KG signal potential B-complex deficiencies. Low activity of this enzyme has been shown to produce elevated excretion of α-KG [39].

The compounds that make up the CAC can be derived from amino acids as discussed above for aspartic acid. This would explain the energy-boosting effect people often report when they take free-form amino acid supplements. The effect is due to the conversion of specific amino acids directly into depleted CAC intermediates needed for the energy-producing cycle. α-Ketoglutarate also is also formed in the catabolic breakdown of glutamic acid, histidine, arginine, proline, and glutamine. The fatigue-reducing effect of supplementation of aspartate salts and ketoglutaric acid has been attributed to such a mechanism [40, 41]. The importance of α-KG in cell energetics is shown by the effectiveness of oral α-KG in cyanide poisoning where the availability of oxygen is the primary deficit [42].

Levels of α-KG can also serve to mark an aspect of the carbohydrate and fat metabolic relationship. When metabolic conditions, such as insulin intolerance, act to stimulate fatty acid synthesis, the rising levels of palmitoyl-CoA cause an effective inhibition of glutamate dehydrogenase. Since the dominant pathway for α-KG formation is production from glutamate by the activity of glutamate dehydrogenase, mitochondrial α-KG is depleted. [43]. To state the corollary, low α-KG is a marker for up-regulated fatty acid synthesis, increased palmitic acid in plasma and cell membranes, and increased serum triglycerides.

Succinate

Succinate cannot be used as an energy source when coenzyme Q_{10} (CoQ10) is inadequate. Clinical signs of CoQ_{10} and riboflavin deficiency include fatigue, lassitude, and myocardial and neurological degeneration. CoQ_{10} depletion can be tissue specific. For example, a 4-year old boy who presented with progressive muscle weakness, seizures, and cerebellar syndrome had greatly reduced muscle CoQ_{10} without corresponding CoQ_{10} deficits in lymphoblasts or skin fibroblasts [31]. His muscle mitochondria had severely restricted ability to utilize succinate.

Elevated succinate excretion is a marker for deficiency of CoQ_{10} and riboflavin. The reaction in which succinate is converted into fumarate depends on the presence of flavin adenine dinucleotide (FAD) derived from riboflavin. Riboflavin administration has been shown to produce dramatic regression of neurological impairment in cases of deficits in the complex II respiratory chain succinate dehydrogenase enzyme (Figure 6-4) that forms the first link in the transport of electrons from succinate to oxygen [13, 44].

When succinate levels drop below normal, leucine and isoleucine are effective precursors that are converted into succinate to assure functioning of the CAC. Since this conversion requires adequacy of vitamin B_{12}, functional adequacy of B_{12} should be assured when amino acids are used to raise succinate. Individuals with inherited disorders in which the final steps of amino acid conversion to succinate are blocked exhibit serious neurologic symptoms [45].

FIGURE 6-4. Coenzyme Q_{10} Function Coenzyme Q_{10} controls the rate of citric acid cycle processing in the oxidation of succinate to fumarate. Reduced coenzyme Q_{10} (HQ_{10}) accepts hydrogen atoms one at a time from reduced FAD ($FADH_2$).

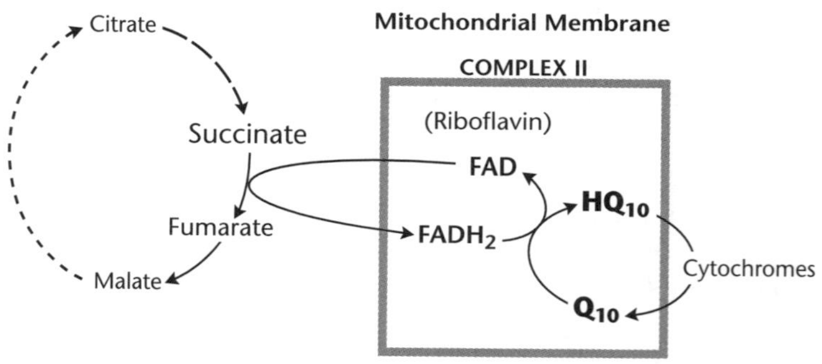

Fumarate and Malate

The removal of electrons from succinate in the reaction shown in Figure 6-4 generates fumarate. Fumarate is then converted to malate. Because of the sequential nature of these reactions in the mitochondrial matrix, fumarate and malate elevations serve as additional markers to substantiate the extent of metabolic interference from coenzyme Q_{10} deficiency or cytochrome oxidase defects [21]. Elevated malate also could be the result of stimulation of fatty acid synthesis. The fatty acid synthesis product palmitoyl-CoA is known to inhibit the malate dehydrogenase enzyme [43]. This sign of metabolic resistance to the oxidation of stored fat is seen in patients on low fat, high carbohydrate diets or with the hyperinsulinism known as syndrome X.

Low fumarate and malate concentrations in urine are a sign that the refilling reactions for the CAC are not keeping up with losses of intermediates to other pathways. Amino acid catabolism supplies a major portion of compounds that re-supply of the CAC (Figure 6-1). The addition of tyrosine or phenylalanine can help supply fumarate when this intermediate is low.

B-Complex Vitamin Markers

All of the compounds that we have placed in this category are metabolic intermediates in the degradation of amino acids (Figure 6-5). Once again, vitamin insufficiencies result in slower rates of the required reactions. Because of the specificity of individual enzymes utilizing the vitamin-derived coenzymes, discrete patterns of elevated excretion of specific intermediates signal functional deficiencies of individual vitamins.

FIGURE 6-5. B-Complex Vitamin Markers Elevations of specific compounds appearing in urine indicate deficiencies of nutrients required in their catabolic pathways. Deficiencies of the B-complex vitamins cause blocked pathways shown as:

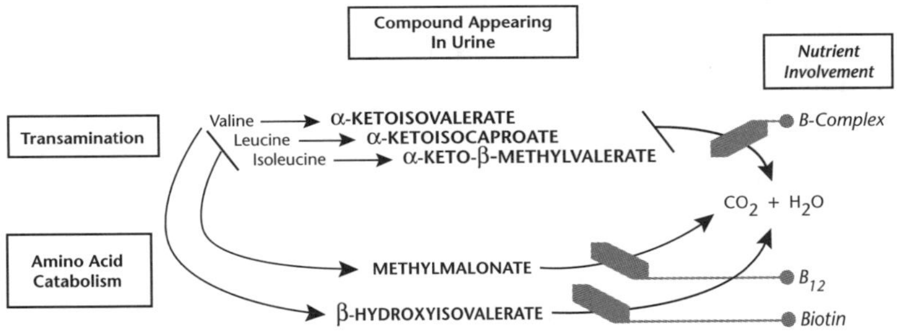

α-Ketoisovalerate, α-Ketoisocaproate, and α-Keto-β-methylvalerate

When hepatic or small intestinal enzymes remove the amino groups from the branched chain amino acids valine, leucine, and isoleucine the products formed are α-ketoisovalerate, α-keto-isocaproate, and α-keto-β-methylvalerate, respectively. In most current biochemical literature, these compounds are named 2-oxo, rather than α-keto. In medical discussions, the older Greek letter designation still prevails.

These compounds have been discussed previously. Although that discussion noted the specific response to thiamin, the reactions require all of the B-complex cofactors: vitamins B1, B2, B3, B5, and lipoic acid. Elevations of the branched chain keto acids provide a functional assessment of the sufficiency of these vitamins, especially thiamin [10].

Patients taking the hypolipidemic agent, clofibrate, may have a diminished sensitivity for these markers of vitamin deficiency because the drug stimulates their metabolism by the branched chain keto acid dehydrogenase complex [46].

Methylmalonate

Methylmalonyl-CoA is produced in the multi-step process of breaking down the amino acid valine into energy-yielding products. The intermediate is converted into succinic acid using a vitamin B_{12}-dependent enzyme, methylmalonyl CoA mutase (see Chapter 2). The lack of vitamin B_{12} impairs this conversion as well as the one in which methionine is recovered from homocysteine, leading to accumulation of both homocysteine and methylmalonate. Such biochemical impairment helps explain the fatigue resulting from vitamin B_{12} deficiency and the rapid improvement of symptoms vitamin B_{12}-deficient patients experience with administration of intramuscular vitamin B_{12}.

A differential diagnosis of vitamin B_{12}, folate, and vitamin B_6 deficiencies can be made with simultaneous assay of methylmalonate and the amino acid, homocysteine (Table 6-3). If only methylmalonate is elevated with normal homocysteine, only B_{12} supplementation is needed. Vitamin B_{12} deficiency has been demonstrated to be present in over 12% of the free-living elderly population in the United States [47] and is especially common in the early stages of HIV infection [48].

TABLE 6-3. Differentiation B_{12}, Folate, and B_6 Deficiency

Elevated Marker	Vitamin Specified
Homocysteine	B12, Folate, or B6
Methylmalonate	B12
Homocysteine with normal Methylmalonate	B6 or Folate
Both Homocysteine and Methylmalonate	B12, Folate, and B6

β-Hydroxyisovalerate

For many years, students of nutrition were taught that there is no evidence of biotin deficiency in adult humans [49]. The discovery of biotin production by some organisms that inhabit the human gut added impetus to the attitude that, although biotin is clearly not made in human tissues, biotin deficiency was of little concern. It was thought that only certain disorders of infancy, such as Leiner's disease (desquamative erythroderma) and other forms of seborrheic dermatitis, are clinical manifestations of biotin deficiency. Recent findings are shifting the clinical outlook for biotin deficiency in adults.

The biochemical function of biotin-requiring enzymes is the insertion of carboxyl groups to allow modification of metabolic intermediates. The catabolism of the amino acid isoleucine is a high-flux process that offers a biochemical marker of biotin deficiency. The product formed after the first three steps of the pathway, β-methylcrotonyl-CoA, requires a biotin-dependent carboxylation to allow the flow to continue. This compound accumulates if biotin is deficient, and its hydrated product, β-hydroxyisovalerate, spills in urine. β-Hydroxyisovaleric aciduria appears early in people who are made biotin deficient by consuming the biotin-binding protein, avidin. After starting avidin administration, elevated β-hydroxy-isovalerate appears at the third day while serum biotin concentrations remain in the normal range until the tenth day (Figure 6-6). The effects are completely reversible.

FIGURE 6-6. Human Biotin Deficiency and β-Hydroxyisovalerate Excretion

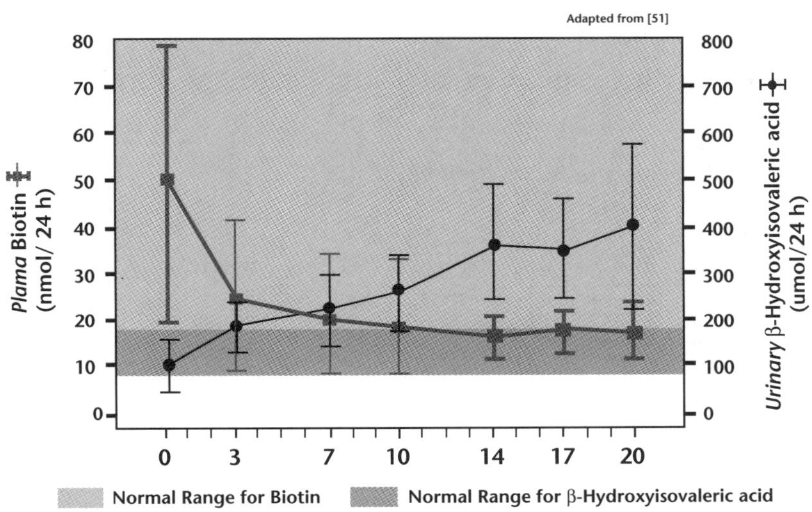

Heritable disorders of biotin metabolism lead to the condition called multiple carboxylase deficiency in which the activities of enzymes that have absolute requirements for biotin to carry out carboxylation reactions are deficient [50]. β-Hydroxyisovalerate is a compound that is elevated in biotin deficiency and multiple carboxylase insufficiency [51]. Biotin deficiencies of various degrees have been shown to develop in normal pregnancies [52] and in patients on long-term anticonvulsant therapy [53]. When β-hydroxyisovalerate was used to assess biotin in pregnant women, biotin status was found to decrease during pregnancy, and 9 out of 13 women studied were biotin depleted even in early pregnancy [54]. Symptoms of biotin deficiency include alopecia, skin rash, Candida dermatitis, unusual odor to the urine, immune deficiencies, and muscle weakness.

The enzymes that use biotin as a cofactor are called carboxylases because they use carbon dioxide to insert carboxyl groups into substrates. Biotin from food or from intestinal microbial synthesis is absorbed in the upper small intestine and transported to tissues bound to several blood proteins. Cellular biotin must be incorporated into the carboxylase enzymes by the action of other enzymes called synthases (Figure 6-7). If these enzymes are not fully active, higher biotin concentrations can increase enzymatic activity and enhance the reaction rate. The carboxylase enzymes have critical roles in major pathways for the utilization of energy from amino acids (where β-hydroxyisovalerate is formed), the synthesis of fatty acids for cell membrane replacement, and the maintenance of blood glucose via gluconeogenesis. At cell death, the biotin may be recovered if there is sufficient activity of the enzyme biotinidase, which acts on the biotin-peptide fragments called "biocytin." All five of the major factors (boxed text) in Figure 6-7 contribute to the maintenance of active carboxylase enzymes. Biotin deficiency can be caused by lack of biotin-rich foods or genetic variations in the enzymes shown in Figure 6-7. Antibiotic overuse can contribute to biotin insufficiency by lowering the population of biotin-producing organisms and

FIGURE 6-7. Defects of Biotin Pathways

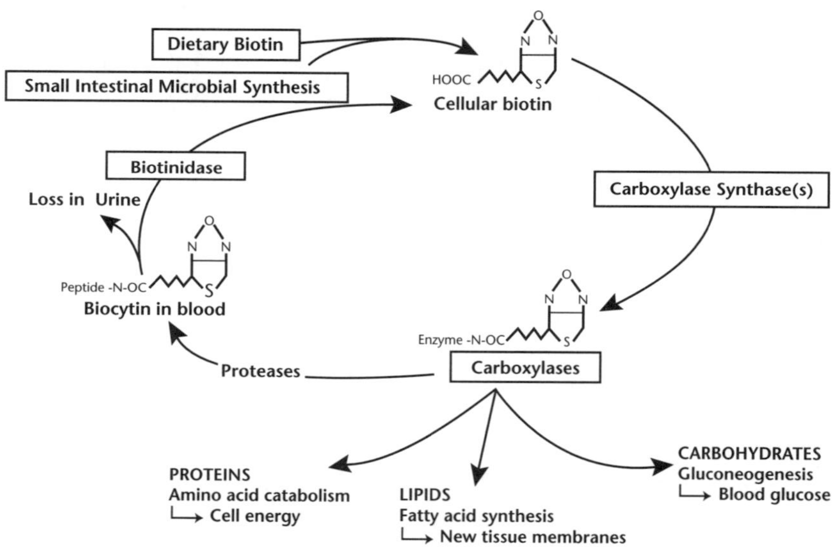

favoring the overgrowth of non-biotin producing species.

A case involving biotinidase deficiency illustrates the use of biotin replacement. A two-year-old male patient presented with ataxia, seizures, and a history of normal development for 19 months, and then experienced loss of language and developmental milestones and progressive gait disturbances [55]. Upon physical examination, the patient was obtunded with eczematous dermatitis, truncal ataxia, and a fine volitional tremor on the arms. Symptoms of ataxia, tremors, loss of language development, gait disturbances, and eczematous dermatitis cleared only after increasing biotin intake to 30 mg/day. The data presented in Table 6-4 are from this report. After one year of high dose biotin, the symptoms did not return and there was no further progression of hearing loss but some attention disorder with hyperactivity.

In addition to increased β-hydroxyisovalerate, elevations of lactate and alanine in urine and accumulations of odd-chain fatty acids in plasma or red cell membranes have also been found in biotin deficiency [50].

TABLE 6-4. Biotin Doses for 2 yr. Old Male with Biotinidase Deficiency

Biotin Dose	Clearing Signs/Symptoms
6 mg	Truncal ataxia Convulsions Alertness
20 mg	Organic acids Leukocyte carboxylase
30 mg	Dermatitis (Returned at 25 mg)

Hydroxymethylglutarate

Hydroxymethylglutarate (HMG) is the metabolic precursor of both cholesterol and coenzyme Q_{10} (CoQ_{10}) (Figure 6-8). Low levels of HMG may reflect inadequate synthesis and possible deficiency of CoQ_{10}. Drugs that are used to lower serum cholesterol by inhibiting the HMG-CoA reductase simultaneously inhibit the endogenous

FIGURE 6-8. Coenzyme Q_{10} Pathways

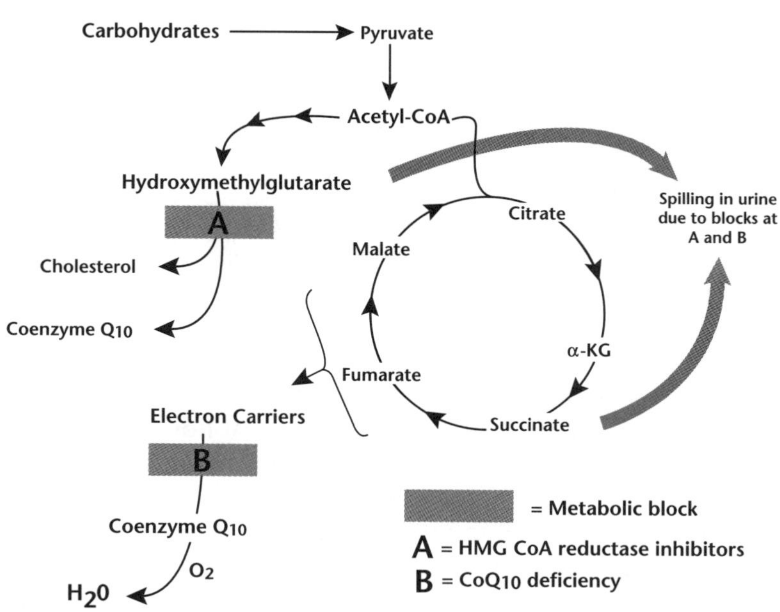

synthesis of CoQ_{10} and cause accumulation of the HMG intermediate in the pathway [56]. Therefore, high levels of HMG reveal inhibition in the synthesis of CoQ_{10}. CoQ_{10} is utilized in the mitochondrial oxidative phosphorylation pathway for ATP synthesis and is a potent anti-oxidant. It has been used extensively as a cardiovascular protective agent [57]. Although it may seem paradoxical, cases where simultaneous low serum CoQ_{10} levels were found with low urinary HMG suggest that CoQ_{10} inadequacy may be associated both low and high levels of HMG. The rationale for this association is that a block earlier in the pathway inhibits the formation of HMG producing low levels in urine, while a block subsequent to the formation of HMG produces high levels.

Neurotransmitter Metabolism Markers

Vanilmandelate and Homovanillate

Vanilmandelate, also known as vanilmandelic or vanillylmandelic acid (VMA), is the main metabolite of the catecholamines, epinephrine, and norepinephrine. Homovanillate (HVA) is the main metabolite of dopamine that appears in urine (Figure 6-9). Low urinary levels of VMA and HVA have been associated with low CNS levels of these neurotransmitters [58]. Symptoms associated with this condition include depression, sleep disturbances, anxiety, and fatigue. Treatments aimed at improving protein digestion and supplementation of the amino acid precursor (L-tyrosine) can normalize neurotransmitter levels in the CNS [59]. The relationship of tyrosine to epinephrine also gives rise to the antihypertensive role of tyrosine [60].

FIGURE 6-9. Neurotransmitter Metabolism Markers

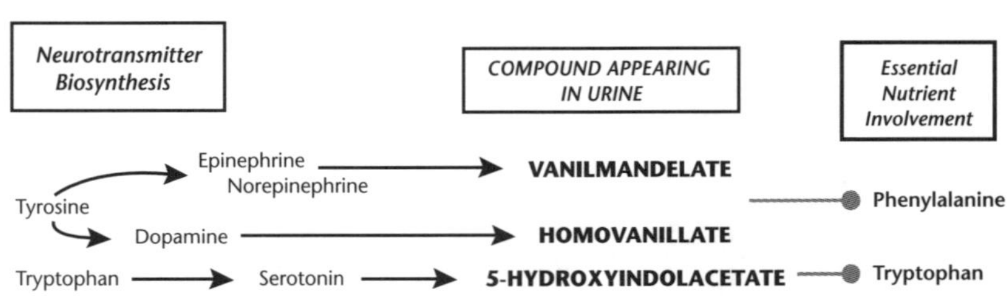

Elevated levels of VMA and HVA signal an increased rate of synthesis and degradation in normal tissue or abnormal production by tumor tissue. Neuroblastic tumors frequently cause a profound elevation in VMA, which may be expressed as the VMA/HVA ratio [61]. The incidence of elevated values appears to increase as a function of tumor size, and small tumors are not likely to result in positive urinary measurements [62]. Elevation of VMA along with specific elevation of cardiac troponin I are, together, precursors to myocardial injury [63].

In the absence of such disease processes, increased catecholamine synthesis results from the synergism of pituitary adrenocorticotropic hormone and adrenal cortisol. This constitutes the widely recognized chronic stress response that manifests as heightened sympathetic reactions to stress.

5-Hydroxyindoleacetate

Catabolic breakdown of serotonin leads to excretion of 5-hydroxyindoleacetate (5-HIA) (Figure 6-9). Abnormal high levels of this metabolite result from the use of serotonin-specific re-uptake inhibitor (SSRI) drugs or increased release of serotonin from any of three primary sites: central nervous system, intestinal argentaffin cells, or platelets. Carcinoid tumors composed of chromaffin tissue also can release large amounts of serotonin. Because of the magnitude of total body serotonin synthesis, increased rates of serotonin turnover indicated by elevated urinary 5-HIA can lead to depletion of the essential amino acid precursor, L-tryptophan. A very high 5-HIA result calls attention to potential deficiency of tryptophan. Plasma amino acid analysis provides a useful, direct measure of this amino acid (see Chapter 4 – Amino Acids).

Serotonin is required for control of gut motility as it activates smooth muscle activity. Inadequate production of serotonin leads to constipation. Low levels of serotonin have also been associated with depression, fatigue, insomnia, suicide, and attention deficit and behavioral disorders. Ethanol consumption also causes lowered 5-HIA excretion because the biochemical pathway that catalyzes serotonin metabolism is extremely sensitive to ethanol [64]. When 5-HIA levels are low, increased consumption of foods high in tryptophan including turkey, bananas, low fat milk, lentils, and eggs can minimize the need for oral tryptophan or 5-hydroxytryptophan.

Detoxification Markers

The compounds discussed here are also presented in Chapter 8, which elucidates their relationship to other tests for detoxification. They are organic acids that are significant markers of exogenous and endogenous toxin accumulation. Glucarate is included in this set even though it may not be reported as part of a profile. The ability of the laboratory to measure glucarate is determined by the type of sample preparation chosen. Glucarate is extremely water-soluble and does not move into the organic solvent layer when solvent extraction (the standard method) is used.

FIGURE 6-10. Orotate and Ammonia

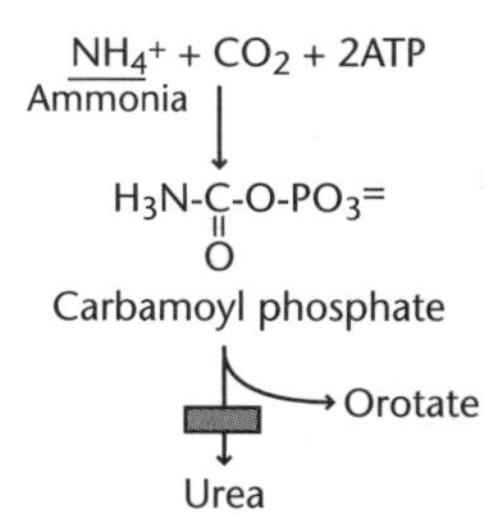

Orotate

When there is insufficient capacity for detoxifying the load of ammonia via the urea cycle, carbamoyl phosphate leaves the mitochondria and stimulates the synthesis of orotate [65] (Figure 6-10). Increased orotic acid (orotate) production is a sensitive indicator of arginine deficiency. Most of the symptoms that develop following arginine deprivation can largely be accounted for by a decreased efficiency of ammonia detoxification and reduced formation of nitric oxide. Increased orotic biosynthesis is observed with increasing ammonia concentrations in rat, mouse, and human liver and is reduced by *in vitro* arginine supplementation [37, 66]. In addition, orotate requires magnesium for its metabolism. While a normal level in a urinary organic acid panel may not necessarily indicate magnesium sufficiency, high levels should alert one to a significant possibility of intracellular magnesium insufficiency.

Oral glutamine supplementation at levels above 10 gm/day in adults can result in elevation of urinary orotate through the mechanism described above [67].

p-Hydroxyphenyllactate

p-Hydroxyphenyllactate (HPLA) is a metabolite of tyrosine that acts as part of the cell signaling mechanism for apoptosis. The methyl ester, methyl p-hydroxyphenyllactate (MeHPLA), is an important cell growth-regulating agent. Tumor cells contain esterase activities that hydrolyze the compound to the free acid, HPLA. MeHPLA binds strongly to nuclear sites, blocks estradiol stimulation of uterine growth, and inhibits growth of human breast cancer cells [68]. HPLA, on the other hand, binds more weakly to the same nuclear site and does not block uterotropic responses to estradiol or inhibit growth of breast cancer cells. Thus, HPLA is an important regulator of normal and malignant cell growth, and it appears to mediate the cancer promoting effects of estrogen. MeHPLA is deficient in a variety of rat and mouse mammary tumors and human breast cancer preparations, and this deficiency correlates with the loss of regulatory control [68]. Other estradiol-potentiating effects are mediated by toxins such as pertussis [69] and cholera toxin [70]. These effects are worsened by high HPLA.

Elevated levels of HPLA result in a dramatic decrease of ascorbic acid (vitamin C) concentration in the liver, adrenal glands, and blood [71]. High doses of ascorbic acid (100 mg/kg body weight daily) were shown to arrest or significantly inhibit the excretion of HPLA in patients with hemoblastoses and nephroblastoma [72]. Thus, high urinary HPLA signals an increased tissue growth response that may be associated with neoplastic disease or increased growth of normal tissue. The dramatic responses to ascorbic acid suggest its involvement in mediating this level of cell growth. Tissue saturating levels of vitamin C should be considered in patients with elevated HPLA.

FIGURE 6-11. Metabolism of Xylene to 2-Methylhippurate

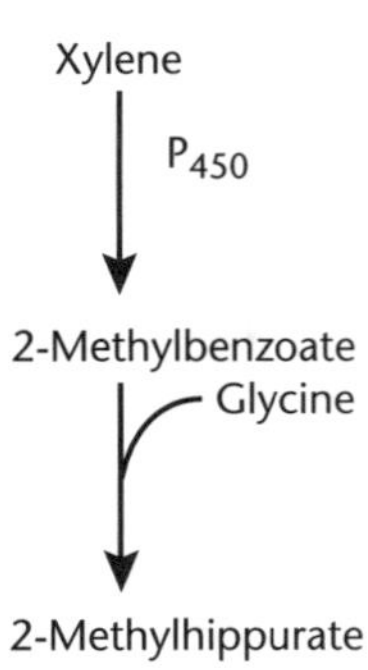

2-Methylhippurate

This compound is a byproduct from the detoxification of the common solvent, xylene. Xylene is oxidized via hepatic p450 oxidase enzymes, and the 2-methylbenzoate product is conjugated with glycine to form 2-methylhippurate (Figure 6-11). Spray-painting workers show elevated methylhippurate, indicating recent exposure to xylene [73]. Patient counseling in avoidance is indicated. Xylene is a very common solvent found in paint thinners and building products, fuel and exhaust fumes, and industrial degreasers and solvents.

Glucarate

Glucaric acid (glucarate) is an oxidation product of glucuronic acid. It is a by-product of the predominant liver Phase I detoxification reactions involving cytochrome p450. This oxidation step converts glucose into glucuronic acid, which is then used in Phase II conjugation reactions. Decreased glucarate is an indicator of reduced overall hepatic function. A great variety of drugs, food components, and products of gut microbial metabolism are prepared for excretion by glucuronidation. Glucarate elevation is an indication of enzyme induction due to such potentially toxic exposures [74-76]. Any exposure that results in stimulation of hepatic P-450 activity will result in increased excretion of glucarate. Urinary D-glucaric acid, for example, was elevated in pesticide-exposed groups [77]. Glucarate measurements have been advanced as useful biomarkers to xenobiotic exposure, being particularly useful as a screening tool for xenobiotic exposure in reproductive epidemiology [78].

Pyroglutamate

Small amounts of pyroglutamate are always present in overnight urine because it is produced as an intermediate in a cycle used in the active transport of amino acids in renal tubules [79]. This process utilizes glutathione as a carrier. With each molecule of amino acid recovered, glutathione is split and then reformed by an ATP-requiring enzymatic step (Figure 6-12). When the cycle is impaired, the

FIGURE 6-12. Pyroglutamate and Glutathione An important in the ATP-required step is shown and the resulting shunt pathway R represented by thick, grey arrows.

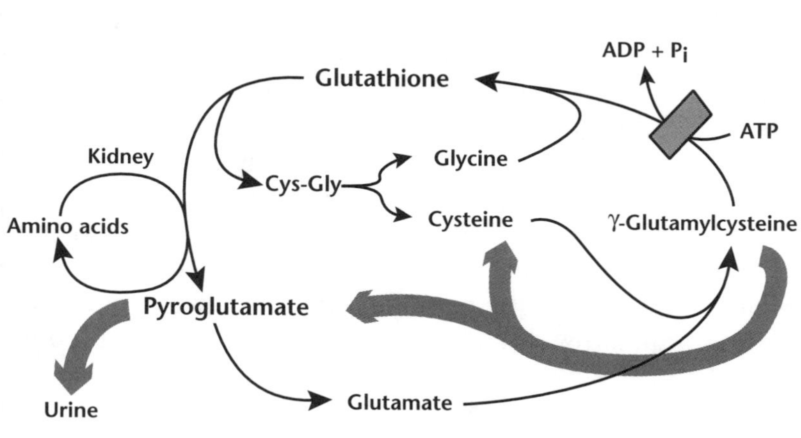

glutamic acid portion of glutathione is converted to pyroglutamate, which is excreted. This shunt pathway conserves amino acids at the expense of glutathione. Up to one third of the glutathione circulating in blood may be used in this amino acid recovery process. Figure 6-13 shows actual results for urinary pyroglutamate in 1500 sequential cases submitted to a clinical laboratory. N-acetylcysteine (NAC) is an effective oral agent for rebuilding total body glutathione, and oral taurine spares sulfur amino acids while providing an effective antioxidant.

FIGURE 6-13. Distribution of Pyroglutamate Results
The cutoff of 80 ug/mg creatinine is indicated by the shaded bar. A large number of individuals have some degree of wasting of glutathione and some individuals are in a state of severe depletion, which can effect detoxification and protection against free radical damage.

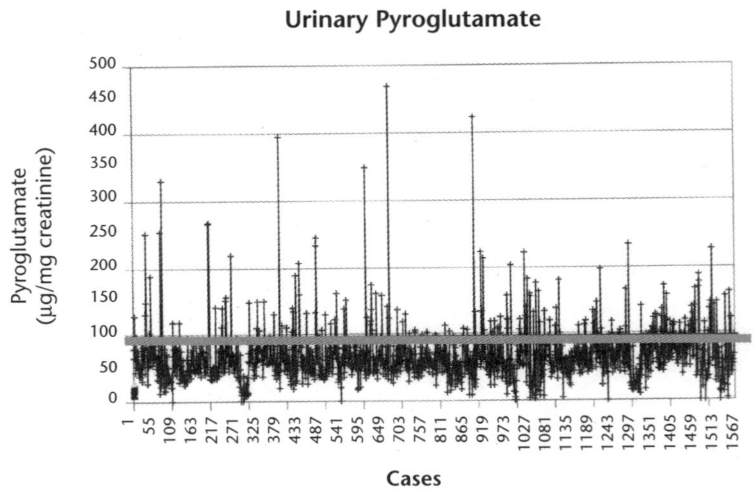

Sulfate

The sulfation pathway is used in Phase II liver detoxification for biotransformation of many drugs, steroid hormones, phenolic compounds, and other compounds. The addition of a sulfate group increases water solubility of hydrophobic compounds in preparation for their excretion in urine (see Chapter 8). The ratio of sulfate to creatinine has been used to assess the body's reserve of sulfur-containing compounds (especially glutathione) used in Phase II pathways [80]. When the ratio of sulfate to creatinine is low, these stores need replenishment. Glutathione administration with oral cysteine, taurine, and salts of sulfate are used in combinations to replenish sulfur pathways and restore the hepatic supply of inorganic sulfate [81].

Since both pyroglutamate and sulfate are markers of sulfur compound depletion, explanation of the interpretation of various combinations of abnormalities is needed. Severe depletion of organic sulfur sources will cause simultaneous high pyroglutamate and low sulfate excretion. High pyroglutamate with normal sulfate indicates inadequate organic sulfur sources for production of cysteine required for glutathione synthesis. Only organic sulfur in the form of compounds such as N-acetylcysteine or methionine will restore normal glutathione levels. Normal urinary pyroglutamate with low sulfate levels can occur in individuals with impaired sulfate activation. In these cases, rapid replenishment of hepatic sulfate may be accomplished with either sulfur donors like N-acetylcysteine or inorganic sulfate such as sodium sulfate [80].

Intestinal Dysbiosis Markers

Intestinal dysbiosis markers are discussed in Chapter 7 since they are related to gastrointestinal function. These organic acids provide an entirely different type of information from that discussed so far. They are reported as a part of the full profiling of organic acids in urine because the specimen set up and analytical methods are the same as for the other organic acids. In addition to their coverage in Chapter 7, we will review their interpretation here in order to complete the coverage of organic acids.

The abnormal overgrowth of unfavorable microflora in the small and large intestine is sometimes referred to as "gut dysbiosis" in order to distinguish this clinical condition from that of infection. Dysbiosis has been related to a wide variety of symptoms due to pathogenic toxins produced by the populations of microflora. Some compounds that appear in urine are unique metabolic products of the microbes that inhabit the lumen of gut (Figure 6-14). The compounds

TABLE 6-5. Origins of Dysbiosis

- Widespread use of antacids
- Prescription drugs that block stomach acid production
- Repeated use of antibiotics
- Food allergies and sensitivities
- Use of NSAIDs
- Consumption of infected foods

FIGURE 6-14. Dysbiosis Markers
Amino acids and sugars that are not assimilated can be used by gut microbes for growth. The specific compounds that are formed depend on the starting material and the species of organism present. In additon to bacterial formation from phenylalanine, dietary intake of benzoate may add to urinary levels of benzoate and hippurate. The other compounds are of microbial origin.

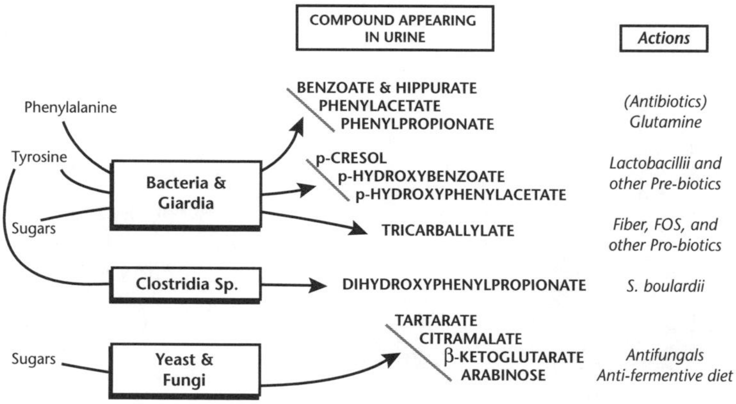

have a wide range of relative toxicities, with cresol and tartarate at the upper extreme and hippurate and benzoate at the lower end. Among the factors that lead to dysbiosis, importance of mealtime should not be overlooked because of the power to restore intestinal health by allowing normal stomach emptying rate and adequate flow of gastric and pancreaticobiliary fluids. The condition of microbial overgrowth is usually chronic, and the toxic products can affect multiple organs.

FIGURE 6-15. Conversion of Tyrosine to p-Cresol by Gut Microbes

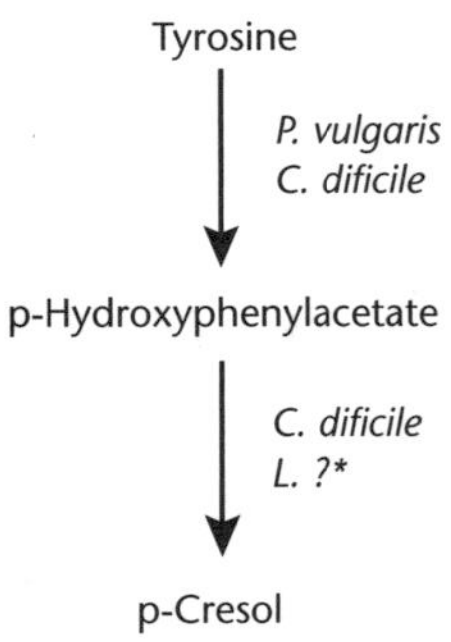

**Lactobacillus species other than L. acidophilus, leichmanii, delbrukii, plantarum, brevis, minutis, or fermentus.*

The high predictive value of urinary markers has been established from several lines of investigation. No false negative results and only 2% false positive results for small bowel disease and bacterial overgrowth syndrome were found in a study of p-hydroxyphenylacetate in 360 acutely ill infants and children [82]. Three of the patients who tested positive for this compound were diagnosed with *Giardia lamblia* infection.

No other species has a digestive tract exactly like humans. The one that has the closest resemblance is swine. Studies in newly weaned pigs [83] have revealed microbes that carry out tyrosine degradation (Figure 6-15). Both the transamination to form p-hydroxyphenylacetate (HPA) and the decarboxylation to p-cresol are carried out by *Clostridium dificile*. Since *Proteus vulgaris* can do only the first of these steps, HPA will increase in urine if *Proteus vulgaris* is the predominant organism. When *P. vulgaris* is accompanied by overgrowth of a newly identified strain of lactobacillus, however, p-cresol will be the major product to accumulate. The lactobacillus was not given species identification, but the characteristics eliminate any of the species listed in the footnote of Figure 6-15. Such studies illustrate the potential for more specific identifications based on patterns of products appearing in urine.

HPA is elevated in a wide variety of conditions involving direct intestinal pathology or digestive organ failure (Figure 6-16), which are obvious candidates for dysbiosis. Although treatments vary greatly

FIGURE 6-16. Response of HPA to Treatment
The dramatic lowering of HPA after treatments commensurate with six different clinical presentations demonstrates the removal of microbial overgrowth in every case, whether the cause was lactose intolerance or giardiasis.

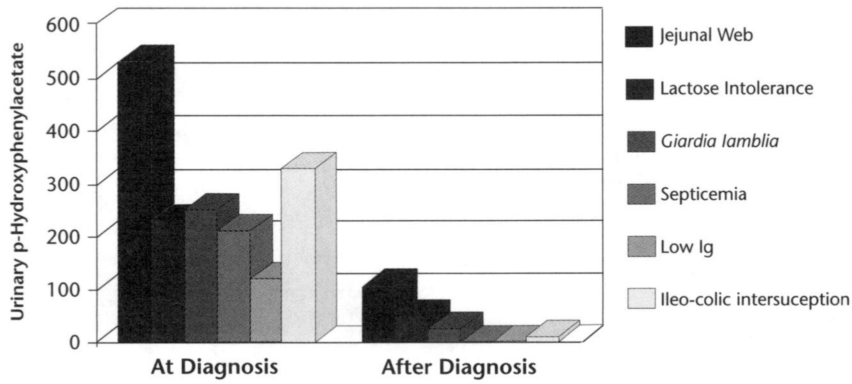

depending on the nature of the disorder, the lowering of urinary HPA reveals a normalized intestinal bacterial population. Some microbial compounds are absorbed and enter the detoxification pathways of the liver to be excreted as modified products that can serve as indicators of gastrointestinal activities. For example, bacterial amines are converted to piperidine, a sensitive biochemical index of gastrointestinal flora changes in celiac disease [84]. Other compounds appear due to genetic traits that affect how bacterial products are metabolized. The anaerobic bacterial product, 3-phenylpropionate, for example, is normally converted to common hippuric acid but is excreted as 3-phenylpropionylglycine in individuals with a relatively common inborn error of fatty acid oxidation [85]. Some compounds excreted in these instances are not organic acids, so they must be analyzed in separate assays to enhance the interpretation of origins for microbial compounds in urine.

Benzoate and Hippurate

Bacterial deamination of the amino acid phenylalanine produces benzoate, which is conjugated with glycine in the liver to form hippurate. Benzoic acid is also a common food component. It is used as a preservative in packaged foods such as pickles and lunch meats, and it occurs naturally in cranberries. This should be taken into account when interpreting elevated hippurate levels in urine. Whether the source is dietary intake or jejunal bacterial production, benzoate is usually converted to hippurate by conjugation with glycine in the liver. Glycine and pantothenic acid are the limiting factors in this process. Therefore, elevated benzoate is a marker of inadequate levels of these nutrients [86, 87].

Phenylacetate and Phenylpropionate

Intestinal bacterial action on phenylalanine also causes the appearance of phenylacetate and phenylpropionate in urine. Excretion of these compounds in urine is markedly increased after the gastrointestinal tracts of germ-free rats are inoculated with fecal microorganisms, indicating their microbial origin [9]. For individuals with normal, healthy intestinal function, phenylacetate and phenylpropionate should not appear at more than background concentrations in urine. They are products of unidentified, specific strains of bacteria, marking a state of bacterial overgrowth when they appear elevated in urine.

p-Cresol, p-Hydroxybenzoate, and p-Hydroxyphenylacetate

Tyrosine from dietary protein is the parent compound from which p-cresol, p-hydroxybenzoate, and p-hydroxyphenylacetate are formed. The compounds are not products of normal human metabolism but are produced by bacteria and protozoa that can populate the gut. Cresol has a chemical structure very similar to phenol and is highly toxic. Cresol excretion is not affected by dietary protein intake, suggesting that the bacteria responsible reside in the lower portions of the small intestine and colon. They produce cresol from intestinal secretions rather than from dietary sources [88]. Almost all adult celiac disease patients excrete unusually high amounts of p-cresol [89]. The variable responses of these patients to short-term antibiotic therapy indicate the variety of bacterial populations that produce p-cresol. Due to the loss of renal function, uremic patients accumulate cresol. The resultant increase in serum cresol can be prevented by the use of oral sorbents [90], demonstrating that the origin of the p-cresol is the bowel. Finely powdered, activated charcoal is a widely available sorbent, but newer synthetic compounds such as AST120 may be more effective.

p-Hydroxyphenylacetic aciduria has been found useful in detecting small bowel disease associated with *Giardia lamblia* infestation, ileal resection with blind loop, and other diseases of the small intestine associated with anaerobic bacterial overgrowth [82]. Use of antibiotics that act primarily against aerobic bacteria (such as neomycin) can encourage the growth of protozoa and anaerobic bacteria that then produce greater amounts of these compounds [91].

Tricarballylate

Tricarballylate (tricarb) is produced by a strain of aerobic bacteria that quickly repopulates in the gut of germ-free animals [92]. As its name implies, tricarb contains three carboxylic acid groups that are ionized at physiological pH to give a small molecule with three negative charges akin to the structure of EDTA. Magnesium is bound so tightly by tricarb that magnesium deficiency results from overgrowth of tricarb-producing intestinal bacteria in ruminants [93]. This condition, known as "grass tetany," is also accompanied by lower levels of calcium and zinc, all of which can form divalent ion complexes with tricarb (Figure 6-17).

FIGURE 6-17. The Magnesium-tricarballylate Complex

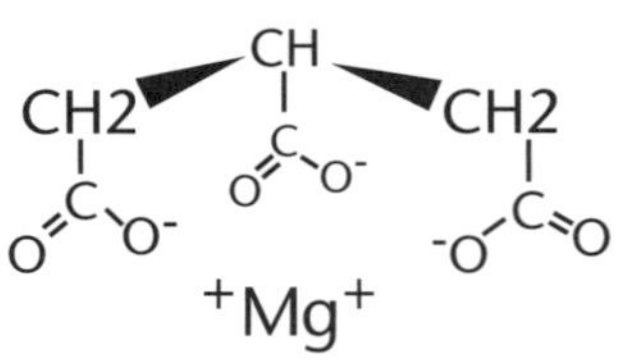

Dihydroxyphenylpropionate

Cases of confirmed *Clostridium* overgrowth show elevated levels of dihydroxyphenylpropionate that fall to baseline with Flagyl but are unaffected by Nystatin [94]. While other organisms may produce this compound, clostridia is the most commonly encountered genera among those susceptible to Flagyl. Various compounds closely related to dihydroxyphenylpropionate also are produced by the genus Clostridium [95].

FIGURE 6-18. From Ear Infection to Autism
The progression from normal development to autism in two brothers was apparently mediated by opportunistic intestinal overgrowth of yeast. In the process that has been proposed by recurrent otitis media is the initiating event and elaboration of toxic fungal products is the critical point at which treatment should be directed.

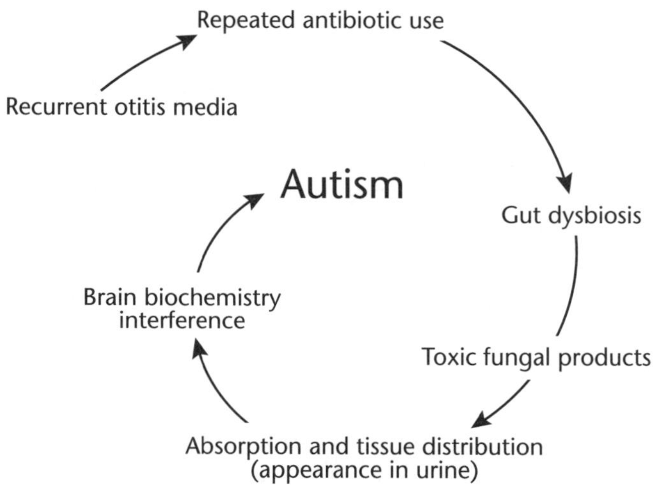

Citramalate, β-Ketoglutarate, Tartarate, and Arabinose

The structures of these compounds are closely related to those of intermediates of normal human metabolism, and they are sometimes called anti-metabolites because they block metabolic pathways through molecular mimicry. Their structures are very close to those of normal metabolic intermediates. The pathways that they impact are those of central energy production. They have been reported to be elevated in children who began to display autistic traits after repeated treatment with broad-spectrum antibiotics [96]. Patterns of elevations of any of the compounds in this category are

FIGURE 6-19. **Nutrient Deficiency Due to Gut Fermentation**

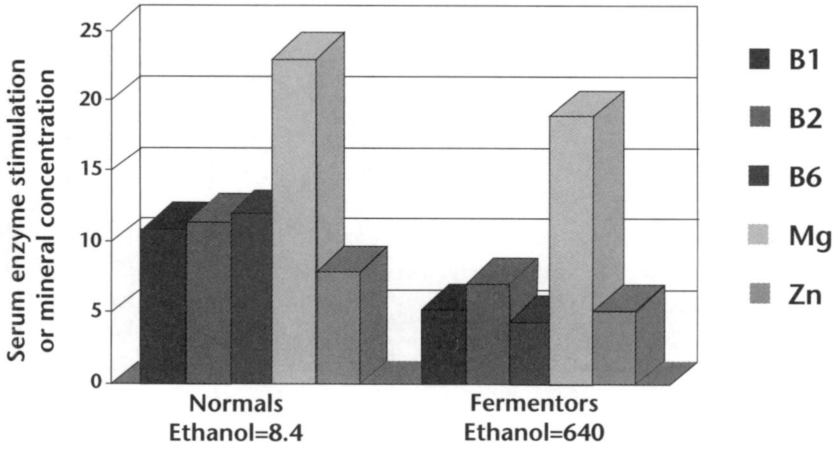

often normalized with anti-fungal treatment and introduction of beneficial bacteria (such as lactobacilli) to keep yeast growth in check (Figure 6-18). It also is noteworthy that there are multiple negative impacts on nutrients because of dysbiosis. Lowered levels of vitamins B_1, B_2, B_6, magnesium, and zinc are found in patients with elevated nascent production of ethanol due to gut fermentation [97] (Figure 6-19).

Actions to Consider for Elevated Dysbiosis Markers

Although species identification is not possible from the limited number of compounds currently detected, and no antibiotic sensitivity testing can be done, the information available is sufficient for clinical decisions about appropriate interventions. First, whenever possible, potent wide spectrum antibiotics are to be avoided because of the obvious encouragement of the cycle of opportunistic overgrowth of pathogens. The susceptibilities of various bacteria known to reside in the upper jejunum to five antibiotics are shown in Table 6-6. In many cases, the less challenging approach of bacteriostatic herbal preparations and compounds that inhibit growth of yeast, such as undecylenic acid, may be sufficient. A summary of interventions commonly used is presented in Table 6-7. Additional insight may be gained from studies of the antibiotic sensitivities of organisms present in cases of small intestinal bacterial

TABLE 6-6. Upper Jejunal Bacterial Strains in Small Intestinal Bacterial Overgrowth [98]

		Antibiotic Sensitivity % of Susceptible Strains (See footnote for explanation)				
Bacteria	Prevalence (%)	1	2	3	4	5
Microaerophilic	100					
Streptococcus	71	85	85	80	57	43
E. coli	69	63	76	76	?	?
Straphylococcus	25	38	62	69	46	100
Micrococcus	22	75	75	75	87	50
Klebsiella	20	0	100	100	?	100
Nisseria	16	100	100	100	100	30
Proteus	11	100	100	100	?	100
Acinetobacter	9	0	0	0	?	100
Enterobacter	7	0	0	0	?	100
Anaerobic	93					
Lactobacillus	75	89	100	?	72	28
Bacteroides	29	16	100	16	42	63
Clostridium	25	90	100	?	52	52
Veillonella	25	89	92	?	13	61
Fusobacterium	13	80	100	?	73	20
Peptostreptococcus	13	80	90	?	60	5

1: Amoxicillin
2: Amoxicillin-Clavulanic Acid (Frequent digestive intolerance)
3: Cephalotin
4: Erythromycin
5: Trimethoprim-Sulfamethoxazol (Significant side effects)

overgrowth syndrome. Since the specimen used for the data in Table 6-6 was taken in the upper jejunum [98] the data represent only the proximal part of the small bowel where microaerobic species are most likely to persist. The more distal regions of the jejunum and the entire ileum are less favorable for aerobes.

Since the microbial compounds that appear in urine originate in the lumen of the gut, oral sorbents that bind them and prevent their absorption can be used. This approach is especially helpful for those compounds of high toxicity like cresol. The absorbent, AST-120 (3-7 gm/day), has been shown to effectively lower p-cresol accumulation in uremic patients [90, 99]. Uremic coma and bleeding tendency is attributed to the accumulation of p-cresol, and p-cresol is a cocarcinogen in the mouse skin test. The fact that an oral sorbent lowers levels of cresol in uremia demonstrates that the gut is a significant source of the toxin. Activated charcoal may be used for this purpose, but there is greater tendency for constipation than with AST-120.

TABLE 6-7. Intestinal Overgrowth Interventions

Class	Examples
General	Encourage high fiber diet, remove mucosal irritants such as allergenic or IgG-positive foods, alcohol, and NSAIDs
Antibacterial	Pharmaceutical: Amoxycilin/Clavulanic Acid Natural: Goldenseal or other berberine-containing herbs, citrus seed extract, garlic, uva ursi, aloe vera, glycyrrhiza, olive leaf extract, garlic
Anti-fungal	Pharmaceutical: Nystatin Natural: See natural antibacterials plus oil of oregano, citrus seed extract, garlic, uva ursi, undecylenic acid
Anti-protozoal	Pharmaceutical: Flagyl Natural: See natural antibacterials
Probiotic	*L. acidophilus, L. sporogenes, S. boulardii,* favorable soil organisms
Prebiotic	Fructo-oligosaccharide (FOS), increase use of raw and cooked vegetables
Mucosal regeneration	Glutamine, pantothenic acid, deglycyrrhizanated licorice, slippery elm, oligopeptide preparations

It is worth noting that fasting is an effective way to reduce microbial populations in the gut. If the resident microbes do not receive a relatively constant infusion of substrate for growth, such as carbohydrates or amino acids, they die. A discussion of the various clinical ramifications of fasting is beyond the scope of this book. Although many patients may be candidates for fasting [100], the vast majority is reluctant to apply the discipline required to properly execute severe food restriction.

Regular oral dosing with organisms favorable to the human gut (pro-biotics) and use of substrates that encourage their growth such as dietary fiber and fructo-oligosaccharide (pre-biotics) are generally implemented to achieve long term control of intestinal microbes [101]. When aggressive intervention is warranted, use of amoxicillin-clavulanic acid is a suitable candidate because offending microaerophilic bacteria are very susceptible to its antibacterial action, and the anaerobic populations decline as well, possibly due to the more oxygen rich environment produced by declining populations of the oxygen-consuming species.

References

[1] Sweetman, L. and J.C. Williams, *Branched chain organic acidurias, in The Metabolic and Molecular Bases of Inherited Disease*, C. Scriver, *et al.*, Editors. 1995, McGraw-Hill, Inc.: New York. p. 1387-1422.

[2] Tanake, K., *et al.*, Gas-chromatographic method of analysis for urinary organic acids. I. retention indices of 155 metabolically important compounds. *Clin Chem*, 26((13)). 1839-46, 1980.

[3] Scriver, C., *et al.*, *The metabolic and molecular bases of inherited disease*. 7th ed. Vol. 1 - 3. 1995, New York: McGraw-Hill.

[4] Scriver, C. and L. Rosenberg, Amino Acid Metabolism and It's Disorders. *Major problems in clinical pediatrics*, ed. A. Schaffer. Vol. X. 1973, Philadelphia: W. B. Saunders. 491.

[5] Nyhan, W., *Abnormalities in amino acid metabolism in clinical medicine*. 1984, Norwald: Appleton-Century-Crofts. 463.

[6] Lord, R. and J. Bralley, Organics in urine: Assessment of gut dysbiosis, nutrient deficiencies and toxemia. *Nutr Pers*, 20(4). 25-31, 1997.

[7] Bralley, J. and R. Lord, *Urinary organic acids profiling, in Textbook of Natural Medicine*, P.a. Murray, Editor. 1998, Churchill Livingstone: Edinburgh. p. 229 - 237.

[8] Ong, C.N., *et al.*, Elevated levels of benzene-related compounds in the urine of cigarette smokers. *Int J Cancer,* 59(2). 177-80, 1994.

[9] Goodwin, B.L., C.R. Ruthven, and M. Sandler, Gut flora and the origin of some urinary aromatic phenolic compounds. *Biochem Pharmacol*, 47(12). 2294-7, 1994.

[10] Sweetman, L., Qualitative and quantitative analysis of organic acids in physiologic fluids for diagnosis of the organic acidurias., in *Abnormalities in amino acid metabolism in clinical medicine*, W. Nyhan, Editor. 1984, Appleton-Century-Crofts: Norwalk. p. 419-53.

[11] Duez, P., A. Kumps, and Y. Mardens, GC-MS profilng of urinary organic acids evaluated as a quantitative method. *Clin Chem*, 42(10). 1609-1615, 1996.

[12] Feller, A.G. and D. Rudman, Role of carnitine in human nutrition. *J Nutr*, 118(5). 541-547, 1988.

[13] Scholte, H.R., *et al.*, Riboflavin-responsive complex I deficiency. *Biochim Biophys Acta*, 1271(1). 75-83, 1995.

[14] Trauner, D.A., E. Horvath, and L.E. Davis, Inhibition of fatty acid beta oxidation by influenza B virus and salicylic acid in mice: implications for Reye's syndrome. *Neurology*, 38(2). 239-41, 1988.

[15] Wallace, D.C., Mitochondrial genetics: a paradigm for aging and degenerative diseases? *Science*, 256(5057). 628-32, 1992.

[16] Bates, C.J., Liberation of 14CO2 from [14C]adipic acid and [14C]octanoic acid by adult rats during riboflavin deficiency and its reversal. *Br J Nutr*, 63(3). 553-62, 1990.

[17] Oshida, Y., *et al.*, Effect of insulin on intramuscular 3-hydroxybutyrate levels in diabetic rats. *Horm Metab Res*, 30(2). 70-1, 1998.

[18] Anderson, R.A., *et al.*, Elevated intakes of supplemental chromium improve glucose and insulin variables in individuals with type 2 diabetes. *Diabetes*, 46(11). 1786-91, 1997.

[19] Halberstam, M., *et al.*, Oral vanadyl sulfate improves insulin sensitivity in NIDDM but not in obese nondiabetic subjects [published erratum appears in Diabetes 1996 Sep;45(9):1285]. *Diabetes*, 45(5). 659-66, 1996.

[20] Pandey, S.K., M.B. Anand-Srivastava, and A.K. Srivastava, Vanadyl sulfate-stimulated glycogen synthesis is associated with activation of phosphatidylinositol 3-kinase and is independent of insulin receptor tyrosine phosphorylation. *Biochemistry*, 37(19). 7006-14, 1998.

[21] Bardosi, A., W. Cruetzfeldt, and e. al., Lactate, hydroxybutyrate, and fumarate elevation in an adult with defective cytochrome c oxidase. *Acta Neuropathol*, 74(3). 248-58, 1987.

[22] Dissmann, R., T. Linderer, and R. Schroder, Estimation of enzymatic infarct size: direct comparison of the marker enzymes creatine kinase and alpha-hydroxybutyrate dehydrogenase. *Am Heart J*, 135(1). 1-9, 1998.

[23] Imaki, M., *et al.*, Evaluation of the effects of various factors on the serum alpha hydroxybutyrate dehydrogenase activity in young females. *Appl Human Sci*, 14(6). 297-302, 1995.

[24] Kano, K. and T. Ichimura, Increased alpha-hydroxybutyrate dehydrogenase in serum from children with measles. *Clin Chem*, 38(5). 624-7, 1992.

[25] Naito, E., *et al.*, Thiamine-responsive lactic acidaemia: role of pyruvate dehydrogenase complex. *Eur J Pediatr*, 157(8). 648-52, 1998.

[26] Gries, C.L. and M.L. Scott, The pathology of thiamin, riboflavin, pantothenic acid and niacin deficiencies in the chick. *J Nutr,* 102(10). 1269-85, 1972.

[27] Munujos, P., *et al.*, Brain pyruvate oxidation in experimental thiamin-deficiency encephalopathy. *Clin Chim Acta*, 255(1). 13-25, 1996.

[28] Rao, G.A., D.E. Riley, and E.C. Larkin, Fatty liver caused by chronic alcohol ingestion is prevented by dietary supplementation with pyruvate or glycerol. *Lipids*, 19(8). 583-8, 1984.

[29] Chen, R.S., C.C. Huang, and N.S. Chu, Coenzyme Q_{10} treatment in mitochondrial encephalomyopathies. Short-term double-blind, crossover study. *Eur Neurol*, 37(4). 212-8, 1997.

[30] Sobreira, C., *et al.*, Mitochondrial encephalomyopathy with coenzyme Q_{10} deficiency. *Neurology*, 48(5). 1238-43, 1997.

[31] Boitier, E., *et al.*, A case of mitochondrial encephalomyopathy associated with a muscle coenzyme Q_{10} deficiency. *J Neurol Sci*, 156(1). 41-6, 1998.

[32] Fouty, B., F. Frerman, and R. Reves, Riboflavin to treat nucleoside analogue-induced lactic acidosis [letter]. *Lancet*, 352(9124). 291-2, 1998.

[33] Santos, J.L., *et al.*, Nonsynergic effect of ethanol and lead on heme metabolism in rats. *Ecotoxicol Environ Saf*, 43(1). 98-102, 1999.

[34] Bardosi, A., *et al.*, Myo-, neuro-, gastrointestinal encephalopathy (MNGIE syndrome) due to partial deficiency of cytochrome-c-oxidase. A new mitochondrial multisystem disorder. *Acta Neuropathol*, 74(3). 248-58, 1987.

[35] Matsumoto, I., *et al.*, Urinary organic acid profile studies in a patient with cytochrome c oxidase deficiency. *39th ASMS Conf.*, 1991.

[36] Hamm, L.L. and E.E. Simon, Roles and mechanisms of urinary buffer excretion. *Am J Physiol*, 253(4 Pt 2). F595-605, 1987.

[37] Milner, J.A., Metabolic aberrations associated with arginine deficiency. *J Nutr*, 115(4). 516-23, 1985.

[38] Prior, R.L., A. Zimber, and W.J. Visek, Citric, orotic, and other organic acids in rats injected with active or inactive urease. *Am J Physiol*, 228(3). 828-33, 1975.

[39] al Aqeel, A., *et al.*, A new patient with alpha-ketoglutaric aciduria and progressive extrapyramidal tract disease. *Brain Dev*, 16. 33-7, 1994.

[40] Shaw, D.e.a., Management of fatigue: A physiological approach. *Am J Med Sci*. 243:758, 1962.

[41] Hicks, J., Treatment of fatigue in general practice: A double blind study. *Clin Med*. 85, 1964 (Jan.).

[42] Hume, A.S., *et al.*, Antidotal efficacy of alpha-ketoglutaric acid and sodium thiosulfate in cyanide poisoning. *J Toxicol Clin Toxicol*, 33(6). 721-4, 1995.

[43] Kawaguchi, A. and K. Bloch, Inhibition of glutamate dehydrogenase and malate dehydrogenases by palmitoyl coenzyme A. *J Biol Chem*, 251(5). 1406-12, 1976.

[44] Pinard, J.M., *et al.*, Leigh syndrome and leukodystrophy due to partial succinate dehydrogenase deficiency: regression with riboflavin. *Arch Pediatr*, 6(4). 421-6, 1999.

[45] Nowaczyk, M.J., *et al.*, Ethylmalonic and methylsuccinic aciduria in ethylmalonic encephalopathy arise from abnormal isoleucine metabolism. *Metabolism*, 47(7). 836-9, 1998.

[46] Zhao, Y., J. Jaskiewicz, and R.A. Harris, Effects of clofibric acid on the activity and activity state of the hepatic branched-chain 2-oxo acid dehydrogenase complex. *Biochem J*, 285(Pt 1). 167-72, 1992.

[47] Lindenbaum, J., *et al.*, Prevalence of cobalamin deficiency in the Framingham elderly population [see comments]. *Am J Clin Nutr*, 60(1). 2-11, 1994.

[48] Rule, S.A., *et al.*, Serum vitamin B12 and transcobalamin levels in early HIV disease. *Am J Hematol*, 47(3). 167-71, 1994.

[49] Guthrie, H., *Introductory Nutrition*. Fourth ed. 1979, St. Louis: C.V. Mosby.

[50] Nyhan, W.L., Inborn errors of biotin metabolism. *Arch Dermatol*, 123(12). 1696-1698a, 1987.

[51] Mock, N., *et al.*, Increased urinary excretion of 3-hydroxyisovaleric acid and decreased urinary excretion of biotin are sensitive early indicators of decreased biotin status in experimental biotin deficiency. *Am J Clin Nutr*, 65. 951-8, 1997.

[52] Dostalova, L., Vitamin status during puerperium and lactation. *Ann Nutr Metab*, 28(6). 385-408, 1984.

[53] Krause, K.H., *et al.*, Vitamin status in patients on chronic anticonvulsant therapy. *Int J Vitam Nutr Res*, 52(4). 375-85, 1982.

[54] Mock, D.M., *et al.*, Biotin status assessed longitudinally in pregnant women. *J Nutr*, 127(5). 710-6, 1997.

[55] Riudor, E., *et al.*, Requirement of high biotin doses in a case of biotinidase deficiency. *J Inherit Metab Dis*, 12(3). 338-9, 1989.

[56] Folkers, K., *et al.*, Lovastatin decreases coenzyme Q levels in humans. *Proc Natl Acad Sci U S A*, 87(22). 8931-4, 1990.

[57] Mortensen, S.A., *et al.*, Coenzyme Q_{10}: clinical benefits with biochemical correlates suggesting a scientific breakthrough in the management of chronic heart failure. *Int J Tissue React*, 12(3). 155-62, 1990.

[58] Orgacka, H. and Z. Zbytniewski, Excretion of vanillic acid and homovanillic acid and tissue distribution of catecholamines and their metabolites in mice with various levels of pigmentation. *Endokrynol Pol*, 42(3). 471-9, 1991.

[59] Braverman, E. and C. Pfeiffer, Tyrosine: The antidepressant, in *The Healing Nutrients Within*. 1987, Keats: New Canaan. p. 44-58.

[60] Mauron, J., Tyrosine and hypertension. *Bibl Nutr Dieta*, 38. 209-18, 1986.

[61] Akimaru, K., *et al.*, Amplification of N-myc gene and increase of urinary VMA and HVA in patients with neuroblastic tumors. *Nippon Ika Daigaku Zasshi*, 61(2). 148-53, 1994.

[62] Lucas, K., *et al.*, Catecholamine metabolites in ganglioneuroma. *Med Pediatr Oncol*, 22(4). 240-3, 1994.

[63] Parekh, N., *et al.*, Cardiac troponin I predicts myocardial dysfunction in aneurysmal subarachnoid hemorrhage. *J Am Coll Cardiol*, 36(4). 1328-35, 2000.

[64] Keung, W.M., *et al.*, Volitional ethanol consumption affects overall serotonin metabolism in Syrian golden hamsters (Mesocricetus auratus). *Biochem Biophys Res Commun*, 271(3). 823-30, 2000.

[65] Visek, W.J., Nitrogen-stimulated orotic acid synthesis and nucleotide imbalance. *Cancer Res*, 52(7 Suppl). 2082s-2084s, 1992.

[66] Milner, J.A. and W.J. Visek, Orotate, citrate, and urea excretion in rats fed various levels of arginine. *Proc Soc Exp Biol Med*, 147(3). 754-9, 1974.

[67] Nelson, J., *et al.*, Regulation of orotic acid biosynthesis and excretion induced by oral glutamine administration in mice. *Biochem Med Metab Biol*, 49(3). 338-50, 1993.

[68] Markaverich, B.M., *et al.*, Methyl p-hydroxyphenyllactate. An inhibitor of cell growth and proliferation and an endogenous ligand for nuclear type-II binding sites. *J Biol Chem*, 263(15). 7203-10, 1988.

[69] Markaverich, B.M., *et al.*, Methyl p-hydroxyphenyllactate and nuclear type II binding sites in malignant cells: metabolic fate and mammary tumor growth. *Cancer Res*, 50(5). 1470-8, 1990.

[70] Sheffield, L.G. and L.C. Kotolski, Epidermal growth factor modulates cholera toxin induced mammary gland development. *Endocr Res*, 19(4). 259-71, 1993.

[71] Levchuk, A.A., *et al.*, Effect of the carcinogenic tyrosine metabolite p-hydroxyphenyllactic acid on the ascorbic acid concentration in the organs and blood of mice. *Biull Eksp Biol Med*, 102(10). 462-3, 1986.

[72] Baikova, V.N., *et al.*, Congenital disorders of tyrosine metabolism and their correction in children with tumors. *Vopr Onkol*, 33(11). 42-8, 1987.

[73] Triebig, G., K.H. Schaller, and D. Weltle, Neurotoxicity of solvent mixtures in spray painters. I. Study design, workplace exposure, and questionnaire. Int Arch Occup Environ *Health*, 64(5). 353-9, 1992.

[74] Hunter, J. and et al, Urinary D-glucaric acid excretion as a test for hepatic enzyme induction in man. *Lancet*, (March 20). 572-575, 1971.

[75] Hunter, J., *et al.*, Urinary D-glucaric acid excretion and total liver content of cytochrome P-450 in guinea-pigs: relationship during enzyme induction and following inhibition of protein synthesis. *Biochem Pharmacol*, 22(6). 743-7, 1973.

[76] Sandle, L.N. and J.M. Braganza, An evaluation of the low-pH enzymatic assay of urinary D-glucaric acid, and its use as a marker of enzyme induction in exocrine pancreatic disease [published erratum appears in Clin Chim Acta 1990 May;188(3):279]. *Clin Chim Acta*, 162(3). 245-56, 1987.

[77] Edwards, J.W. and B.G. Priestly, Effect of occupational exposure to aldrin on urinary D-glucaric acid, plasma dieldrin, and lymphocyte sister chromatid exchange. *Int Arch Occup Environ Health*, 66(4). 229-34, 1994.

[78] Hogue, C.J. and M.A. Brewster, The potential of exposure biomarkers in epidemiologic studies of reproductive health. *Environ Health Perspect*, 90. 261-9, 1991.

[79] Chalmers, R. and A. Lawson, Organic acidurias due to disorders in other metabolic pathways, in *Organic Acids in Man*. 1982, Chapman & Hall: London. p. 350 - 381.

[80] Levy, G., Sulfate conjugation in drug metabolism: role of inorganic sulfate. *Fed Proc*, 45(8). 2235-40, 1986.

[81] Gregus, Z., *et al.*, Effect of glutathione depletion on sulfate activation and sulfate ester formation in rats. *Biochem Pharmacol*, 37(22). 4307-12, 1988.

[82] Chalmers, R.A., H.B. Valman, and M.M. Liberman, Measurement of 4-hydroxyphenylacetic aciduria as a screening test for small-bowel disease. *Clin Chem*, 25(10). 1791-4, 1979.

[83] Ward, L.A., *et al.*, Isolation from swine feces of a bacterium which decarboxylates p-hydroxyphenylacetic acid to 4-methylphenol (p-cresol). *Appl Environ Microbiol*, 53(1). 189-92, 1987.

[84] Lindblad, B.S., *et al.*, Absorption of biological amines of bacterial origin in normal and sick infants. *Ciba Found Symp*, (70). 281-91, 1979.

[85] Bennett, M.J., *et al.*, When do gut flora in the newborn produce 3-phenylpropionic acid? Implications for early diagnosis of medium-chain acyl-CoA dehydrogenase deficiency. *Clin Chem*, 38(2). 278-81, 1992.

[86] Temellini, A., *et al.*, Conjugation of benzoic acid with glycine in human liver and kidney: a study on the interindividual variability. *Xenobiotica*, 23(12). 1427-33, 1993.

[87] Quick, A., The study of benzoic acid conjugation in the dog with a direct quantitative method fro hippuric acid. *J Biol Chem*. 477-90, 1934.

[88] Bures, J., *et al.*, Excretion of phenol and p-cresol in the urine in fasting obese individuals and in persons treated with total enteral nutrition. *Cas Lek Cesk*, 129(37). 1166-71, 1990.

[89] Tamm, A.O., Biochemical activity of intestinal microflora in adult coeliac disease. *Nahrung*,

28(6-7). 711-5, 1984.

[90] Niwa, T., *et al.*, Suppressive effect of an oral sorbent on the accumulation of p-cresol in the serum of experimental uremic rats. *Nephron*, 65(1). 82-7, 1993.

[91] Fellaman, J.H., N.R. Buist, and N.G. Kennaway, Pitfalls in metabolic studies: the origin of urinary p-tyramine. *Clin Biochem*, 10(5). 168-70, 1977.

[92] McDevitt, J. and P. Goldman, Effect of the intestinal flora on the urinary organic acid profile of rats ingesting a chemically simplified diet. *Food Chem Toxicol*, 29(2). 107-13, 1991.

[93] Schwartz, R., M. Topley, and J.B. Russell, Effect of tricarballylic acid, a nonmetabolizable rumen fermentation product of trans-aconitic acid, on Mg, Ca and Zn utilization of rats. *J Nutr*, 118(2). 183-8, 1988.

[94] Shaw, W. Experience with organic acid testing to evaluate abnormal microbial metabolites in the urine of children with autism, in Presented at the Conference of American College for the Advancement of Medicine. 1997. Orlando, Florida.

[95] Elsden, S.R., M.G. Hilton, and J.M. Waller, The end products of the metabolism of aromatic amino acids by Clostridia. *Arch Microbiol*, 107(3). 283-8, 1976.

[96] Shaw, W., E. Kassen, and E. Chaves, Increased urinary excretion of analogs of Krebs cycle metabolites and arabinose in two brothers with autistic features. *Clin Chem*, 41(8 Pt 1). 1094-1104, 1995.

[97] Eaton, K.K., *et al.*, Abnormal gut fermation: Laboratory studies reveal deficiency of B vitamins, zinc, and magnesium. *J Nutr Biochem*, 4. 635-638, 1993.

[98] Bouhnik, Y., *et al.*, Bacterial populations contaminating the upper gut in patients with small intestinal bacterial overgrowth syndrome. *Am J Gastroenterol*, 94(5). 1327-31, 1999.

[99] Sanaka, T., *et al.*, Therapeutic effects of oral sorbent in undialyzed uremia. *Am J Kidney Dis*, 12(2). 97-103, 1988.

[100] Furhman, J., *Fasting and ating for health*. 1995, New York: St. Martin's Griffin.

[101] Bouhnik, Y., *et al.*, Effects of fructooligo-saccharides ingestion on fecal bifidobacteria and selected metabolic indexes of colon carcinogenesis in healthy humans. *Nutr Cancer*, 26(1). 21-9, 1996.

Chapter 7

GI Function

TABLE 7-1. Summary of laboratory Evaluations for Gastrointestinal Function

GI Aspect	Function	Testing	Abnormal	Intervention
Immune Barrier	Not-self elimination	Serum IgE Salivary or Fecal IgA	++ Foods ↓ IgA	•Eliminate all •Immune-support nutrients
Physical Barrier	Regulate nutrient admission and restrict toxicant and microbial access	Serum IgG	Many ++ Foods Few ++ Foods Urinary	•Eliminate reactive foods (Rotation Diet) •Medical foods, Zinc, B_5 •Free-form amino acids •Eliminate all + foods
		Lactulose/Mannitol challenge	↑ Lactulose ↓ Mannitol	•Mucosal restoration
Microbial Populations	Normal – nutrient support Pathogen – toxin production	Urinary metabolic markers	↑ Bacterial markers ↑ Protozoal markers ↑ Yeast markers	•Herbal or pharmaceutical bacteriostatic agents (antibiotics), pre-biotics and pro-biotics •Restrict simple sugars •Antifungals
		Hydrogen / methane breath test	↑ Expired gases	•Herbal or pharmaceutical bacteriostatic agents
		Stool culture & sensitivity	↑ Growth	•Specific antibiotics
Stomach	Gastric acid, Pepsin	Heidleberg capsule Direct pH readings Indirect indicators	↓ pH ↓ Trace elements or amino acids	•Betaine-HCl, free-form amino acids, B-vitamins, minerals, chloride
Pancreas	Protease Ribonuclease Lipase	Fecal chymotrypsin PABA test Plasma fatty acids Fecal fats	↓ Activity ↓ Index ↓ PUFA ↑ Fat	•Pancreatic replacement enzymes and essential fatty acids
Liver/Gall Bladder	Bile acid secretion	Fecal fatty acids	↑ Fatty acids	•Ox bile and essential fatty acids
Small Intestine	Absorption	Schilling test	↓ Urinary B_{12}	•B_{12} by injection or sublingual
		Mannitol challenge	↓ Urinary mannitol	•Mucosal restoration
Colon	Water resorption, microbial containment	Fecal butyrate or other SCFA Fecal pH	↓ Butyrate ↑ Isobutyrate	•Increase dietary fiber •Butyrate enemas

Introduction

Proper gastrointestinal function is critical to adequate nutritional status and can impact all aspects of body function. Every disorder that pertains to the gut has the potential to manifest as a factor in nutritional deficiencies. The tests described in this chapter reveal the functional status of the organs required for the digestion and assimilation of food and some of the toxicologic and immunologic consequences of failure to properly carry out those functions. The subjects of leaky gut and the associated immunologic challenge are discussed with special reference to ways of testing for clinically relevant factors. Testing also is discussed relevant to stomach, pancreatic, hepatic, and intestinal function. The detection of abnormal microbial growth within the gut and the impact of the various toxins produced by their growth is an important and relatively new aspect, especially in the analysis of urinary compounds that reveal the presence of various classes of microbes.

Topologically speaking, the human body's fundamental form is that of a donut. From our embryologic beginnings, the inside surface of our mouth, esophagus, stomach, and intestines – everything we call our digestive tract (*i.e.* the donut hole) – is contiguous with the outside surface of our bodies that we call our skin. Stretch the donut longitudinally and it becomes a tube. Granted, the tube is a closer geometric representation of our digestive tract; nonetheless, it retains the topological characteristics of the donut. This simple lesson in geometry teaches that, anatomical differences aside, a common attribute shared by the epithelial surface of our skin and digestive tract is they both face outward from the body. Nature also imposed upon this arrangement an important function: both serve foremost as barriers to the outside world. A breach of these barriers, especially the gastrointestinal barrier, is an underlying factor in chronic illness.

The proper function of the GI tract depends to a large degree upon a myriad of autonomic processes that regulate rhythmic contractions of smooth muscle, coordinated with the opening and closing of the appropriate sphincter valves. These actions control the unidirectional movement of food through the gut at a pace slow enough to allow absorption but fast enough to continuously bring new nutrients as needed by the body. Digestive secretions, formed principally in response to the presence of food in the GI tract, vary according to the amount and type of food present. The amount of each digestive secretion must be accurately modulated to provide just the amount required for proper digestion. The nervous system that regulates these activities is largely independent of the brain, although it responds to signals from the sensory organs and the immune system.

Typical daily food consumption includes kilograms of carbohydrates, 0.5–1 kg of fat, 500–700 g of amino acids, and 20 or more liters of water. Establishing a clear picture of functional adequacy of the gastrointestinal tract for achieving the delivery of nutrients from such a mass of food is a difficult clinical challenge for which laboratory evaluations offer only partial assistance.

Because of the difficulty of direct sampling of intestinal contents and the remoteness of specimens like peripheral blood, urine, and feces from the intestinal sites where most of these phenomena occur, routine laboratory evaluations using such specimens give partial answers. Due to the critical role of the gastrointestinal tract in nutrient supply, even partial answers can be of great value. Invasive procedures for observation or luminal specimen retrieval can be highly informative, but they can only be performed by trained specialists. Such procedures are beyond the scope of this book. This chapter will deal with non-invasive laboratory evaluations of gastrointestinal function.

The Immune Barrier

We begin with the immune barrier aspect of gastrointestinal health because of the paramount importance of evaluating adverse immune reactions to food. The importance derives from the predictive value for overall intestinal health and the practical information regarding food choices that enhance health restoration in the patient with compromised GI function.

It is difficult to grasp the magnitude of the challenge presented to the immune system in the gut. The enzymes of human tissues are designed to handle a flow of substances presented one molecule at a time. Each day, the act of eating introduces an astronomical number of molecules in the form of polymers unsuitable for cell processing, along with toxins and microbes. Whenever the digestive process fails to work in a timely and efficient manner, the undigested or partially digested food molecules provide an antigen-rich slurry that passes into a microbial growth chamber with constant temperature, neutral pH, and steady flow – the lumen of the small and large intestines. Microbial action yields another host of antigenic molecules. Cells of the immune system must constantly survey every region of the gut for entry of foreign molecules or microbes, monitor signals for increasing defensive barriers, and, when necessary, mount aggressive removal strategies while keeping destructive forces in check so that normal tissues can continue the never-ending process of nutrient assimilation, repair, and maintenance.

The field of immunology is still in a state of controversy over just how adverse reactions to foods should be classified and how best to evaluate the immune function of the GI system. We will present a brief overview of certain details to allow the reader to see the intricacies of this system. Excellent reviews of the cellular and molecular aspects of immunology are available [1]. The principal laboratory tests currently available for evaluation of the state of the gastrointestinal immune barrier function are the measurements of serum levels of antibodies to specific foods and salivary or fecal

"In A Nutshell"

1. The gut processes large quantities of food daily and interacts with the hostile environment in the lumen. (Imagine the liquid mixture from a meal being spread out over a surface the size of two tennis courts.

2. Undigested food molecules present a challenge recognized by the immune system as "bad" molecules. The physical barrier must keep "bad" out while admitting the "good" nutrients.

3. Every microscopic section of the huge surface area must be protected from "foreign" entry. Therefore, the gut contains a very large immune component. Undigested food causes immune responses. The response is greatly exaggerated when the barrier has "holes" that allow much greater rates of "foreign" entry.

secretory antibodies.

Throughout the entire length of the small intestine are patches of lymphoid tissue, collectively called gut-associated lymphoid tissue (GALT). Because these cells are in the mucosal layer, the term mucosal-associated lymphoid tissue (MALT) is sometimes used as well. In terms of the total cell mass of the immune system, the gut is by far the largest immune organ, containing over 60% of the mass. Intraepithelial lymphocytes (IELs) are interspersed among the epithelial cells (enterocytes) lining the intestine. The IELs collectively represent a pool of cells comparable in size to that of all peripheral lymphocytes in the spleen. There has been an explosion of data from research laboratories about the intercellular communication that regulates the activities of this part of the immune system. Cytokines, nuclear factors, interferons, and integrins regulate CD4 T-lymphocytes and CD8 lymphocytes [2]. The constant challenge of food ingestion described above frequently leads to abnormal growth. Neoplastic colorectal lesions, for example, contain cytotoxic IELs. Because of the destructive power of activated T-cell populations, there are mechanisms to control their numbers. Repeated stimulation leads to activation-induced cell death mediated by yet other cell regulators [3].

One of the results of hyperstimulation of intestinal cellular immune responses is an increase in the permeability of the cellular lining of the small intestine. Tall columnar epithelial cells that form the physical boundary are constantly replaced in the normal small intestine. The barrier function depends on tight junctions between the cells. Any breach in cell-to-cell junctions presents an opening for antigen flow and further immune response. Antibody responses to food antigens result in clinical food allergy symptoms.

Interactions between food macromolecules or the fragments released from them by partial digestion and the GALT produce symptoms in distant tissues, especially the skin [4]. The reactions can also cause lesions in brain white matter associated with inflammatory bowel disease [5]. There is growing evidence that food or microbe-initiated inflammatory products from the GALT can trigger reactions that are part of the initiation of degenerative diseases, including arthritis, atherosclerosis, and dementia [6]. Reduced production of such inflammatory signals may be achieved by normalizing function throughout the gastrointestinal tract.

There is a complex relationship between intestinal hyper-permeability and adverse food reactions. Children and adults with eczema, urticaria, or asthma triggered by atopic food allergy exhibit elevated intestinal permeability. The increase in intestinal permeability likely leads to the development of food sensitivity, and it may be a result of food allergy [7-9].

Secretory IgA

The lymphoid tissue of the intestinal mucosa secretes immunoglobulin A into the lumen of the GI tract as a first line of defense against microbes and antigens. Secretory IgA (sIgA) forms immune complexes with pathogens and allergens, which prevents them from binding to and penetrating the intestinal mucosa [20]. Individuals with selective IgA deficiency have high titers of antibodies against various food antigens [21]. A high antigenic load can result in depressed sIgA even in healthy, asymptomatic individuals. With time, this decrease in resistance can lead to dysbiosis and an increased risk for infection and allergy [22, 23].

The secretory IgA function may be evaluated by measurement of salivary or fecal IgA levels. Salivary IgA is a predictor of the release of IgA at intestinal surfaces [24]. Failure of the immune barrier can lead to elevated serum IgA and increased intestinal permeability to toxic gluten lectins, which is improved in 64% of IgA nephropathy patients by a gluten-free diet [25]. Fecal sIgA correlates with salivary IgA, and both are predictors of partial mucosal IgA deficiency as defined by serum IgA > 0.05 g/l but 2 SD below normal levels [26]. Detection of specific fecal IgAs directed toward Rotavirus and Hepatitis A is useful in diagnosis of early stages of these infections [27, 28].

Food-Directed Antibody Testing

Food antibody tests measure levels of specific immunoglobulins in serum. Elevated levels can cause allergic food responses involving the anaphylactic response or sensitivity reactions due to food antigen-antibody complex deposition (Table 7-2). Such reactions can lead to a wide variety of symptoms, which can be as diverse as ill-defined malaise and fatigue to digestive disorders, skin problems, aching joints, or back problems. In children, recurrent ear infections are often associated with cow milk allergies, and hyperactivity

TABLE 7-2. Types of Adverse Immune Reactions To Food

Immunoglobulin E-mediated (atopic) reactions
• Result from infrequent exposure
• Anaphylaxis is inherent in IgE reactions
• Symptoms of IgE-mediated allergies are very predictable, and the offending food is usually obvious
• Treatment involves permanent avoidance and immunotherapy
• *In-vitro* testing for serum IgE is appropriate for confirmation
• Incidence is relatively low
Immunoglobulin G-mediated (hypersensitivity) reactions
• Result from frequent exposure
• The result is serum sickness with widely diverse symptoms
• Symptoms are varied, and the offending food is usually not suspected
• Treatment with temporary avoidance can improve digestion and build healthy gut tissue
• *In-vitro* testing for serum IgG shows food offenders and extent of gut permeability
• Incidence is relatively high

disorders have shown systematic improvement when foods causing antibody complex formation are eliminated. In many people, undetected food reactions can be a significant factor in overall health. The challenge is to clearly define which foods may be causing adverse reactions. The ultimate proof of an adverse food reaction is via an elimination-provocation process where the suspected food is removed from the diet and then later reintroduced. If symptoms subside when the food is avoided and return upon reintroduction of the food, this is good evidence of food intolerance. This approach is time consuming but may be used for immediate reactions where the symptoms may be assessed relatively easily. If watery eyes and headache occur whenever shrimp are eaten, then elimination of shrimp from the diet may prevent the symptoms. If they occur again when a trial of consuming a few shrimp is done, then some antigen in shrimp is causing the problem.

The obviously tedious process of elimination/provocation can be expedited with the use of *in vitro* food-specific antibody tests. These tests measure the amounts of antibodies that cause food reactions. While there are several ways foods can cause adverse reactions, the antibody-mediated types are the most common, and these tests provide the clinician and patient useful data to design appropriate diets that avoid the major offending foods.

IgE Food Antibodies

Individuals prone to developing strong immediate food allergic responses are called "atopic" and are said to suffer from allergies. The IgE antibody is the cause of allergic reactions most frequently seen as inhalant allergies, (*e.g.*, hay fever). Foods can also cause a build-up of IgE in the blood during the early stages of allergic food reactions. The sequence of events in the GALT involves activation of a class of lymphocytes known as Th2 helper T cells. These cells secrete cytokines that instruct B-lymphocytes to produce IgE and stimulate mast cells, basophils, and eosinophils [1].

In most cases, people will easily recognize an IgE mediated food reaction because the allergic response is immediate – a person eats a food and has a noticeable, immediate reaction. One then learns to avoid these foods in the future. However, very low levels of IgE to a food can also cause reactions that are not as noticeable and can contribute to chronic illnesses not necessarily associated with the allergic food response. These low-level IgE responses can constitute a "hidden food allergy," causing chronic symptoms unless the person identifies the foods and eliminates them from the diet.

Another cardinal feature of the atopic individual is that the offending food causes characteristic symptoms of sneezing, runny nose, and itching eyes. Other responses may be present, but the appearance of the characteristic symptoms within minutes of exposure to a food is evidence of IgE-mediated allergy. When allergy first developed as a medical specialty, allergists thought the story ended here. Recent scientific insight acknowledges that there are other mechanisms of "allergy" which produce symptoms related to gut-immune responses. One mechanism that is measurable by laboratory evaluation is food sensitivity reactions from the IgG class of antibody directed at food antigens.

IgG Food Antibodies

Measurement of serum levels of IgG-class antibodies reveals an entirely different type of immune system-mediated adverse response to food. G-class immunoglobulins have been called blocking antibodies because they moderate IgE reactions by interacting with food antigens [10]. Such "blocking" interactions afford protection from relatively massive antigen exposure, in contrast to the high sensitivity of the IgE-mediated food allergies. Adverse reactions from IgG-food antigen reactions are due to the accumulation of large antigen-antibody complexes that tend to form deposits and cause strong inflammatory reactions. Elevated IgG causes stimulation of the complement fixation sequence [11], which elicits many of the effector functions of the humoral (immunoglobulin) immunity and inflammation.

The behavior of circulating IgG-antibody complexes in the presence of varying concentrations of antibody and antigen can explain the immune component of the phenomenon of food addiction [12]. People with no elevated IgG levels to foods have a combination of favorable factors. The immune and physical barriers in the small intestine are intact, and there is adequate capacity for lymphocyte uptake and removal of circulating complexes. Lymphocytes may accomplish a large degree of circulating food antigen-IgG antibody complex removal. In this role, the immune system is acting as an adjunct digestive system. When the capacity for such "digestion" is exceeded, however, with continuing antigen exposure, increasingly large complexes are formed (Figure 7-1). Such complexes have been found deposited in glomerular basement membranes of the kidney in experimental serum sickness [13]. Serum sickness involves relatively massive antigen-antibody levels because of the direct introduction of a relatively large mass of antigen. Patients with bone marrow failure have been treated in this way with horse serum containing anti-thrombocyte antibodies [14]. The treatment helps to suppress the immune response to the transplant, but intravenous administration of several cubic centimeters of horse serum introduces massive numbers of IgG molecules that form large immune complexes. Other instances of serum sickness have been due to antepartum Rh immunoprophylaxis with repeated administration of Rh immunoglobulin [15]. The mechanism involved in these immune responses is far different from the exquisite sensitivity operative in atopic response discussed above.

The treatment frequently used for IgE-mediated food allergies is repeated injection of antigen in order to elicit IgG blocking antibody formation in response to the high levels of circulating antigen. The IgG class antibodies form in response to direct introduction of

FIGURE 7-1. Food Addiction/Withdrawl Phenomenon

Early exposure to food antigens results in the formation of small complexes because of the relatively low concentration of antibodies (**A**). The small complexes may be removed by lymphocytes. As IgG concentrations rise, the formation of large complexes increases. (**B**) (These large comlexes form since) Relatively similar quantities of antigen and antibody exist allowing multiple cross links to form large aggregate complexes. Symptoms are caused when these complexes deposit in tissues and elicit inflammatory reactions. Eating more of the IgG positive foods can temporarily return the system to a preponderance of antigen and fewer large complexes (**A**). Only long-term removal of the offending food lowers antigen levels to the point where they do not cause large comples formation, even in the presence of elevated antibody levels (**C**). This mechanism explains why people become "addicted" to the foods they are allergic to, *i.e.*, they feel better when they eat them and worse when they don't.

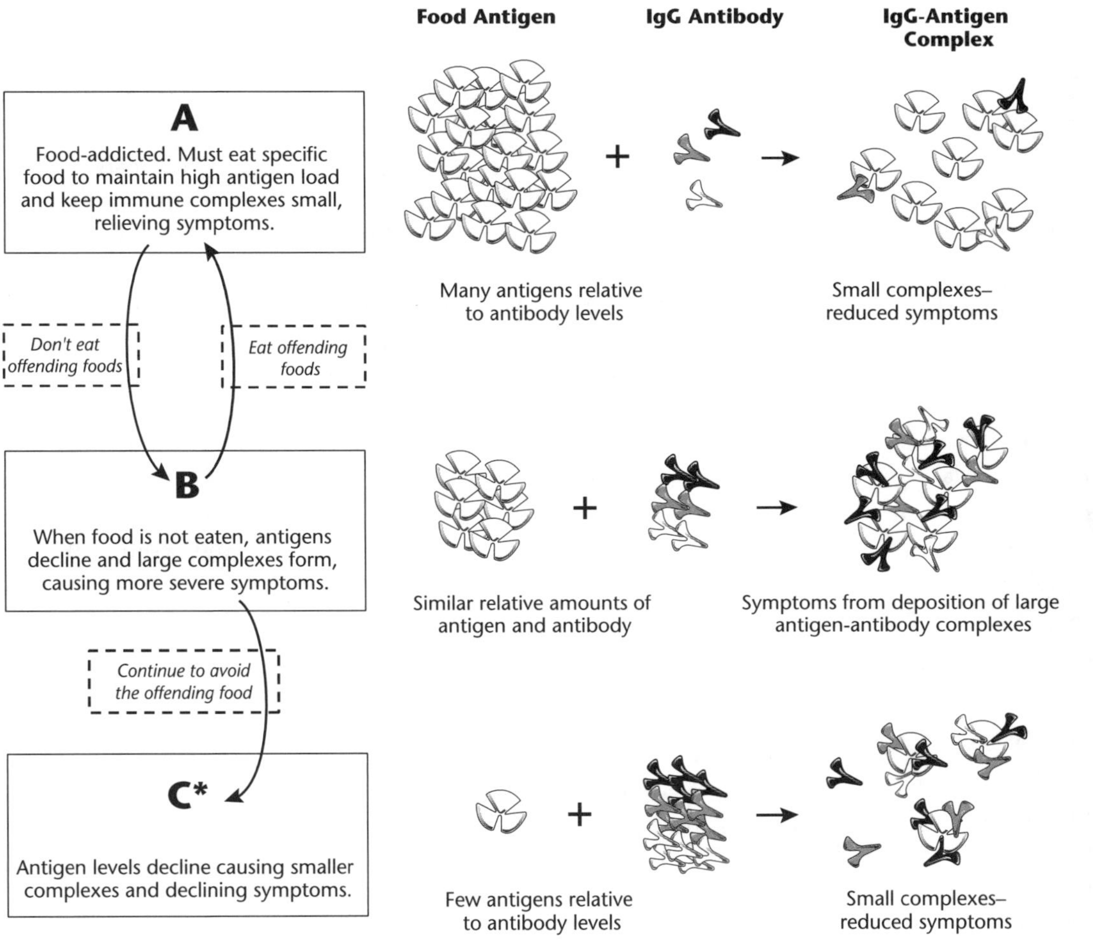

* At this point (**C**), eating the offending food can cause the situation to return to point "**B**", with increasing symptoms. This situation usually lasts 4 to 6 weeks. Continued elimination and restoration of normal gut function results in lower antibody levels as well as lower antigen levels, and the patient can become tolerant of the previously offending food.

antigen into the blood. Any failure of the barriers in the gut will result in such introduction. In patients with any degree of hyperpermeability ("leaky gut"), foods that are frequently consumed will cause the production of specific IgG antibodies directed at those food antigens. Removal of foods causing high IgG reactions results in lowering of the food-specific IgG levels in serum and significant clinical improvement over one year of observation [16].

The IgG1 and IgG4 antibodies (IgG1&4) are two sub-classes of the G-type immunoglobulins that are produced in delayed hypersensitivity reactions. These reactions are the most difficult to detect. There may be no obvious association of food consumption with an adverse reaction, which can occur 1 to 7 days later. IgG4 levels to foods increase in the blood due to the presence of the food antigen in the bloodstream. Elevated levels, therefore, are seen in response to the most commonly eaten foods. The antibody can combine with the specific food antigen to form a food immune complex. These complexes are thought to be the active agents for the delayed allergic response (Types II & III) since they can cause inflammatory reactions at various sites in the body, and these reactions can have a variable time course. Commonly reported symptoms for patients with delayed food sensitivities involve skin and GI, as well as nose and throat (Table 7-3). Treatment based on elimination of IgG-positive foods resulted in a 75% improvement of symptoms in 71% of a population of 80 patients studied. Nine of eleven patients with chronic, disabling symptoms had 80% improvement after elimination of positive foods and 20% of this group obtained 100% relief. All of these eleven patients had negative IgE RAST results for inhalants, but positive histories for delayed food allergy [17].

TABLE 7-3. Commonly Reported Symptoms in Patients With Delayed Food Sensitivities

- Nasal Congestion
- Nasal Drainage
- Sinus Headaches
- Fatigue After Meals
- Throat Clearing
- Gas
- Chronic Fatigue
- Dizziness
- Dry Cough
- Sneezing
- Hoarseness
- Diarrhea
- Migraine Headaches
- Itchy Eyes
- Ringing in Ears
- Ear Fullness
- Nausea
- Cervical Headache
- Watery Eyes
- Itchy Skin
- Asthma
- Burping
- Cramps
- Rash
- Ear Popping

Testing of IgG antibodies is done using a semi-quantitative ELISA assay procedure for semi-quantitative measures of antibody concentrations to go or move foods. Each food is given a numerical score and ranked by relative concentration. High concentrations of IgG1&4 to a particular food are ranked as "severe," and low levels receive a "mild" ranking. However, it is possible to have an adverse reaction to a food even if there is a low level of IgG1&4 present to that food. This reaction will be due to factors other than IgG1&4 levels. In general, three different types of results are documented. A discussion of each one follows.

Many IgG1&4 Reactions. If more than twenty foods have elevations of IgG1&4, an intestinal permeability problem is indicated. This "leaky gut" phenomenon causes numerous food antigens to permeate into the bloodstream and present to the immune system's lymphocytes that then respond by generating IgG1&4 antibodies. High levels may or may not reflect adverse symptoms for the patient. In general, consuming these foods puts the patient under stress. Avoiding the reactive foods may not completely solve the health problem in this case since the leaky gut problem will still exist, and the person will simply make new antibodies to the new foods leaking into the blood from the gut. A better approach is to have the patient eliminate the 3 to 6 most reactive foods and rotate consumption of the others on a four day rotation plan; that is, don't eat a food any more frequently than once every four days. By rotating out the most offending foods, a diminished IgG response can provide an interval of relative calm from inflammatory propagation of the gut pathology.

Since increased intestinal permeability is the apparent cause of this problem, some type of therapeutic program should be started to improve this condition. Nutrients that support the development of intestinal integrity include: 5 to 10 grams daily of glutamine‡ , 1 to 3 grams of pantothenic acid daily, and 25 to 50 mg of zinc daily. The tall columnar epithelial cells of the small intestine derive most of their energy from oxidation of glutamine, and the pantothenic acid and zinc provide assurance of support for increased cell mass. It should not be necessary to maintain this level of support for more than 60 days.

Moderate Number of IgG1&4 Reactions. Three to fifteen elevations may also indicate an intestinal permeability problem, and steps should be taken as indicated above. In many cases, total elimination of all of the reacting foods for one month coupled with nutritional support will ameliorate this condition. Food rotation can also be beneficial as symptoms improve. If symptoms return after a resumption of normal eating patterns, the rotation diet and nutritional support should be continued.

Few IgG1&4 Reactions. If there are one to five reactions to foods, these should again be eliminated from the diet to test for involvement in patient symptoms. In many cases, there will be significant improvement. Other etiological factors may also be involved in patient health problems.

No IgG1&4 Reactions. The absence of reactions in this test may be due to several possibilities. First, steroidal anti-inflammatory drugs are known to suppress the immune response and IgG4 production. Next, some individuals do not produce IgG1&4 antibodies to foods as most people do, and they may compensate for this by producing other IgG subclass antibodies. Finally, since this test measures the immune response to exposure of the system to food antigens, the test will not show a positive reaction if the patient is not consuming a food for reasons of known intolerance. Without antigenic stimulation, IgG will not be produced. This could give a false negative result for those foods not being consumed. The patient should be counseled that a negative result on this test for the food does not necessarily mean they can eat the food now without causing a reaction if they are already on an elimination diet due to known food reactions. Reintroduction of known food allergens should be done with caution.

These effects illustrate an important point about the clinical utility of this test. That is, it is best used on the naive, food allergic patient. It can be very useful in screening for potential foods the person is eating on a regular basis that may be causing adverse physical reactions.

Treatment involves: removal of offending foods; digestive aid supplements of betaine HCl and pancreatic enzymes; oral pro-biotics

‡ The amount of glutamine required may be reduced to 1.5 gm doses by adding 10 mg of pyridoxine and 10 mg of alpha-ketoglutaric acid with each dose. This combination appears to aid the efficiency of epithelial cell utilization of glutamine [18]

and pre-biotics for establishing favorable bacterial populations (see Intestinal Wellness Options at the end of this chapter); oral glutamine, the primary energy source for intestinal epithelial cells; and supportive nutrients as described above for attenuation of gut permeability [19]. Patients can usually start to reintroduce foods that had been positive after six to nine months when other factors are absent.

The Physical Barrier

The lining of the intestinal tract must endure considerable physical, chemical, and biological challenge. In response to the entry of harsh gastric juices, the duodenum releases large quantities of mucus that serves to coat and protect the intestinal mucosa from digestion by gastric juice and pancreatic digestive enzymes. The mucus barrier constantly is being eroded, posing challenge to the underlying cells. Turnover of intestinal epithelial cells is among the highest in the body as new cells must constantly replace those lost to luminal degradation. New cells form at the base of the crypts in the layer of tall columnar epithelial cells that are the outer cellular barrier. The newly formed cells must establish tight junctions to seal off the barrier from passage of large amounts of macromolecular components. The susceptibility of the cellular junctions to penetration by large molecules normally results in about two percent of food proteins getting past the mucosa in a form the immune system can recognize [29].

Testing for serum IgG concentrations also provides evidence of the integrity of the physical barrier because significantly elevated IgG levels are found only when food antigens can penetrate the physical barrier at an abnormally high rate. The finding of significant IgG elevations toward food antibodies is evidence of failure to maintain adequate barrier function because it addresses the central question: "Are large, undigested molecules (usually proteins) able to pass into the sub-mucosal layers?"

Intestinal Hyperpermeability (Leaky Gut Syndrome)

Leaky gut syndrome is a clinical disorder associated with increased intestinal permeability, subsequent to the loss of intestinal mucosa integrity. Patients with this condition have more than the normal one to two percent "leakiness" to large molecules. Degradation of the physical barrier often is due to exposure to toxic substances within the intestinal lumen that, directly or indirectly, damage the "tight junctions" between intestinal epithelial cells, leading to an increase in passive para-cellular absorption [30, 31]. Common causes of intestinal hyperpermeability are ethanol [32], nonsteroidal anti-inflammatory drugs (NSAIDs), and viral, bacterial, yeast, and protozoan agents [33-36]. Also, elevated levels of reactive oxygen species coming from a variety of sources such as bile, food, cytotoxic drugs [37], or inflammatory cells [38, 39] can increase paracellular permeability.

Macromolecules that penetrate the intestinal barrier are carried in the portal vein to the liver where they are degraded by phagocytic Kupffer cells. Xenobiotics and microbial toxins must be processed by the liver's Phase 1 and Phase 2 detoxification systems for excretion into bile or urine (see Chapter 8). In the colon where byproduct production from bacterial growth is high, the colonocytes also carry out most of the detoxification reactions [40]. Even for the intact GI tract, the cost of detoxification is high; antioxidants and conjugates like glutathione are consumed and must be replenished. Compromise of the intestinal barrier can lead to excess production of reactive oxygen species by the hepatic mixed-function oxidase (cytochrome P-450) system. Increased hepatic detoxification causes increased biliary excretion of reactive and carcinogenic compounds. Reflux of this "toxic" bile may be an important cause of chronic pancreatic disease [41, 42] and leaky gut syndrome. The syndrome also causes disorders by immunologic mechanisms. Increased permeability can stimulate hypersensitivity responses to food and components of the gut flora. Dietary gluten and bacterial endotoxin may cause nonspecific activation of the immune system's inflammatory pathways mediated by complement and cytokines. It has been shown experimentally that chronic low-grade endotoxemia leads to autoimmune disorders [43-45].

Intestinal hyperpermeability is found in all chronic inflammatory bowel diseases where it may play an etiologic role, or it may be a secondary consequence due to the vicious cycle involving immune activation, hepatic dysfunction, and pancreatic insufficiency [46]. The role of leaky gut syndrome in many diseases is often missed. Immune activation and toxin effects can manifest as diverse symptoms. The availability of noninvasive and affordable methods for measuring intestinal hyperpermeability makes it possible for clinicians to watch for leaky gut syndrome in their patients and to objectively assess the efficacy of treatment.

Lactulose/Mannitol Intestinal Permeability Challenge Test

The lactulose/mannitol protocol was developed to measure intestinal hyperpermeability that could lead to food sensitivity [7]. While the presence of elevated IgG levels to many foods in the test described above indicates hyperpermeability to macromolecular antigenic molecules, the lactulose/mannitol test provides a direct measure of hyperpermeability to the disaccharide lactulose. Lactulose/mannitol is a challenge test in which patients with suspected hyperpermeability ingest 5 grams each of the metabolically inert sugars, lactulose and mannitol. Since these sugars are not metabolized, any absorbed sugar is fully excreted in the urine within 6 hours. The urine is collected, and concentrations of the two sugars are measured. Percent absorptions are calculated using the formula below.

$$\% \text{ Compound Absorption} = \frac{\text{Compound concentration (mg/ml)} * \text{Urine volume (ml)} * 100}{5 \text{ gm}}$$

Mannitol, a monosaccharide, is passively absorbed through the intestinal mucosa. In contrast, lactulose, a disaccharide, is normally not absorbed unless the mucosal barrier is compromised.

In the healthy intestine the mean absorption of mannitol is 14% of the administered dose, while the mean absorption of lactulose is less than 1%. The normal ratio of lactulose/mannitol recovered in urine is <0.03. An elevated ratio means that excessive lactulose was absorbed, indicating leaky gut syndrome. It is best to have the patient perform the lactulose/mannitol test twice – first in the fasting state, then again after ingestion of a test meal [47]. The lactulose/mannitol ratio was found to be an accurate predictor of relapse when measured in patients with Crohn's disease who were clinically in remission [48]. Recent intake of high carbohydrate meals and prolonged administration of the challenge solutions diminish responses to this test.

Lactulose/Mannitol Testing Protocol

- Swallow a solution of 5 g mannitol and 5 g lactulose
- Collect urine for 6 hours
- Assay for total lactulose and mannitol
- Calculate recoveries
    - < 14% Mannitol = Carbohydrate malabsorption
    - > 1% Lactulose = Disaccharide hyperpermeability

(Lactulose vs. Mannitol Correlation Graph may be provided)

The Microbial Mass

After the partially digested food mass (chyme) leaves the stomach, it moves through the small intestine past the ileocecal valve and into the colon, carrying nutrient rich solutions at neutral pH and at a constant, warm temperature – conditions perfect for the growth of microbial populations. Oxygen, another growth factor, is present at very low concentrations. Thus, the populations that predominate are anaerobic. Normal microbial population densities should be viewed as a continuum, rising from about 10^4 per ml in the upper jejunum to over 10^{12} per ml in the colon. Stomach acid and digestive enzymes are major factors controlling species selection and growth rates in chyme. The approximate fifteen pounds of bacteria resident in a normal healthy colon constitute a mass that exceeds all organs in the body except skeletal muscle.

The dynamically perpetuated mass of microbes can have many beneficial effects. Their presence stimulates the maintenance of the well-developed villus and its supporting vasculature and musculature [49]. The feces normally are about three-quarters water and one-fourth solid matter composed of about 30% bacteria, 10–20% fat, 10–20% inorganic matter, 2–3% protein, and about 30% undigested food roughage [50]. This composition varies as a function of the amount and type of fiber in the diet. The fat in feces is mostly fat synthesized by bacteria or fat from sloughed off epithelial cells, though some can come from unabsorbed dietary fat. The brown color of feces is caused by derivatives of bilirubin. Products of bacterial action that include indole, skatole, mercaptans, and hydrogen sulfide cause the odor; these can vary from person to person depending on their diet and colonic bacterial flora [51].

While colonic bacteria are capable of digesting small amounts of cellulose, the amount of energy derived from this digested fiber is negligible. Other substances formed as byproducts of bacterial activity are short-chain fatty acids, vitamin K, biotin, vitamin B_{12}, thiamin, and riboflavin. Bacterial synthesis of vitamin B_{12} is of little significance because it is poorly absorbed in the colon. On the other hand, bacterial-formed vitamin K is nutritionally relevant. In the absence of vitamin K-forming bacteria, vitamin K status declines because dietary vitamin K is normally insufficient.

Many of the biotransformation products coming from bacterial metabolism are toxic. The colon has limited ability to detoxify these xenobiotics before passing them to the blood [52]. Colonocytes, like hepatocytes in the liver, contain cytochrome P-450 [53] and are

FIGURE 7-2. Production, Absorption, and Distribution of Microbial Metabolites

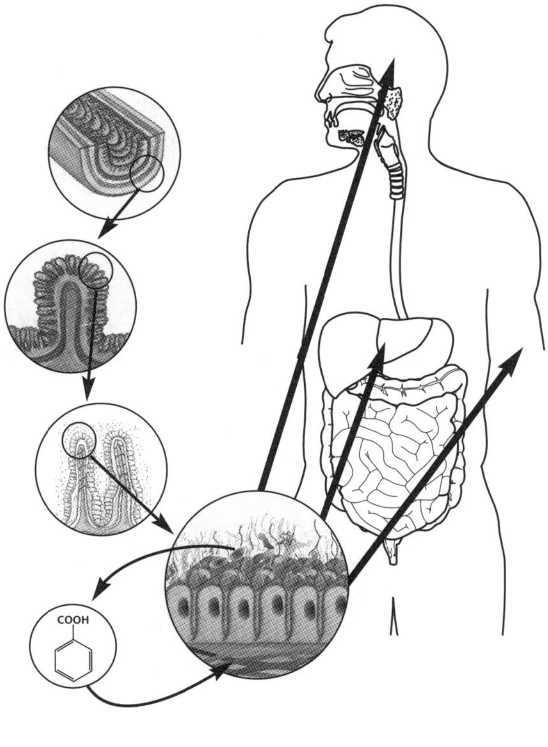

capable of reduction and hydrolysis (Phase I reactions). They can also transform toxic compounds by conjugation with donor substrates (Phase II reactions). Failure of the colon to properly detoxify xenobiotics and drugs results in epithelial cell damage, which has major implications for diseases of the colon, including cancer [40]. Once absorbed, the toxins will present challenges to many tissues. The sequence of stages from microbial metabolic release and intestinal absorption to systemic distribution is illustrated in Figure 7-2.

The integration of immunologic and microbiologic concepts can lead to new insight about the origins of many disorders related to the gut. In many cases, reactions that have been referred to as food allergies may be non-IgE reactions to antigens that arise from bacterial action on dietary proteins and are better described as enterometabolic disorders [54].

Microbial Populations

The intestinal flora is a complex ecosystem consisting of over 400 bacterial species that greatly outnumber the total number of cells making up the entire human body [55]. These metabolically active bacteria reside close to the absorptive mucosal surface and are capable of a remarkable repertoire of transforming chemical reactions. Any compound taken orally, entering the intestine through the biliary tract or by secretion directly into the lumen, is a potential substrate for bacterial transformation.

Anaerobic bacteria are the predominant microorganisms in the human GI tract, outnumbering aerobes by a factor of 10,000 to 1. The most abundant anaerobes are beneficial or benign: *Bifidobacterium, Bacteroides, Fusobacterium, Clostridium, Eubacterium, Peptococcus, and Peptostreptococcus*. *Bifidobacterium* can comprise up to 25% of the total flora in a healthy adult. A great many other species are present but in lesser numbers [56]. An imbalance in proportion and numbers of these species can be induced by broad-spectrum antibiotic use. This leads to the dominance of other bacterial species, including *Pseudomonas, Enterobacter, Serratia, Klebsiella, Citrobacter, Proteus, Providencia*, and fungi, especially yeasts such as Candida. Colonization of the gut by these undesirable microorganisms alters the metabolic or immunologic status of the host [134]. When this state leads to disease or dysfunction, it is termed "dysbiosis" [57].

In health, the upper GI tract is sparsely populated with microorganisms. The vast majority of bacteria washed along with saliva from the oral cavity are destroyed in the stomach by gastric juice. Those that survive (if sufficiently resistant to gastric secretions and bile acids) tend to be gram-positive, facultative forms such as *Streptococcus, Staphylococcus*, and *Lactobacillus* [58]. The small intestine constitutes a zone of transition between the sparsely populated stomach and the luxuriant bacterial flora of the colon. In the distal ileum, the concentrations of bacteria increase to 10^6–10^7 colony-forming units per milliliter. Here, gram-negative bacteria outnumber the gram-positive species. Coliforms predominate and anaerobic bacteria, such as *Bacteroides, Bifidobacterium, Fusobacterium,* and *Clostridium,* are found in large numbers. Beyond the ileocecal valve, the bacterial concentration increases steeply. Colonic bacteria number between 10^{11} and 10^{12} colony-forming units per milliliter of fecal material. Multiple dramatic shifts in populations of species occur between the ileocecal valve and the rectum. By the time they are passed from the body in stools the large majority of the bacteria are no longer viable.

Intestinal Pathogens, Symbiotes, and Parasites

The spectrum of identifiable diseases of intestinal microbial overgrowth ranges from those causing mild, transient diarrhea to those causing pandemic diarrhea resulting in thousands of deaths. Diarrheal episodes result in 2 to 3.5 million physician visits and more than 200,000 hospitalizations per year in the US alone in 1996 [59]. The treated cases are the acute cases that are easily recognized. The difficult cases are those with mild, chronic GI symptoms associated with excess microbial production of toxins in the intestinal lumen. Microbial overgrowth can be bacterial or fungal. In some cases, GI symptoms may be so minor as to go without notice, while neuromuscular effects due to toxic microbial products may persist for years [60]. The incidence of bacterial overgrowth of the small intestine increases steeply after age 70 [34]. Elderly patients develop malabsorption syndrome secondary to bacterial overgrowth, placing them at risk for clinically significant nutritional deficiencies [61].

The clinical perspective regarding gut ecology has shifted with modern cultural changes. In the early 1900s, the Nobel laureate E. E. Metchnikoff put forth his theory of orthobiosis [62] – a state of health dependent on the correct balance of life forms inhabiting the gastrointestinal tract. It was a commonly held belief that eating up to three pounds of meat per day was good practice. Such dietary habits lead to a condition that he termed "putrefactive dysbiosis." Diets high in meat and low in fiber can result in long transit times with undigested protein reaching the colon.

Modern studies give credence to Metchnikoff's hypothesis, showing that numerous toxic chemicals are generated by the action of colonic bacteria on undigested protein [63]. The enzyme urease induced in *Bacteroides, Proteus,* and *Klebsiella* species by a high meat diet hydrolyzes urea to ammonia. This raises stool pH, which is associated with an increased risk for colon cancer [64]. Decarboxylation of amino acids by gut bacteria yields vasoactive and neurotoxic amines, particularly histamine, octopamine, tyramine, and tryptamine, which are absorbed and transported to the liver to be deaminated. In severe cirrhosis, these amines enter systemic circulation and contribute to encephalopathy and hypotension of hepatic failure [65]. The bacterial enzyme tryptophanase, also induced by a high meat diet, degrades tryptophan to carcinogenic phenols [66]. Bacterial β-glucuronidase deconjugates estrogens, increases enterohepatic recirculation of these steroids, and decreases their rate of clearance from the body, effectively raising blood estrogen levels and the risk for breast cancer [67]. β-Glucuronidase and other hydrolytic bacterial enzymes also deconjugate bile acids. Deconjugated bile acids are toxic to the colonic epithelium and cause diarrhea; they have been implicated as etiological agents in the development of colon cancer [68] and ulcerative colitis [69]. The provocation of joint inflammation by bowel infection with *Salmonella, Shigella*, and *Yersinia* suggests an etiologic link between colonic bacteria and arthritis [63]. Not all bacterial activity in the colon is harmful to the host – in fact, much of it is beneficial. For example, *Bifidobacteria* species ferment soluble fiber to yield butyric acid that nourishes colonic cells. As discussed earlier, bacterial activity produces much of the vitamin K required by the body.

Evidence of shifts in microbial populations from symbiotes to pathogens may be found in urinary markers, exhaled gases, or fecal markers. All such evidence is indirectly related to the growth of organisms in the gut. Direct evidence requires sampling of intestinal content without perturbation of the local environment and performing strictly anaerobic culture growth on a fresh specimen. Such techniques are not used in routine clinical practice.

Due to the predominance of anaerobic bacteria in the colon, isolating colony-forming anaerobes from stool more than a few hours old is difficult in the laboratory and often yields misleading results. The detection of bacteria and yeast overgrowth is best done by measuring the levels of specific by-products of microbial activity in urine, which serves as a type of "finger-print" for intestinal overgrowth.

The advantage of colony counts on stool specimens is the direct evidence of microbial populations that persist until the stool is passed and can be processed in the laboratory. From those colonies isolated, sensitivities to antimicrobial agents may be determined, guiding interventions to the most efficacious agents. The hope in such treatments is that the species tested for sensitivity are the same as those causing the clinical problems. Urinary analysis for compounds produced by microbes in the gut provides indirect evidence of overgrowth. Accurate identification of species from compounds found in urine is generally not possible. By knowing the relative toxicities of the compounds found elevated, however, the additional information of metabolic impact from the overgrowth can be weighed as explained (in the following section).

Bacterial and Protozoal Metabolic Markers

The intestinal flora can be thought of as a chemical factory with massive levels of active enzymes. All rapidly growing bacterial species in the small intestine produce metabolic by-products that can be absorbed. These by-products may be helpful, benign, or toxic. Some of the absorbed products are utilized for energy immediately in the epithelial cells of the gut; others may be acted upon by the detoxification systems in the liver; while others are passed into urine unchanged.

The patterns of specific compounds that are detected in urine allow some deductions about the nature of the originating organisms. As previously mentioned, the organisms that are active under the conditions in the gut lumen are anaerobic. Many of them are obligatory anaerobes that are killed by contact with oxygen. This factor is a major detriment to obtaining accurate microbial population data from direct sampling techniques. The patient preparation and invasive nature of the procedure generally leads to altered conditions of microbial growth.

The indirect nature of examining microbial products in urine means that the growth conditions are completely unaltered by the assay. Bacteria much more actively metabolize amino acids than yeast, and not all bacteria carry out the same reactions. When they grow under anaerobic conditions, yeast produce unique products, allowing discrimination of their presence. Many of the microbial products that appear in urine are organic acids that have been further discussed in Chapter 5 – Fatty Acids. Direct assay of the microbial mass by stool culture and sensitivity testing is discussed below under the topic of the colon.

Urinary Indican

Bacteria in the upper bowel produce the enzymes that catalyze the conversion of tryptophan to indole (Figure 7-3). Absorbed indole is converted in the liver to indoxyl, which is then sulfated to allow urinary excretion. Indoxyl sulfate (also known as "indican") can be measured colorimetrically by conversion to colored oxidation products or directly by liquid chromatography with a UV absorption detector.

Because the upper bowel is sparsely populated with bacteria, indican is normally present in urine at low levels (in health). An elevated level of urinary indican is an indication of upper bowel bacterial overgrowth. Such an overgrowth was reported for some patients with adult celiac disease [70]. Bacterial overgrowth was demonstrated with indican in 8 out of 12 patients following jejuno-ileal bypass surgery [71].

Oral, unabsorbed antibiotics reduce indican excretion [72]. Indican excretion is also reduced when the gut is populated with strains of *Lactobacillus* at levels above 10^5 organisms/g [73]. *L. salivarius, L. plantarum, and L. casei* were more effective in achieving reduced indican than were two strains of *L. acidophilus*.

Indican testing can differentiate pancreatic insufficiency from biliary stasis as the cause of steatorrhoea (fatty stools). Patients with steatorrhoea due to pancreatic insufficiency show a rise of indican from low values to above normal when they are treated with pancreatic enzyme extract [74]. Urinary indican did not rise in patients with steatorrhoea not due to pancreatic insufficiency nor in the normal subjects who received pancreatic enzymes. This scenario

FIGURE 7-3. Intestinal Tryptophan Conversion to Urinary Indican

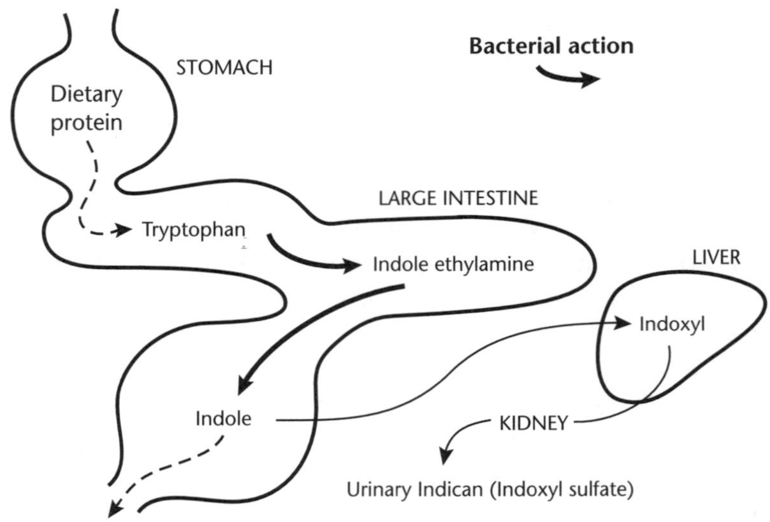

demonstrates how bacterial populations respond to increased concentrations of luminal amino acids. Large shifts in bacterial populations induced by the artificial sweetener saccharin have also been demonstrated by changes in indican excretion [75].

No age adjustment for reference limits is necessary since excretion is constant for young and elderly control subjects [76]. The test may be performed after oral loading of 5 g tryptophan [77]. Tryptophan loading, which can be monitored by measuring periodic levels of urinary indican, results in neuropsychiatric manifestations due to products of intestinal bacterial conversion of the amino acid [72]. Including elevations of phenol and p-cresol with that of indican as criteria of abnormal bacterial colonization of the small intestine reduces the number of false positives [78].

The use of indican results is complicated by impaired protein digestion, which increases the tryptophan available for bacterial action. The relationship between incomplete indican and incomplete digestion might be utilized as a measure of protein digestive adequacy. Increased urinary indican has been shown to correlate with enteric protein loss [79]. Indican elevation has revealed that impaired protein digestion and increased bacterial conversion of tryptophan is a complication of cirrhosis of the liver [80]. Some degree of malabsorption was found in 30% of an elderly population by combinations of indican with the Shilling test as well as other tests described below [81].

Urinary Benzoate, Phenylacetate, p-Cresol, p-Hydroxybenzoate, and p-Hydroxyphenylacetate

Each species of microbe expresses unique patterns of enzyme activities, yielding one or more of the above products reported by a urinary dysbiosis test. When bacteria metabolize the amino acids phenylalanine and tyrosine, the major products are benzoate and p-cresol, respectively. Normal levels of these two compounds are much greater than the typical levels of p-hydroxyphenylacetate and phenylacetate. These compounds, normally present at lower concentrations, are by-products of species that have less efficient degradation pathways. The same compounds may become greatly elevated, of course, when the populations of the less efficient organisms increase such as with clinical conditions of overgrowth.

Intestinal bacteria that contain L-amino acid decarboxylase enzymes degrade tyrosine to tyramine. The tyramine is then deaminated and oxidized to p-hydroxyphenylacetate. This product is excreted unchanged and unconjugated in urine. It was used to identify small bowel disease and bacterial overgrowth syndromes in 360 randomly selected, acutely ill infants and children. In this study, no false negative and only 2% false positive results were found [82]. Among those with high levels of p-hydroxyphenylacetate were children ranging from 11 months to 8 years old with confirmed Giardiasis. Treatment with metronidazole or mepacrine eliminated the p-hydroxyphenylaceturia. Although p-hydroxyphenylacetate can be produced in the liver, abnormally high levels in urine are of bacterial origin.

Elevated urinary phenylacetate comes from similar degradation reactions starting with phenylalanine. An increase in dietary protein leads to altered products of colonic metabolism. Urinary p-cresol increases along with an increase in fecal ammonia and fecal volatile sulfur substances. The urinary output of p-cresol is probably due to the action of colonic bacteria on protein; this is based on observed urinary p-cresol increases following increases in fecal short chain fatty acids and ammonia in people on a high protein diet [83].

In addition to serving as markers of intestinal dysbiosis, these compounds have important physiological impact. Phenylacetate is one of the compounds that accumulates in the genetic disorder phenylketonuria (PKU). The neurotoxicity of phenylacetate is probably due to very strong inhibition of synaptic choline acetyltransferase [84]. Phenylacetate elevation due to dysbiosis has the same metabolic effects as elevation due to PKU, where it is linked to behavior and learning disabilities [85].

Urinary Benzoate and Hippurate

High urinary benzoate and hippurate have been associated with intestinal bacterial overgrowth [86], and they can convey additional information. Using a major hepatic detoxification pathway, benzoate is conjugated with glycine to form hippurate. This is a two-step process that can be limited by deficiency of pantothenic acid required for coenzyme A production. Individuals with high requirement for hepatic detoxification pathways are frequently depleted in glycine and pantothenic acid. This situation will be reflected as an elevation of benzoate without concurrent elevation of hippurate. Glycine and pantothenic acid supplementation can relieve the congestion in this hepatic pathway. Intestinal dysbiosis with weakened mucosal epithelium is a common cause of toxemia and the resulting up-regulation of the hepatic pathways.

Urinary Tricarballylate

Some bacteria convert the citric acid cycle intermediate aconitate into tricarballylate. Extensive studies in animals have shown that tricarballylate in urine comes from such bacteria in the intestines [87]. Species that produce this compound frequently infect the foregut of ruminant animals like cows. Tricarballylate binds magnesium very tightly and prevents absorption, leading to magnesium deficiency [88].

Urinary Dihydroxyphenylpropionate

Clostridium species constitute a major portion of the bacteria of the lower ileum and colon. Being strictly anaerobic, they are difficult to culture from stool and are not reported on most stool cultures. *Clostridium difficile* is associated with enterocolonopathies. Cases of confirmed clostridium overgrowth show elevated levels of dihydroxyphenylpropionate, which fall to baseline with Flagyl treatment but are unaffected by Nystatin [89]. While other organisms may produce dihydroxyphenylpropionate, clostridia are the most commonly encountered genera among those susceptible to Flagyl. Compounds closely related to dihydroxyphenylpropionate are produced by the genus *Clostridium* [90].

Simultaneous elevations of multiple bacterial and protozoal by-products are consistent with the condition of putrefactive dysbiosis that is also characterized by lowered ratio of Bifidobacteria to *Bacteroides*, the major genera of the anaerobic organisms. Fecal pH may be elevated, and β-glucuronidase increases. This scenario has been linked to increased occurrence of colon and breast cancer and hepatic encephalopathy [91].

Breath Hydrogen and Methane

This test offers reasonable sensitivity, and, it is very convenient to administer, although it will sometimes give false-negative results. For greater predictive value, it is best to also obtain a baseline breath sample from the patient before consumption of the challenge solution. The fasting patient drinks a challenge dose of lactulose (10 g) or glucose (75 g) solution. Breath samples should be collected every 15 minutes for up to 2 hours. If bacteria exist in the small intestine, they will ferment the sugar and release hydrogen and methane, which can be detected in the breath. The typical fasting breath sample contains less than 10 ppm hydrogen or methane; levels higher than 20 ppm indicates a high probability of bacterial overgrowth. Following a lactulose or glucose challenge, a two-phase response may be seen. The first rise in breath hydrogen occurs within 30 to 60 minutes when lactulose contacts the small intestine; a second, more pronounced rise occurs about two hours later when the sugar enters the large intestine. A rapid and prolonged first-phase response is frequently due to small intestine bacterial overgrowth [92]. Interpretation of results is complicated by the large number of false-positive findings as compared to results obtained from bacterial cultures of intestinal lumen aspirates [93, 94]. Combining the findings of elevated fasting breath hydrogen (>20 ppm) and raising the limit for the post-lactulose challenge increase in breath hydrogen and methane to greater than 15 ppm reduce the chance of false-positive responses [95].

[^{14}C]Xylose Breath Test

For this test, the patient drinks a solution containing [^{14}C]xylose, which is normally absorbed and excreted in the urine. In the presence of excess bacteria in the small intestine, [^{14}C]xylose is metabolized by bacteria to form $^{14}CO_2$, which is absorbed and eliminated through the lungs. Excess breath $^{14}CO_2$ is diagnostic for bacterial overgrowth of the small intestine. This test has limited usefulness for diagnosing overgrowth in the small intestine because it is prone to giving excessive numbers of false-positive and false-negative results, especially in patients with intestinal motility problems [96, 97].

Yeast Overgrowth (Fermentative Dysbiosis)

As elaborated above, Metchnikoff's ideas of putrefactive dysbiosis have been recognized for several decades. More recently, the prevalence of a new type of dysbiosis has gained recognition. In contrast to putrefactive dysbiosis in the colon, fermentative dysbiosis can be present in the small intestine. While colonic fermentation is desirable, the proliferation of yeast, fungi, and fermentative bacteria in the small intestine has detrimental effects. The production of toxic by-products can overload the detoxification pathways in the liver, and fungal enzymes can damage the mucosal epithelium.

Opportunities for fermentative dysbiosis are most apparent following jejunal or ileal resection or jejuno-ileal bypass surgery, where aerobic and/or anaerobic colonic organisms were found in jejunal aspirates from 8 out of 12 patients [71]. Abnormal overgrowth of microbes occurs in surgical blind loop, intestinal stasis, gastric hypochlorhydria, and immune and nutritional deficiencies. Bacterial consumption of cobalamin lowers blood levels of vitamin B_{12}. Bacterial dehydroxylation of bile salts causes impairment of micelle formation that can lead to steatorrhoea [69]. Some of the damage resulting from small intestinal overgrowth is attributable to the action of bacterial proteases that degrade pancreatic and intestinal brush border enzymes. This causes pancreatic insufficiency and mucosal damage leading to malabsorption. Candida colonization of the GI tract impairs mucosal barrier defense against gram-negative bacteria [98]. Endotoxemia resulting from bacterial overgrowth contributes to hepatic damage [99].

Regular use of antacids reduces the killing of yeast and bacteria in the stomach, and repeated courses of antibiotics cause sudden, dramatic reductions in populations of beneficial bacteria. These conditions set the stage for increased growth of microbes that produce toxins instead of nutrients. What was once healthy fermentation can become unfavorable dysbiotic overgrowth in the upper bowel [31]. Clinicians should be watchful for antacid abuse which (in conjunction with antibiotic treatment) can lead to the development of disparate symptomatology secondary to intestinal dysbiosis (Table 7-4). This is particularly true for such patients consuming a diet rich in simple fruit sugars. Responses vary widely depending on factors of mucosal integrity, immune competence, and metabolic function.

Carbohydrate intolerance may be the only symptom of bacterial overgrowth, making it indistinguishable from intestinal candidiasis; in either case, dietary sugars can be fermented to produce endogenous ethanol [100]. The production of ethanol following an oral glucose load was found in 61% of chronically ill patients [101]. The symptoms include abdominal distension, carbohydrate intolerance, fatigue, and impaired cognitive function. The majority of cases are thought to be due to yeast overgrowth. Another product of bacterial fermentation of sugar is D-lactic acid. Although D-lactic acidosis is usually a complication of short-bowel syndrome or of jejuno-ileal bypass surgery (colonic bacteria being the source of acidosis), elevated levels of D-lactate were found in blood samples of 13 out of 470 randomly selected hospitalized patients [102]. Small intestine fermentation was a likely cause of D-lactic acidosis in these 13 patients, 60 percent of whom had a history of gastrointestinal surgery or disease.

TABLE 7-4. Factors, Symptoms, and Diseases Associated with Dysbiosis and Intestinal Hyperpermeability

Contributing Factors
Alcohol abuse
Corticosteroid use
NSAID use
Excessive Stress
Nutrient insufficiencies
Gastrointestinal infections
Fodd reactions
Improper fasting
Symptoms
Abdominal distention
Diarrhea
Constipation
Abdominal pain
Food intolerances
Skin rashes
Poor exercise tolerance
Shortness of breath
Cognitive deficits
Fatigue and malaise
Arthralgia
Myalgia
Fevers of unknown origin
Diseases
Inflammatory bowel disease
Irritable bowel syndrome
Celiac disease
Infectious enterocolitis
Cystic fibrosis
Chronic fatigue immune Deficiency Syndrome
Acne
Eczema
Psoriasis
Urticaria
Dematitis herpetiformis
Autism
Childhood hyperactivity
Spondyloarthropathies
Pancreatic insufficiency
HIV infection
Neoplasia treated with cytototoxic drugs
Hepatic dysfunction
Environmental illness

Yeast and Fungal Metabolic Markers in Urine

Tartarate, Arabinitol, Arabinose, Citramalate, and β-Ketoglutarate

Products of intestinal yeast and fungi that have been detected in urine include tartarate, arabinitol, arabinose, citramalate, and β-ketoglutarate. Some of these compounds may cause metabolic interference because they are structurally related to known intermediates in central energy metabolism. Tartarate has known toxicity related to metabolic interference [103]. Arabinose is an aldopentose that reacts with the amino acid side chains of proteins to produce cross-linked polymers called pentosidines [104]. Measuring serum D-arabinitol allows prompt diagnosis of invasive candidiasis [105]. The ratio of D- to L-arabinitol in serum reveals the presence of disseminated candidiasis in immunosuppressed patients [106]. Elevated arabinose in urine was found in women with Candida overgrowth of vulvovaginitis [107]. Citramalate has been reported to be associated with fermentative intestinal overgrowth and is elevated in cerebrospinal fluid of patients with bacterial meningitis [108].

The normal flora is kept in check by competition and symbiosis among the hundreds of microbial species. Repeated use of broad-spectrum antibiotics sets up conditions for opportunistic overgrowth of organisms that are not affected by the drugs and are able to re-colonize rapidly when treatment has ended. Fungal infection is especially prevalent in intensive care wards where combinations of antibacterials and immune suppression frequently lead to disseminated candidiasis [109]. The question of mucocutaneous yeast overgrowth has been a point of considerable controversy. Clinicians who have treated suspected mild mucocutaneous candidiasis based on symptomatic evidence have been convinced of a significant incidence of a "yeast syndrome" [110]. Yeast has been etiologically associated with at least 6% of diarrhea in children and 19% in adults [111].

The greatest problem with unequivocal demonstration of intestinal yeast and fungi overgrowth is the difficulty of sampling upper intestinal contents under normal conditions. The ease with which yeast and fungi can persist in the anaerobic environment of the gut is most easily demonstrated by the universal occurrence of these organisms in stool specimens of individuals not on antifungal therapy. Largely anaerobic proliferation of yeast is evident in cases of severe immune suppression such as with HIV-infected patients [112]. Disseminated candidiasis is characterized by candidemia, endophthalmitis, and multiple organ involvement. The appearance of yeast and fungal metabolic products in urine is the strongest existing physical evidence of intestinal overgrowth of these organisms. The disappearance of tartarate and arabinose from urine upon antifungal treatment argues that the organisms susceptible to such treatment are the source of these compounds [60].

Upper bowel fermentative dysbiosis is associated with intolerance for carbohydrate and soluble fiber. There may be increased concentrations of short chain fatty acids in stools and fecal pH may fall. Analysis of breath hydrogen and methane following a standard carbohydrate load is one method of detection of fermentative bacterial overgrowth, but the breath test does not reveal the presence of yeast overgrowth; yeast produce alcohol rather than hydrogen and methane.

Evaluating Digestive Function

The Stomach

Gastric secretion of hydrochloric acid (HCl) is essential for proper digestion of protein (proteolytic activity of pepsin requires a pH less than 5) and for adequate sterilization of stomach contents. Low pH is also required for proper release of vitamin B_{12} from food sources. When stimulated, the parietal cells of the stomach secrete an isotonic solution containing enough HCl to make the gastric fluid pH less than one. At this extreme acidity, the hydrogen ion concentration of gastric fluid is about three million times greater than that of arterial blood. It takes a tremendous amount of cellular energy to maintain this gradient because of the high activity of membrane pumps utilizing ATP. Perhaps it is not surprising that problems of intestinal absorption are often due to inadequate secretion of HCl [113].

Depletion of B-complex vitamin may be the precipitating event leading to reduced energy production in cells of the gastric pits. Patients over 50 years of age, who are particularly prone to inadequate secretion of HCl (hypochlorhydria) and intrinsic factor because of degenerative gastric disorders (*e.g.*, atrophic gastritis), are frequently vitamin B_{12} deficient [114].

Symptoms of inadequate stomach acid are similar to those of excess acid and are frequently mistaken. Upper abdominal burning sensations associated with mealtime are reported in both cases. Excessive acidification of stomach contents leads to burning sensation from exposure of the lower esophagus to highly acidic chyme. With inadequate stomach acid production, regurgitation of chyme into the lower esophagus causes such burning pain. In these cases, advising antacids is contraindicated and counterproductive because they neutralize the acid that is produced, further interfering with digestion.

Stomach acid secretion is a principle line of defense against infection of the GI tract [115]. The low pH and proteolytic enzyme activity in normal gastric secretions kills most of the bacteria and parasites that contaminate food. Chronic hypochlorhydria, whether induced by habitual use of antacids (*i.e.,* "H-2 blockers") or due to gastric disorders, increases the risk for infection and intestinal overgrowth [116]. The risk of infection is multiplied when hypochlorhydria simultaneously reduces zinc and other trace elements critical for immune system function. Patients with malabsorptive disease, in particular, are at risk for various mineral deficiencies. Inhibition of gastric acid secretion reduces absorption of most trace elements [117] because gastric acid serves to release minerals from their complexed native forms in the stomach [118]. Variations in stomach acid production can explain why the absorption efficiency for dietary calcium varies widely from individual to individual, ranging from 15% to 45% even with adequate calcium intake. Thus, some individuals seem inherently to be better absorbers of calcium [119].

Heidleberg Test

The Heidleberg test is not considered a laboratory evaluation because the test must be administered with the patient as an active participant. It will be described here because only this type of test can give definitive answers about the adequacy of gastric acid secretion. The test uses a tiny plastic encapsulated pH probe that is swallowed by the patient. The capsule, small enough to safely pass the circuitous course of the GI tract, contains a miniature radio transmitter that continuously measures gastrointestinal pH and transmits the data to a waistband antenna connected to a bedside receiver. The pH readings are recorded for a permanent record. The capsule can either be tied to a thin string for retrieval or swallowed untethered. The latter approach allows additional measurements of upper intestinal pH to be gathered. After swallowing the capsule, pH readings typically start around 7.0 and then drop towards 1.0 as the capsule settles toward the stomach bottom. The patient then drinks a challenge solution consisting of concentrated sodium bicarbonate (baking soda), which has strong buffering capacity. Within half a minute, the pH will normally rise to approximately 7. If acid secretion is normal, the pH will fall again, returning to between 1.0 and 2.0 within 20 minutes. The challenge solution is given again and repeated up to four times, as long as the pH response time is less than 20 minutes. Hypochlorhydria is indicated for a patient requiring more than 20 minutes to re-acidify. With achlorhydria, the patient's stomach secretes little acid, and the pH will not fall below 4.0, even on the first challenge [120]. Experienced technologists must administer this test because of factors, such as the timing of bicarbonate solutions, that are critical for accurate, reproducible results.

Other Indirect Stomach Acid Indicators

Other laboratory abnormalities are frequently related to or caused by inadequate stomach acid production. Multiple trace element deficiencies are sometimes found in individuals who have apparent adequate dietary intake. Because of the acid dependency of trace elements, evidence of multiple trace element deficiencies has been used as an indirect measure of adequacy of gastric acid flow. For example, simultaneous low levels of iron, zinc, copper, and manganese in serum, erythrocytes, or hair is often due to gastric acid inadequacy, especially when intake of trace elements is normal. The critical function of low pH in the stomach is required to set up mineral absorption (see Chapter 3 – Minerals). High levels of compounds produced by bacterial action on amino acids are even more directly associated with inadequate hydrochloric acid. The loss of bactericidal action and the failure to digest protein due to low stomach acid simultaneously lead to higher bacterial populations and greater availability of unassimilated amino acids for bacterial conversion.

The Pancreas and Gall Bladder

Pancreaticobiliary fluid composition can be highly variable. The pancreas contributes to digestion by secreting alkaline bicarbonate and a variety of digestive enzymes (Table 7-5). Secretion of pancreatic fluid is controlled in part by vagus nerve stimulation. A more important regulatory mechanism of pancreatic secretion is the control exerted by the hormones secretin and cholecystokinin. Both are synthesized in the duodenum (upper small intestine) and secreted in response to the presence of acidified chyme in the small intestine. Additionally, cholecystokinin will stimulate the contraction of the gall bladder, causing the release of bile into the duodenum. Secretin stimulates the flow of bicarbonate-rich pancreatic fluid that serves to raise the pH of normal chyme from below four to above seven, allowing trypsin and other pancreatic digestive enzymes to reach their maximal activities.

Chyme in the upper portions of the small intestine is also a stimulus for the release of bile necessary for proper digestion of fat. Bile is synthesized in the liver and stored in the gall bladder until released. Bile, which contains no digestive enzyme activity, is important for digestion due to its high content of bile salts. Bile salts emulsify fat into smaller-sized globules that can be digested by pancreatic lipase.

TABLE 7-5. Pancreatic Solutions

Secretion	Breakdown of	To Yield
Bicarbonate	Stomach acid	Neutralization
Trypsin, chymotrypsin, and carboxypeptidase	Protein	Peptides and amino acids
Ribonuclease and deoxyribonuclease	DNA and RNA	Nucleotides
Amylase	Carbohydrate	Glucose
Lipase	Fat	Fatty acids + glycerol
Cholesterol esterase	Fatty acylcholesterols	Cholesterol + fatty acids

Fecal Chymotrypsin

The fecal chymotrypsin test is a useful non-invasive test to determine chronic pancreatic insufficiency [121]. False-positive results up to 10% have been reported in normal individuals. Even greater reliability for diagnosing chronic pancreatitis can be obtained by performing the fecal chymotrypsin test in combination with the bentiromide (N-benzoyl-L-tyrosyl-p-aminobenzoic acid) test [122]. The bentiromide test measures the amount of p-aminobenzoic acid (PABA) appearing in urine following an oral loading with the test compound, bentiromide. PABA excretion rates are lower in patients with compromised pancreatic function than in healthy control subjects [123].

Fecal Fat

Fat absorption is a passive process. Bile and pancreatic lipases greatly facilitate the rate of fat absorption. Brush-border lipase activity and simple diffusion allow for the absorption of a considerable proportion of fat in the healthy intestine even in the absence of bile. Oils containing unsaturated fatty acids are absorbed more quickly than saturated fats. Within the intestinal cell, fatty acids are packaged and released into lymphatic circulation as chylomicron particles. Since the lymphatic vessels drain into the arterial system, dietary fat is presented directly to extrahepatic tissues. However, medium chain-length fatty acids (6–12 carbons) are water-soluble and require minimal lipase activity and less bile salt for solubilization. Therefore, they enter portal circulation rather than the lymphatic drainage of the small intestine.

Steatorrhoea, defined as the presence of excess fat in the stool, is established by a total fecal fat test. Normal fecal excretion of fat is less than 6 g/day. However, this test does not distinguish between fat maldigestion or malabsorption. Instead, tests for fecal triglycerides and long-chain, free fatty acids help to differentiate between these two disorders. Since most of the dietary fat is composed of triglycerides, excess fecal triglyceride levels indicate incomplete fat hydrolysis (maldigestion) possibly due to pancreatic insufficiency. On the other hand, if fecal triglycerides are normal but the feces contains excess long-chain free fatty acids (carbon length >12C), malabsorption is indicated. This is because long-chain, free fatty acids are readily absorbed in the presence of a fully functioning intestinal mucosa.

The usefulness of these tests is illustrated by an experimental study that differentiated fat maldigestion from malabsorption in patients fed fat containing radio-labeled triglyceride. It was found that significantly greater amounts of radioactivity were present in the triglyceride fraction of feces from those patients with pancreatic insufficiency compared to controls. On the other hand, in patients with short bowel syndrome, almost the entire radioactivity was associated with the free fatty acid fraction of the feces [124]. Maldigestion is a result of inadequate secretion or activation of pancreatic lipase, which causes excessive, unhydrolyzed triglycerides to be excreted in the feces, as shown by other studies [125]. On the other hand, excess fecal content of long-chain, free fatty acids is evidence of malabsorption.

Fecal Steroids

Fecal steroids come from dietary cholesterol, intestinal mucosal sloughing, and bile acids and cholesterol contained in bile. Fecal cholesterol content can range from 75 to 200 mg/day, most of which comes from bile. The wide variation in fecal steroid content is due to many factors. Especially interesting is the effect of a high dietary polyunsaturated to saturated fatty acid ratio resulting in increased fecal steroids [126, 127]. These data demonstrate that increasing intake of dietary fat containing polyunsaturated fatty acids is effective for lowering total serum cholesterol without reducing HDL-cholesterol. The mechanism involves reduced absorption of cholesterol. Additional significant benefits from reduced sterol absorption may be obtained by increasing dietary fiber. Several reports have demonstrated the ability of various dietary fibers to modulate fecal steroid content [128-131].

Fecal Fibers

Microscopic inspection of stool can reveal the presence of meat and vegetable fibers. The increase in the amount of these fibers that occurs with impaired digestion is an indirect indicator of hypochlorhydria or insufficient output of pancreatic enzymes [132, 133].

The Small Intestines – Digestion and Absorption

The intestinal brush-border contains various membrane-associated enzymes (*e.g.*, sucrase, lactase, dextrinase, maltase, isomaltase, trehalase) that cleave numerous polysaccharides and disaccharides (sucrose and lactose) to absorbable monosaccharides (glucose and fructose), the products of luminal digestion of dietary starches and fiber. Simple sugars (glucose, fructose, and galactose) are principally absorbed by saturable active-transport processes. Impairment of brush-border enzyme activity causes carbohydrate malabsorption, which leads to excess bacterial fermentation of unabsorbed disaccharides and an increase in osmolarity that is accompanied by diarrhea and flatulence [133]. Although most people consume too little dietary fiber, excessive dietary fiber (> 50 gm/day), which is resistant to the action of human digestive enzymes but is broken down by gut bacteria, can lead to diarrhea and gas as well.

Note: As much as a third to half of the protein in the lumen of the small intestine is not of dietary origin but is protein from sloughed off epithelial cells and digestive enzymes [16].

The amount of mannitol appearing in urine after a standard oral load of this monosaccharide derivative is used as a measure of small intestinal capacity for monosaccharide (glucose and fructose) absorption (see Lactulose/Mannitol test). The disaccharidase functions cannot be measured by routine laboratory evaluations, but clinical observations have indicated that the disaccharide enzymes play a role in inflammatory bowel conditions such as Celiac disease. Elimination of all disaccharide sources has been helpful in many of these cases [135]. The diet is difficult to maintain because all sources of sucrose, maltose, and isomaltose must be eliminated. Maltose and isomaltose are produced in the earlier phases of digestion of starch, so all starch sources must be restricted. Enteral insulin-like growth factor 1 has shown promise for inducing the formation of intestinal disaccharidase enzymes [136].

Brush-border and cytosolic peptidases hydrolyze small peptides to produce amino acids. Although most proteins are digested to free amino acids prior to absorption, the absorption of large peptides plays a significant role in the assimilation of dietary protein [137]. The transport mechanism for the uptake of peptides by mucosal cells differs from the mechanism for free amino acid uptake, and there is no competition between the two. Nonetheless, because peptide hydrolase enzymes in the brush border and cytosol of mucosal cells hydrolyze small peptides, primarily free amino acids pass into the portal vein and go to the liver [138, 139]. This process is one of the principal dynamic actions affecting the concentrations of amino acids in blood plasma (see Chapter 4 – Amino Acids). Direct utilization by intestinal cells contributes to the overall dynamic action of amino acids obtained from dietary protein. Up to 50% of orally consumed amino acids may be used by intestinal cells, never becoming available for transport into portal circulation.

Schilling Test

The Schilling test primarily assesses the absorptive capacity of the GI tract for vitamin B_{12}. It is used for diagnosing pernicious anemia (intrinsic factor deficiency). An abnormal Schilling test in the absence of pernicious anemia may indicate malabsorption of B_{12} in the distal ileum [140], which may be due to pancreatic output or bacterial overgrowth. The Schilling test also is a very sensitive and useful test for diagnosing bacterial overgrowth of the small intestine proximal to the ileocecal valve.

Typically, the Schilling test involves giving the patient an oral dose of radiolabeled vitamin B_{12}, although a non-radioactive version of the test has been described [141]. A week before the start of the test, the patient is injected with a 1000 mcg loading dose of vitamin B_{12} to saturate tissue receptors. This is done so that the vitamin B_{12} absorbed from an oral dose will be excreted in the urine, instead of going to body stores. Next, the patient takes an oral dose of vitamin B_{12} containing radioactive cobalt so that the amount absorbed can be determined. The dosing is best carried out in three stages: 1) without intrinsic factor, 2) with intrinsic factor and pancreatic enzymes, and 3) after a course of treatment with antibiotics (*e.g.*, metronidazole) or anti-inflammatory drugs (*e.g.*, prednisone). The recovery of radioactive cobalt in the urine gives a measure of vitamin B_{12} absorption efficiency. Normal persons excrete more than 8% of the administered oral dose, while patients with pernicious anemia often excrete only 0 to 3%.

In disorders affecting the distal ileum, such as regional enteritis caused by microbial overgrowth, the first and second stage Schilling tests are frequently abnormal. Ileal receptor sites for intrinsic factor may be damaged by inflammation, and the impaired absorption of vitamin B_{12} will not be corrected by the addition of intrinsic factor or antibiotics. However, the Schilling test may normalize after treatment with an anti-inflammatory drug. If bacterial overgrowth is a problem, the bacteria can actually take up the vitamin B_{12}, which results in impaired absorption of the vitamin. Under these conditions, the first and second stages of the Schilling test frequently give abnormal results. After treatment for bacterial overgrowth, the Schilling test response usually returns to normal.

The Colon

The colon is the recipient of the product from the myriad actions on food by the different parts of the upper digestive tract, from mouth to ileocecal valve. Clinical interventions for maintaining or restoring a state of wellness in the colon must deal with the entire sequence of digestive organ functions, which can be evaluated through laboratory analysis. Laboratory testing for colonic health is primarily done by fecal analysis.

Microbiological examination of stool has been used in the detection of ova and parasites. The high percentage of false negatives is a major disadvantage of the procedure. Similar uncertainty applies to all culture and sensitivity results on stool because of the astronomical numbers of organisms that can easily hide pathogens. In addition, the populations of organisms are known to undergo profound shifts as they pass from the small intestine to the rectum. The most clinically useful information is to determine whether the growth is releasing a significant amount of toxic products in the small intestine where absorption is active.

The examination of stool can still provide valuable clinical information. For example, a positive culture for a pathogenic organism suggests that some degree of upper bowel colonization is likely. Antibiotic sensitivity testing can determine which treatments are effective in vitro. Other markers allow assessment of end product biochemical status from the total bioconversion activity of the microbial mass.

Bacterial Culture of Stool

Reduced numbers of beneficial gut bacteria, particularly *Bifidobacteria* and *Lactobacilli* species, leave the intestine susceptible to invasion by pathogens and the production of carcinogens. Quantitative bacteriological culture of stool is a useful tool for finding imbalances in colonic flora. Cultures can reveal the proportion of beneficial *Bifidobacteria, Lactobacilli,* and *E. coli* populations in relation to potential pathogens. Sero-typing for toxigenic *E. coli*, and *Campylobacter* cultures is useful for diarrheic specimens. Testing of susceptibilities of bacterial colonies to a range of common antibacterial agents is commonly performed when pathogens are found. The antibiotic with the highest potency at the lowest dose is the best treatment.

Mycological Culture of Stool

The presence of fecal yeast is revealed by a mycology culture. *C. albicans, C. tropicalis, Rhodotorula,* and *Geotrichum* are some of the more commonly identified species. A differential culture in the presence of antimycotic agents is particularly useful for identifying agents that will be effective against the colonic yeast overgrowth.

Fecal β-Glucuronidase

Bacterial β-glucuronidase is an enzyme that can effectively reverse detoxification that has taken place in the liver during the Phase II conjugation reactions (see Chapter 8). Bacterial flora may express large amounts of glycosidase enzyme activity, the principal glycosidase being β-glucuronidase.

Glycosides are compounds containing a nonsugar molecule (aglycone) attached to a sugar derivative like glucuronic acid by α- or β-glycoside linkage. Glycosides enter the GI tract through dietary intake or from the liver through bile secretions. Most dietary glycosides, predominately flavonoids, come from vegetables. Glycosides coming from the liver include toxic compounds that are inactivated by beta glucuronide formation and subsequently secreted into the bowel by way of the bile. The intestinal flora can hydrolyze the beta-glucuronide bond leading to release of the toxic compound, which may be carcinogenic. Therefore, excess β-glucuronidase activity in the small bowel is associated with increased risk of cancer, including estrogen-related cancers. The cleavage of glucuronide from estrogen metabolites leads to their increased enterohepatic recirculation [142-146].

Fecal pH

The colonic microbial mass converts food components into organic acids and amines according to the nature of the substrate passing into the colon and the type of organisms that predominate. A high rate of organic acid production gives acidic stools, while low acid formation with increased amines results in higher pH of stool. One of the most important colonic health practices is regular intake of high fiber foods so that beneficial organisms in the colon receive substrate for the production of favorable products such as butyric acid. Direct measurement of fecal pH provides an overall indicator of acid and base balance.

Abnormally acidic or alkaline pH of the stool may be an indicator of poor digestive health. There is increasing evidence that fecal pH can serve as a marker for colon cancer [64, 147-149]. High fecal pH, however, is only indirectly associated with the development of colon cancer and therefore is a secondary, rather than a primary, measure of cancer risk [150]. High stool pH appears to correlate with low levels of short-chain fatty acids (especially butyric acid) [151-154]. Alkalinity and low butyric acid levels in the stool appear to signal inadequate intake or digestion of fiber.

Fecal Short Chain Fatty Acids (SCFAs)

Frequently, disorders of the GI tract are associated with intestinal malabsorption. Many tests are useful in the diagnosis of GI disorders but do not effectively differentiate abnormalities of absorption (mucosal function) from digestion (pancreatic function). The fecal short chain fatty acid test can be helpful in making a differential diagnosis. Short-chain fatty acids (SCFAs) are formed from bacterial fermentation of dietary carbohydrates and amino acids that escape absorption in the small bowel. An alteration in the proportion of various SCFAs, which stay constant in healthy colons, signifies an impaired state of colonic health [155].

Short-chain fatty acids are well absorbed by the colon and are a significant source of energy for colonic cells. The short chain fatty acid n-butyrate plays a notable role for maintaining colonic health – a reduction in n-butyrate oxidation by colonic epithelial cells may be a key factor in the pathogenesis of ulcerative colitis [156, 157]. High acetic acid and low butyric acid in relation to total SCFAs in the feces are found in patients with large bowel adenomas and cancer [158]. The fecal content of butyric acid — formed by the bacterial fermentation of fiber — is particularly critical for colonic health since it is such an important source of energy for the epithelial cells of the colon [49]. Research suggests that inadequate amounts of colonic butyric acid could be a primary factor in the etiology of inflammatory bowel disease, ulcerative colitis, and colon cancer [159]. Therefore, low butyrate indicates an absorptive problem in the colon due to failure to maintain the epithelial lining.

Certain products of bacterial fermentation (isobutyric acid, valeric acid, and isovaleric acid) come principally from undigested protein [160, 161]. These SCFAs are normally present at low concentrations in the healthy colon. However, maldigestion of protein due to pancreatic enzyme insufficiency can result in excess protein entering the colon. In these cases, fecal isobutyric acid, valeric acid, and isovaleric acid will be elevated. Their presence is more likely due to a pancreatic dysfunction rather than an inadequacy of mucosal absorption.

Intestinal Wellness Options

The goal of treatment is to establish intestinal microbial balance (orthobiosis) and healthy mucosal epithelium. Many naturally occurring substances help repair the gut or support the liver when stressed by enteric toxins. Nutritional support includes supplementation with vitamins A, C, E, B_1 through B_{12}, amino acids and peptides such as glutamine and glutathione, and minerals such as zinc, selenium, chromium, molybdenum, manganese, and magnesium. Removal of pathogenic overgrowth may be hastened by the use of the wide array of antibiotics. Bactericidal effects should be as specific as possible to avoid the long-term harmful effects of inducing large shifts in microbial populations.

Multiple protocols are available to avoid recurrent overgrowth and to restore mucosal integrity. A protocol known as the "4R program" has been widely applied with success. The general concepts involved in this protocol are outlined in Table 7-6, and additional details are provided in Table 7-7.

Generally, the first order of implementation is restoration of digestive function with betaine HCL, pancreatin, and bile acids as needed to replace the normal output of stomach acid, pancreatic fluid, and bile. There is growing evidence of the efficacy of re-inoculation with favorable species of *Lactobacilli* [162], especially with the introduction of antibiotic resistant strains [163]. The consumption of non-absorbed, fermentable carbohydrates, like fructo-oligosaccharides and inulin, stimulate growth of the genera *Bifidobacteria* and *Lactobacillus* [164]. These forms of carbohydrate are found in onion, garlic, chicory, Jerusalem artichoke, and wheat.

Insoluble fiber lowers yeast, *Clostridia*, *Staphylococcus*, and *Proteus* in stool cultures and lowers output of ammonia and phenols [165]. Both butyrate enemas and increased dietary soluble fiber have shown potent anti-inflammatory effects both *in vivo* and *in vitro* [166]. Shifting from a conventional diet to an uncooked vegan diet has been shown to reduce urinary p-cresol due to a decrease in bacterial enzymes [167]. Small intestinal mucosal regeneration can be aided by intake of supplemental glutamine, the primary energy source for this tissue. The colon mucosa utilizes butyrate, which can be supplied as

TABLE 7-6. The Four "R" Program

"R"	Object	Amplification
Remove	Microbial overgrowth	Use bacteriostatic or mycostatic agents of sufficient strength to reduce excessive growth rates
Replace	Digestive factors	Support insufficient digestive secretory factors with oral replacements
Reinoculate	Favorable microbes	Employ oral dosing of viable organisms known to help control toxin-producing species
Repair	Tissue and immune system	Support the growth of healthy intestinal mucosal cells, goblet cells, and immune cell responses

an enema. Restoration of immune function is aided by deglycyrrhizinated licorice, a stimulant of goblet cell formation [168]. Essential fatty acid balance is needed to assure adequate prostaglandin formation. Deficiencies of essential fatty acids impair the recovery of intestinal mucosa from injury [169].

The patient should be counseled to avoid NSAID drugs, alcohol, and other enterotoxic substances. Stress has observable effects on intestinal microbiota. Release of ACTH from fear and anger leads to increased jejunal *E. coli*, loss of *Bifidobacteria* and *Lactobacillus* from fecal samples, and increased levels of the pathogenic *Bacteroides fragilis*. Eating balanced, nutrient-dense meals that provide appropriate amounts of vegetable fiber should become the mainstay of the patient's dietary habits. Patients with food allergies or sensitivities will need to learn how to choose a diet low in the offending foods. Specialized treatments for dysbiosis and "leaky gut" utilizing herbal and probiotic therapies are listed in Table 7-7.

Summary

Evaluation of gastrointestinal function includes detection of inadequate physical and immune barrier functions and measures of the digestion and absorption of food. Pathogenic overgrowth of intestinal microbes in the upper gastrointestinal tract can be detected by measuring their unique products in urine. The patterns reflect the type of organisms that are present within broad categories of bacteria, protozoa, or yeast. The information allows discrimination between putrefactive dysbiosis in the colon versus fermentative dysbiosis in the small intestine or combinations of both. The number of compounds involved and the degree of elevations found change in direct proportion to the severity of pathogenic overgrowth and loss of mucosal integrity. Stool profiling yields markers of digestive function and results direct obervation of microbial populations.

TABLE 7-7. Intestinal Wellness Options

Focus	Category	Agents
Digestive Function	Stomach	Betaine hydrochloride
	Pancreas	Pancreatic enzymes
	Hepaticobiliary	Ox bile, Taurine, Olive oil
		Avoid: Antacids, H2-blockers, Proton pump-inhibitors Cholecystectomy Large, rushed meals high in protein and fat
Mucosal Integrity	Nutrients supporting cell growth & protection	Zinc, Vitamins A, C, and E, Folic acid, Pantothenic acid, Oligopeptide mixtures, Free form amino acids
	Energy substrate	Small Intestine: Glutamine Large Intestine: Soluble and insoluble fiber*, Butyrate retention enemas
	Cell membrane	Essential fatty acids
		Avoid: NSAIDs
Intestinal Dysbiosis	Prebiotics	
	Bacteria	*L. acidophilus, L. salivarius, L. plantarum, and L. casei* *Bifidobacteria*
	Yeasts	*Saccharomyces boulardii*
	Probotics	*Soluble and insoluble dietary fiber** Fructo-oligosaccharides, Inulin
	Bacteriostatic agents	
	Bacteria – Severe	Amoxycilin, clavulinic acid, etc.
	Bacteria – Moderate	Berberine-containing herbals such as Goldenseal Citrus seed extract, Olive leaf extract, Aloe vera, Garlic, Glycerriza (licorice)
	Yeast – Severe	Nystatin
	Yeast – Moderate	Capric and undexylenic acids Avoid simple sugars

* Whole grains, beans and peas, fresh fruits, and vegetables

References

[1] Abbas, A., A. Lichtman, and J. Pober, *Cellular and molecular immunology*. 1997, Philadelphia: W.B. Saunders. 494.

[2] Choi, K.D., H.S. Lillehoj, and D.S. Zalenga, Changes in local IFN-gamma and TGF-beta4 mRNA expression and intraepithelial lymphocytes following Eimeria acervulina infection. *Vet Immunol Immunopathol,* 71(3-4). 263-75, 1999.

[3] Rubio, C.A., B. Jacobsson, and E. Castanos-Velez, Cytotoxic intraepithelial lymphocytes in colorectal polyps and carcinomas. *Anticancer Res,* 19(4B). 3221-7, 1999.

[4] Oehling, A., *et al.*, Skin manifestations and immunological parameters in childhood food allergy. *J Investig Allergol Clin Immunol,* 7(3). 155-9, 1997.

[5] Geissler, A., *et al.*, Focal white-matter lesions in brain of patients with inflammatory bowel disease [see comments]. *Lancet,* 345(8954). 897-8, 1995.

[6] Hadjivassiliou, M., *et al.*, Clinical, radiological, neurophysiological, and neuropathological characteristics of gluten ataxia. *Lancet,* 352(9140). 1582-5, 1998.

[7] Andre, C., Food allergy. Objective diagnosis and test of therapeutic efficacy by measuring intestinal permeability [Translated]. *Presse Med,* 15(3). 105-8, 1986.

[8] Andre, C., F. Andre, and L. Colin, Effect of allergen ingestion challenge with and without cromoglycate cover on intestinal permeability in atopic dermatitis, urticaria and other symptoms of food allergy. *Allergy,* 44(Suppl 9). 47-51, 1989.

[9] Van Elburg, R.M., *et al.*, Intestinal permeability in pediatric gastroenterology. *Scand J Gastroenterol Suppl,* 194. 19-24, 1992.

[10] Halpern, G.M. and J.R. Scott, Non-IgE antibody mediated mechanisms in food allergy. *Ann Allergy,* 58(1). 14-27, 1987.

[11] Yamada, H., T. Hashimoto, and T. Nishikawa, IgG subclasses of intercellular and basement membrane zone antibodies: the relationship to the capability of complement fixation. *J Invest Dermatol,* 92(4). 585-7, 1989.

[12] Philpott, W., Maladaptive reactions to frequently used foods and commonly met chemicals as precipitating factors in many chronic physical and chronic emotional illnesses, in *A physician's handbook on orthomolecular medicine,* R. Williams and D. Kalita, Editors. 1977, Pergamon Press: New York. p. 140-150.

[13] Davies, D.R. and A.E. Clark, Demonstration of immunoglobulin containing deposits in glomerular basement membrane in experimental chronic serum sickness using horseradish peroxidase labelled antiserum. *Br J Exp Pathol,* 56(1). 28-33, 1975.

[14] Bielory, L., *et al.*, Human serum sickness: a prospective analysis of 35 patients treated with equine anti-thymocyte globulin for bone marrow failure. *Medicine* (Baltimore), 67(1). 40-57, 1988.

[15] Jones, B.F., P.R. Trevillian, and R.S. Nanra, Serum sickness due to Rh(anti-D) immunoglobulin. Aust N Z J Obstet Gynaecol, 24(1). 49-50, 1984.

[16] Marinkovich, V., Specific IgG antibodies as markers of adverse reactions to foods. *Monogr Allergy,* 32. 221-5, 1996.

[17] Dixon, H.S., Treatment of delayed food allergy based on specific immunoglobulin G RAST testing. *Otolaryngol Head Neck Surg,* 123(1 Pt 1). 48-54, 2000.

[18] Salerno, F., R. Abbiati, and F. Fici, Effect of pyridoxine alha-ketoglutarate (PAK) on ammonia and pyruvic and lactic acid blood levels in patients with cirrhosis. *Int J Clin Pharmacol Res,* 3(1). 21-5, 1983.

[19] Yoshida, S., *et al.*, Effects of glutamine supplements and radiochemotherapy on systemic immune and gut barrier function in patients with advanced esophageal cancer. *Ann Surg,* 227(4). 485-91, 1998.

[20] Brandtzaeg, P., *et al.*, Production and secretion of immunoglobulins in the gastrointestinal tract. *Ann Allergy,* 59(5 Pt 2). 21-39, 1987.

[21] Visakorpi, J.K., The immune response of the intestinal mucosa to foreign proteins. *Acta Paediatr Scand Suppl,* 296. 56-9, 1982.

[22] Schreiber, R.A. and W.A. Walker, Food allergy: facts and fiction. *Mayo Clin Proc,* 64(11). 1381-91, 1989.

[23] Cunningham-Rundles, C., Analysis of the gastrointestinal secretory immune barrier in IgA deficiency. *Ann Allergy,* 57(1). 31-5, 1986.

[24] Externest, D., *et al.*, Correlations between antibody immune responses at different mucosal effector sites are controlled by antigen type and dosage. *Infection and Immunity,* 68(7). 3830-9, 2000.

[25] Coppo, R., The pathogenetic potential of environmental antigens in IgA nephropathy. *Am J Kidney Dis,* 12(5). 420-4, 1988.

[26] Nagao, A.T., *et al.*, Measurement of salivary, urinary and fecal secretory IgA levels in children with partial or total IgA deficiency. *J Investig Allergol Clin Immunol,* 4(5). 234-7., 1994.

[27] Bernstein, D.I., J.M. Ziegler, and R.L. Ward, Rotavirus fecal IgA antibody response in adults challenged with human rotavirus. *J Med Virol,* 20(4). 297-304., 1986.

[28] Yoshizawa, H., *et al.*, Diagnosis of type A hepatitis by fecal IgA antibody against hepatitis A antigen. *Gastroenterology*, 78(1). 114-8., 1980.

[29] DeWitt, R.C. and K.A. Kudsk, The gut's role in metabolism, mucosal barrier function, and gut immunology. *Infect Dis Clin North Am*, 13(2). 465-81, x, 1999.

[30] Fink, M.P., Leaky gut hypothesis: a historical perspective [editorial]. *Crit Care Med*, 18(5). 579-80, 1990.

[31] Galland, L. Fire in the belly: update on gut fermentation. in Am. Coll. Adv. Med. 1996.

[32] Anonymous, The leaky gut of alcoholism. *Nutr Rev*, 43(3). 72-4, 1985.

[33] Batt, R.M., *et al.*, Small intestinal bacterial overgrowth and enhanced intestinal permeability in healthy beagles. *Am J Vet Res*, 53(10). 1935-40, 1992.

[34] Riordan, S.M., *et al.*, Small intestinal bacterial overgrowth in the symptomatic elderly. *Am J Gastroenterol*, 92(1). 47-51, 1997.

[35] Pignata, C., *et al.*, Jejunal bacterial overgrowth and intestinal permeability in children with immunodeficiency syndromes. *Gut*, 31(8). 879-82, 1990.

[36] Serrander, R., K.E. Magnusson, and T. Sundqvist, Acute infections with Giardia lamblia and rotavirus decrease intestinal permeability to low-molecular weight polyethylene glycols (PEG 400). *Scand J Infect Dis*, 16(4). 339-44, 1984.

[37] Lifschitz, C.H. and D.H. Mahoney, Low-dose methotrexate-induced changes in intestinal permeability determined by polyethylene glycol polymers. *J Pediatr Gastroenterol Nutr*, 9(3). 301-6, 1989.

[38] Grisham, M.B., *et al.*, Effects of neutrophil-derived oxidants on intestinal permeability, electrolyte transport, and epithelial cell viability. *Inflammation*, 14(5). 531-42, 1990.

[39] Sundstrom, G.M., *et al.*, Intestinal permeability in patients with acute myeloid leukemia. *Eur J Haematol*, 61(4). 250-4, 1998.

[40] Roediger, W.E. and W. Babidge, Human colonocyte detoxification. *Gut*, 41(6). 731-4, 1997.

[41] Braganza, J.M., Pancreatic disease: a casualty of hepatic "detoxification"? *Lancet*, 2(8357). 1000-3, 1983.

[42] Braganza, J.M., *et al.*, Lipid-peroxidation (free-radical-oxidation) products in bile from patients with pancreatic disease. *Lancet*, 2(8346). 375-9, 1983.

[43] Walker, W.A., *et al.*, Intestinal uptake of macromolecules. III. Studies on the mechanism by which immunization interferes with antigen uptake. *J Immunol*, 115(3). 854-61, 1975.

[44] Walker, W.A., *et al.*, Intestinal uptake of macromolecules. IV.--The effect of pancreatic duct ligation on the breakdown of antigen and antigen-antibody complexes on the intestinal surface. *Gastroenterology*, 69(6). 1223-9, 1975.

[45] Walker, W.A., Antigen absorption from the small intestine and gastrointestinal disease. *Pediatr Clin North Am*, 22(4). 731-46, 1975.

[46] Galland, L., Leaky gut syndromes: breaking the vicious cycle, in *Townsend Letter for Doctors*. 1995. p. 62-67.

[47] Andre, C., *et al.*, Measurement of intestinal permeability to mannitol and lactulose as a means of diagnosing food allergy and evaluating therapeutic effectiveness of disodium cromoglycate. *Ann Allergy*, 59(5 Pt 2). 127-30, 1987.

[48] Wyatt, J., *et al.*, Intestinal permeability and the prediction of relapse in Crohn's disease. *Lancet*, 341(8858). 1437-9, 1993.

[49] McCullough, J.S., *et al.*, Dietary fibre and intestinal microflora: effects on intestinal morphometry and crypt branching. *Gut*, 42(6). 799-806, 1998.

[50] Goldin, B.R., A.H. Lichtenstein, and S.L. Gorbach, Nutritional and metabolic roles of intestinal flora, in *Modern Nutrition in Health and Disease*, M.E. Shils, J.A. Olson, and M. Shike, Editors. 1994, Lea & Febiger: Philadelphia. p. 569-582.

[51] Hentges, D.J., Does diet influence human fecal microflora composition? *Nutr Rev,* 38(10). 329-36, 1980.

[52] Foster, A.S.J., *et al.*, eds. *Intestinal metabolism of xenobiotics*. 1989, Gustav-Fischer Verlag: Stuttgart.

[53] Peters, W.H. and P.G. Kremers, Cytochromes P-450 in the intestinal mucosa of man. *Biochem Pharmacol,* 38(9). 1535-8, 1989.

[54] Hunter, J.O., Food allergy--or enterometabolic disorder? *Lancet*, 338(8765). 495-6, 1991.

[55] Finegold, S.M., H.R. Attebery, and V.L. Sutter, Effect of diet on human fecal flora: comparison of Japanese and American diets. *Am J Clin Nutr*, 27(12). 1456-69, 1974.

[56] Moore, W.E. and L.V. Holdeman, Human fecal flora: the normal flora of 20 Japanese-Hawaiians. *Appl Microbiol*, 27(5). 961-79, 1974.

[57] Galland, L. and S. Barrie, Intestinal dysbiosis and the causes of diseases. *J Advancement Med*, 6. 67-82, 1993.

[58] Draser, B.S. and M.J. Hill, *Human Intestinal Flora.* 1974, New York: Academic Press.

[59] Sack, R.B., Current treatment of infectious diarrhea. *Infect Med*, 13(4). 301-313, 1996.

[60] Shaw, W., E. Kassen, and E. Chaves, Increased urinary excretion of analogs of Krebs cycle metabolites and arabinose in two brothers with autistic features. *Clin Chem*, 41(8 Pt 1). 1094-1104, 1995.

[61] Keusch, G.T. and N.W. Solomons,

Microorganisms, malabsorption, diarrhea and dysnutrition. *J Environ Pathol Toxicol Oncol*, 5(6). 165-209, 1985.

[62] Petrov, R.V. and T.I. Ulyankina, The genius of E. E. Metchnikoff--discoveries over the centuries. *Biosci Rep*, 16(2). 189-205, 1996.

[63] Conway, P., Microbial ecology of the human large intestine, in *Human colonic bacteria: Role in nutrition, physiology, and pathology*, G. Gibson and G. Macfarlane, Editors. 1995, CRC Press: Boca Raton. p. 292.

[64] Malhotra, S.L., Faecal urobilinogen levels and pH of stools in population groups with different incidence of cancer of the colon, and their possible role in its aetiology. *J R Soc Med*, 75(9). 709-14, 1982.

[65] Brown, J.P., Role of gut bacterial flora in nutrition and health: a review of recent advances in bacteriological techniques, metabolism, and factors affecting flora composition. *CRC Crit Rev Food Sci Nutr*, 8(3). 229-336, 1977.

[66] Chung, K.T., G.E. Fulk, and M.W. Slein, Tryptophanase of fecal flora as a possible factor in the etiology of colon cancer. *J Natl Cancer Inst*, 54(5). 1073-8, 1975.

[67] Goldin, B.R., The metabolism of the intestinal microflora and its relationship to dietary fat, colon and breast cancer, in Dietary Fat and Cancer. 1986, Alan R. Liss: New York. p. 655-685.

[68] Hill, M.J., *et al.*, Faecal bile acids, dysplasia, and carcinoma in ulcerative colitis. *Lancet*, 2(8552). 185-6, 1987.

[69] Bennet, J.D., Ulcerative colitis: the result of an altered bacterial metabolism of bile acids or cholesterol. *Med Hypotheses*, 20(2). 125-32, 1986.

[70] Tamm, A.O., Biochemical activity of intestinal microflora in adult coeliac disease. *Nahrung*, 28(6-7). 711-5, 1984.

[71] Powell-Jackson, P.R., *et al.*, Intestinal bacterial metabolism of protein and bile acids: role in pathogenesis of hepatic disease after jejuno-ileal bypass surgery. *Br J Surg*, 66(11). 772-5, 1979.

[72] Yoshida, K. and C. Hirayama, Tryptophan metabolism in liver cirrhosis: influence of oral antibiotics on neuropsychiatric symptoms. *Tohoku J Exp Med*, 142(1). 35-41, 1984.

[73] Tohyama, K., *et al.*, Effect of lactobacilli on urinary indican excretion in gnotobiotic rats and in man. *Microbiol Immunol*, 25(2). 101-12, 1981.

[74] Miloszewski, K., *et al.*, Increase in urinary indican excretion in pancreatic steatorrhoea following replacement therapy. *Scand J Gastroenterol*, 10(5). 481-5, 1975.

[75] Lawrie, C.A., A.G. Renwick, and J. Sims, The urinary excretion of bacterial amino-acid metabolites by rats fed saccharin in the diet. *Food Chem Toxicol*, 23(4-5). 445-50, 1985.

[76] Kirkland, J.L., E. Vargas, and M. Lye, Indican excretion in the elderly. *Postgrad Med J*, 59(697). 717-9, 1983.

[77] Smith, D.F., Effects of age on serum tryptophan and urine indican in adults given a tryptophan load test. *Eur J Drug Metab Pharmacokinet*, 7(1). 55-8, 1982.

[78] Aarbakke, J. and H. Schjonsby, Value of urinary simple phenol and indican determinations in the diagnosis of the stagnant loop syndrome. *Scand J Gastroenterol*, 11(4). 409-14, 1976.

[79] Mayer, P. and W. Beeken, The role of urinary indican as a predictor of bacterial colonization in the human jejunum. *Am J Dig Dis*, 20(11). 1003 - 9, 1975.

[80] Patney, N.L., *et al.*, Urinary indican excretion in cirrhosis of liver. *J Assoc Physicians India*, 24(5). 291-5, 1976.

[81] Montgomery, R.D., *et al.*, The ageing gut: a study of intestinal absorption in relation to nutrition in the elderly. *Q J Med*, 47(186). 197-24, 1978.

[82] Chalmers, R.A., H.B. Valman, and M.M. Liberman, Measurement of 4-hydroxyphenylacetic aciduria as a screening test for small-bowel disease. *Clin Chem*, 25(10). 1791-4, 1979.

[83] Geypens, B., *et al.*, Influence of dietary protein supplements on the formation of bacterial metabolites in the colon. *Gut*, 41(1). 70-6, 1997.

[84] Potempska, A., Y.H. Loo, and H.M. Wisniewski, On the possible mechanism of phenylacetate neurotoxicity: inhibition of choline acetyltransferase by phenylacetyl-CoA. *J Neurochem*, 42(5). 1499-501, 1984.

[85] Michals, K. and R. Matalon, Phenylalanine metabolites, attention span and hyperactivity. *Am J Clin Nutr*, 42(2). 361-5, 1985.

[86] Van der Heiden C, W.E., Ketting D, Duran M, Wadman, K., Gas chromatographic analysis of urinary tyrosine and phenylalanine metabolites in patients with gastrointestinal disorders. *Clinical Chimia Acta*, 34. 289-296, 1971.

[87] McDevitt, J. and P. Goldman, Effect of the intestinal flora on the urinary organic acid profile of rats ingesting a chemically simplified diet. *Food Chem Toxicol*, 29(2). 107-13, 1991.

[88] Schwartz, R., M. Topley, and J.B. Russell, Effect of tricarballylic acid, a nonmetabolizable rumen fermentation product of trans-aconitic acid, on Mg, Ca and Zn utilization of rats. *J Nutr*, 118(2). 183-8, 1988.

[89] Shaw, W. Experience with organic acid testing to evaluate abnormal microbial metabolites in the urine of children with autism. *Conference of American College for the Advancement of Medicine*. 1997. Orlando, Florida.

[90] Elsden, S.R., M.G. Hilton, and J.M. Waller, The end products of the metabolism of aromatic amino acids by Clostridia. *Arch Microbiol*, 107(3). 283-8, 1976.

[91] Rowland, I.R., Toxicology of the colon: Role of the

intestinal microflora, in *Human Colonic Bacteria: Role in Nutrition, Physiology, and Pathology*, G. Gibson and G. MacFarlane, Editors. 1995, CRC Press: Boca Raton. p. 292.

[92] Rhodes, J.M., P. Middleton, and D.P. Jewell, The lactulose hydrogen breath test as a diagnostic test for small-bowel bacterial overgrowth. *Scand J Gastroenterol*, 14(3). 333-6, 1979.

[93] Corazza, G.R., *et al.*, The diagnosis of small bowel bacterial overgrowth. Reliability of jejunal culture and inadequacy of breath hydrogen testing. *Gastroenterology*, 98(2). 302-9, 1990.

[94] Riordan, S.M., *et al.*, The lactulose breath hydrogen test and small intestinal bacterial overgrowth. *Am J Gastroenterol*, 91(9). 1795-803, 1996.

[95] Hamilton, L.H., *Breath testing and gastroenterology*. 1992, QuinTron Division, The EF BrewPer Co.: Menomonee Falls, WI.

[96] Saltzman, J.R., *et al.*, Bacterial overgrowth without clinical malabsorption in elderly hypochlorhydric subjects. *Gastroenterology*, 106(3). 615-23, 1994.

[97] Valdovinos, M.A., *et al.*, Reduced accuracy of 14C-D-xylose breath test for detecting bacterial overgrowth in gastrointestinal motility disorders. *Scand J Gastroenterol*, 28(11). 963-8, 1993.

[98] Diebel, L.N., *et al.*, Synergistic effects of Candida and Escherichia coli on gut barrier function. *J Trauma*, 47(6). 1045-50; discussion 1050-1, 1999.

[99] Lichtman, S.N., *et al.*, Hepatic injury associated with small bowel bacterial overgrowth in rats is prevented by metronidazole and tetracycline. *Gastroenterology*, 100(2). 513-9, 1991.

[100] Bode, J.C., S. Rust, and C. Bode, The effect of cimetidine treatment on ethanol formation in the human stomach. *Scand J Gastroenterol*, 19(6). 853-6, 1984.

[101] Hunnisett, A., J. Howard, and S. Davies, Gut fermentation (or the 'Auto-brewery') syndrome: a new clinical test with initial observations and discussion of clinical and biochemical implications. *J Nut Med*, 1. 33-38, 1990.

[102] Thurn, J.R., *et al.*, D-lactate encephalopathy. *Am J Med*, 79(6). 717-21, 1985.

[103] Robertson, B. and L. Lonnell, Human tartrate nephropathy. Report of a fatal case. *Acta Pathol Microbiol Scand*, 74(3). 305-10, 1968.

[104] Sell, D.R. and V.M. Monnier, Structure elucidation of a senescence cross-link from human extracellular matrix. Implication of pentoses in the aging process. *J Biol Chem*, 264(36). 21597-602, 1989.

[105] Tokunaga, S., *et al.*, Clinical significance of measurement of serum D-arabinitol levels in candiduria patients. *Urol Int*, 48(2). 195-9, 1992.

[106] Roboz, J., Diagnosis and monitoring of disseminated candidiasis based on serum/urine D/L-arabinitol ratios. *Chirality*, 6(2). 51-7, 1994.

[107] Horowitz, B.J., S.W. Edelstein, and L. Lippman, Sugar chromatography studies in recurrent Candida vulvovaginitis. *J Reprod Med*, 29(7). 441-3, 1984.

[108] Perlman, S. and S.A. Carr, Citramalic acid in cerebrospinal fluid of patients with bacterial meningitis. *Clin Chem*, 30(7). 1209-12, 1984.

[109] Marklein, G., H.P. Weil, and K. Rommelsheim, Laboratory diagnostic possibilities in fungus infections in intensive. *Anasth Intensivther Notfallmed*, 24(3). 172-6, 1989.

[110] Crook, W., *The Yeast Connection*. 1983, Jackson: Professional Books. 280.

[111] Mendis, L., *et al.*, Bacteria, viruses, yeasts and protozoans associated with diarrheal disease in Singapore. *Pathology*, 27. 48-52, 1995.

[112] Ampel, N.M., Emerging disease issues and fungal pathogens associated with HIV infection. *Emerg Infect Dis*, 2(2). 109-16, 1996.

[113] Russell, R.M., Implications of gastric atrophy for vitamin and mineral nutriture, in *Nutrition and Aging*, M.L. Hutchinson and H.N. Munro, Editors. 1986, Academic Press: New York. p. 59.

[114] Russell, R.M., Gastric hypochlorhydria and achlorhydria in older adults [letter]. *Jama*, 278(20). 1659-60, 1997.

[115] Giannella, R.A., S.A. Broitman, and N. Zamcheck, Gastric acid barrier to ingested microorganisms in man: studies in vivo and in vitro. *Gut*, 13(4). 251-6, 1972.

[116] Neal, K.R., *et al.*, Omeprazole as a risk factor for campylobacter gastroenteritis: case-control study. *Bmj*, 312(7028). 414-5, 1996.

[117] Sturniolo, G.C., *et al.*, Inhibition of gastric acid secretion reduces zinc absorption in man. *J Am Coll Nutr*, 10(4). 372-5, 1991.

[118] Ivan, M., P. Jui, and Hidiroglou, The effects of nitrilotriacetic acid on solubilities of zinc, copper, manganese, and iron in the stomach of sheep. *Can J Physiol Pharmacol*, 57(4). 369-74, 1979.

[119] Heaney, R.P., *et al.*, Calcium absorptive consistency. *J Bone Miner Res*, 5(11). 1139-42, 1990.

[120] Wright, J., A proposal for standardized challenge testing of gastric acid secretory capacity using the Heidelberg capsule radiotelemetry system. *J John Bastyr Col Nat Med*, 1(2). 3-11, 1979.

[121] Henry, J.P. and W.M. Steinberg, Pancreatic function tests in the rat model of chronic pancreatic insufficiency. *Pancreas*, 8(5). 622-6, 1993.

[122] Kataoka, K., *et al.*, Diagnosis of chronic pancreatitis using noninvasive tests of exocrine pancreatic function--comparison to duodenal intubation tests. *Pancreas*, 15(4). 409-15, 1997.

[123] Gagee, P., *et al.*, The BT-PABA/PAS test in tropical diabetes. *Clin Chim Acta*, 212(3). 103-11, 1992.

[124] Hatakeyama, K. and S. Koyama, Differential

diagnosis of maldigestion and malabsorption of fat. II. Comparison of 131I-triolein with 14C-triolein in normal, pancreatic juice-deficient, short bowel and bile-deficient dogs. *Jpn J Surg*, 11(4). 272-6, 1981.

[125] Thorsgaard Pedersen, N. and H. Halgreen, Simultaneous assessment of fat maldigestion and fat malabsorption by a double-isotope method using fecal radioactivity. *Gastroenterology*, 88(1 Pt 1). 47-54, 1985.

[126] Oh, S.Y. and P.A. Monaco, Effect of dietary cholesterol and degree of fat unsaturation on plasma lipid levels, lipoprotein composition, and fecal steroid excretion in normal young adult men. *Am J Clin Nutr*, 42(3). 399-413, 1985.

[127] Hill, P., B.S. Reddy, and E.L. Wynder, Effect of unsaturated fats and cholesterol on serum and fecal lipids. A study of healthy middle-aged men. *J Am Diet Assoc*, 75(4). 414-20, 1979.

[128] Kesaniemi, Y.A., S. Tarpila, and T.A. Miettinen, Low vs high dietary fiber and serum, biliary, and fecal lipids in middle-aged men. *Am J Clin Nutr*, 51(6). 1007-12, 1990.

[129] Schweizer, T.F., *et al.*, Metabolic effects of dietary fiber from dehulled soybeans in humans. *Am J Clin Nutr*, 38(1). 1-11, 1983.

[130] Salvioli, G., R. Lugli, and J.M. Pradelli, Cholesterol absorption and sterol balance in normal subjects receiving dietary fiber or ursodeoxycholic acid. *Dig Dis Sci*, 30(4). 301-7, 1985.

[131] Ullrich, I.H., *et al.*, Alterations of fecal steroid composition induced by changes in dietary fiber consumption. *Am J Clin Nutr*, 34(10). 2054-60, 1981.

[132] Lankisch, P.G., Exocrine pancreatic function tests. *Gut*, 23(9). 777-98, 1982.

[133] Moore, J.G., *et al.*, Simple fecal tests of absorption. A prospective study and critique. *Am J Dig Dis*, 16(2). 97-105, 1971.

[134] Van Eldere, J., *et al.*, Isolation and identification of intestinal steroid-desulfating bacteria from rats and humans. *Appl Environ Microbiol*, 54(8). 2112-7, 1988.

[135] Gottschall, E., *Breaking the vicious cycle*. 1994, Ontario: Kirkton Press. 164.

[136] Park, Y.K., M.H. Monaco, and S.M. Donovan, Enteral insulin-like growth factor-I augments intestinal disaccharidase activity in piglets receiving total parenteral nutrition. *J Pediatr Gastroenterol Nutr*, 29(2). 198-206, 1999.

[137] Freeman, H.J., M.H. Sleisenger, and Y.S. Kim, Human protein digestion and absorption: normal mechanisms and protein-energy malnutrition. *Clin Gastroenterol*, 12(2). 357-78, 1983.

[138] Munro, H.N., Second Boyd Orr Memorial Lecture. Regulation of body protein metabolism in relation to diet. *Proc Nutr Soc*, 35(3). 297-308, 1976.

[139] Freeman, H.J., Y.S. Kim, and M.H. Sleisenger, Protein digestion and absorption in man. Normal mechanisms and protein-energy malnutrition. *Am J Med*, 67(6). 1030-6, 1979.

[140] Domstad, P.A., *et al.*, Reliability of the dual-isotope Schilling test for the diagnosis of pernicious anemia or malabsorption syndrome. *Am J Clin Pathol*, 75(5). 723-6, 1981.

[141] Magnus, E. and C. Muller, A new, peroral non-radioactive vitamin B12 absorption test compared with the Schilling test. *Eur J Haematol*, 54(2). 117-9, 1995.

[142] Reddy, B.S., *et al.*, Metabolic epidemiology of large bowel cancer: fecal bulk and constituents of high-risk North American and low-risk Finnish population. *Cancer*, 42(6). 2832-8, 1978.

[143] Johansson, G.K., L. Ottova, and J.A. Gustafsson, Shift from a mixed diet to a lactovegetarian diet: influence on some cancer-associated intestinal bacterial enzyme activities. *Nutr Cancer*, 14(3-4). 239-46, 1990.

[144] Bouhnik, Y., *et al.*, Effects of fructo-oligo-saccharides ingestion on fecal bifidobacteria and selected metabolic indexes of colon carcinogenesis in healthy humans. *Nutr Cancer*, 26(1). 21-9, 1996.

[145] Iwasaki, I., *et al.*, Promoting effects of bile acid to intestinal tumorigenesis in gnotobiotic ICR mice. *Acta Pathol Jpn*, 35(6). 1427-33, 1985.

[146] Domellof, L., *et al.*, Fecal sterols and bacterial beta-glucuronidase activity: a preliminary metabolic epidemiology study of healthy volunteers from Umea, Sweden, and metropolitan New York. *Nutr Cancer*, 4(2). 120-7, 1982.

[147] Kashtan, H., *et al.*, Manipulation of fecal pH by dietary means. *Prev Med*, 19(6). 607-13, 1990.

[148] Walker, A.R., B.F. Walker, and A.J. Walker, Faecal pH, dietary fibre intake, and proneness to colon cancer in four South African populations. *Br J Cancer*, 53(4). 489-95, 1986.

[149] Newmark, H.L. and J.R. Lupton, Determinants and consequences of colonic luminal pH: implications for colon cancer. *Nutr Cancer*, 14(3-4). 161-73, 1990.

[150] Kashtan, H., *et al.*, Effects of sodium sulfate on fecal pH and proliferation of colonic mucosa in patients at high risk for colon cancer. *J Natl Cancer Inst*, 82(11). 950-2, 1990.

[151] Zoran, D.L., *et al.*, Wheat bran diet reduces tumor incidence in a rat model of colon cancer independent of effects on distal luminal butyrate concentrations. *J Nutr*, 127(11). 2217-25, 1997.

[152] Segal, I., *et al.*, Fecal short chain fatty acids in South African urban Africans and whites. *Dis Colon Rectum*, 38(7). 732-4, 1995.

[153] Phillips, J., *et al.*, Effect of resistant starch on fecal bulk and fermentation-dependent events in

humans. *Am J Clin Nutr*, 62(1). 121-30, 1995.

[154] Folino, M., A. McIntyre, and G.P. Young, Dietary fibers differ in their effects on large bowel epithelial proliferation and fecal fermentation-dependent events in rats. *J Nutr*, 125(6). 1521-8, 1995.

[155] Hoverstad, R. The normal microflora and short-chain fatty acids. in *Proceedings of the Fifth Bengt E. Gustafsson Symposium*. 1988. Stockholm.

[156] Moore, J.W., *et al.*, Thiolmethyltransferase activity in the human colonic mucosa: implications for ulcerative colitis. *J Gastroenterol Hepatol*, 12(9-10). 678-84, 1997.

[157] Roediger, W.E., J. Moore, and W. Babidge, Colonic sulfide in pathogenesis and treatment of ulcerative colitis. *Dig Dis Sci*, 42(8). 1571-9, 1997.

[158] Latella, G. and R. Caprilli, Metabolism of large bowel mucosa in health and disease. *Int J Colorectal Dis*, 6(2). 127-32, 1991.

[159] Royall, D., T.M. Wolever, and K.N. Jeejeebhoy, Clinical significance of colonic fermentation. *Am J Gastroenterol*, 85(10). 1307-12, 1990.

[160] Rasmussen, H.S., K. Holtug, and P.B. Mortensen, Degradation of amino acids to short-chain fatty acids in humans. An in vitro study. *Scand J Gastroenterol*, 23(2). 178-82, 1988.

[161] Zarling, E.J. and M.A. Ruchim, Protein origin of the volatile fatty acids isobutyrate and isovalerate in human stool. *J Lab Clin Med*, 109(5). 566-70, 1987.

[162] Ling, W.H., *et al.*, Lactobacillus strain GG supplementation decreases colonic hydrolytic. *J Nutr*, 124(1). 18-23, 1994.

[163] Herman, R.E., Antibiotic resistant strain of lactobacillus acidophilus, in *USPTO*. 1993, Bio Techniques Laboratories, Inc.: US. p. 7.

[164] Hudson, M. and P. Marsh, Carbohydrate metabolism in the colon, in *Human colonic bacteria: Role in nutrition, physiology, and pathology*, G. Gibson and G. Macfarlane, Editors. 1995, CRC Press: Boca Raton. p. 292.

[165] Birkett, A., *et al.*, Resistant starch lowers fecal concentrations of ammonia and phenols in. *Am J Clin Nutr*, 63(5). 766-72, 1996.

[166] Andoh, A., T. Bamba, and M. Sasaki, Physiological and anti-inflammatory roles of dietary fiber and butyrate in intestinal functions. *JPEN J Parenter Enteral Nutr*, 23(5 Suppl). S70-3, 1999.

[167] Ling, W.H. and O. Hanninen, Shifting from a conventional diet to an uncooked vegan diet reversibly alters fecal hydrolytic activities in humans. *J Nutr*, 122(4). 924-30, 1992.

[168] Matsumoto, T., *et al.*, Effect of licorice roots on carrageenan-induced decrease in immune complexes clearance in mice. *J Ethnopharmacol*, 53(1). 1-4, 1996.

[169] Vanderhoof, J.A., *et al.*, Effects of dietary lipids on recovery from mucosal injury. *Gastroenterology*, 98(5 Pt 1). 1226-31, 1990.

Chapter 8

Testing for Toxins and Detoxification

TABLE 8-1. Tests for Toxin Exposure, Removal, and Protection

ASSESSMENT OF EXPOSURE

Test	Result	Meaning	Interventions
Toxic metals – blood, urine, or hair	High	Increased body burden	Reduce exposure; oral and i.v. chelation: protective nutrients
Oral cavity mercury vapor	High	Dental amalgam mercury release	Replace dental fillings; i.v. chelation; nutrient support
Organic toxins – blood	High	Recent exposure	Reduce exsposure; detoxification protocols:
Organic toxins – urinary metabolites	High	Recent exposure	Hepatic detoxification support nutrients
Microbial toxins – urine	High	Intestinal dysbiosis	Restore microbial balance (see Chapters 7 & 8)

ASSESSMENT OF DETOXIFICATION CAPACITY

Test	Result	Meaning	Associations	Interventions
Orotate – urine	High	Urea cycle insufficiency	Arginine deficiency	L-Arginine, α-KG
Citrate – urine	High	Ammonurea	Arginine deficiency	L-Arginine, α-KG
Glucaric acid – urine	High	Phase I & II induction	Stimulation of hepatic oxidation and conjugation	Reduce toxin load and use support nutrients
Caffeine clearance – saliva	Low	Impaired Phase I, Slow P450 pathway	Amphetamines, cimetidine, isoniazid, oral contraceptives	Lipoic acid and other hepatic-protective nutrients
Caffeine clearance – saliva	High	P450 induction	Toxin exposure	Antioxidant protection
Salicyluric acid or acetaminophen mercapturate, sulfate, or glucuronide – urine	Low	Low Phase II	Low cysteine, glutathione, and related nutrients	N-Acetylcysteine Olive oil
Cys/Sulfate ratio	High	Impaired sulfoxidation	Neurological disorders	N-Acetylcysteine, $MgSO_4$ (i.v.)
Cysteine – plasma	High	Impaired sulfoxidation	Excessive cysteine	
Sulfate – plasma or urine	Low	Glutathione and sulfate deficiency	Impaired hepatic phase II	N-Acetylcyteine, $MgSO_4$ (i.v.)

TABLE 8-1. Cont.

Test	Result	Meaning	Associations	Interventions
Bilirubin (indirect) – serum	High	Impaired glucuronidation	Slow drug clearance	Lower drug threshold
Benzoate – urine	High	Impaired glycine conjugation	Glycine depletion	Glycine, Vitamin B_5
Hepatic Phase I/Phase II ratio	High	Increased risk of carcinogenisis	Deficiencies of conjugation cofactors	N-Acetylcysteine, Glycine, Lipoate Vitamin B_5, Antioxidants

ASSESSMENT OF MAJOR TOXIN-REMOVAL ORGANS

Test	Result	Meaning	Associations	Interventions
BUN – serum	High	Renal clearance failure	Impaired excretion	Low protein diet
ALT, AST – serum	High	Liver pathology	Impaired hepatic detoxification	Lipoate, amino acids, B-vitamins, Silymarin, Curcumin

ASSESSMENT OF PROTECTIVE NUTRIENTS

Test	Result	Meaning	Associations	Interventions
Met, Cys, Tau – Plasma	Low	Poor sulfur availability	Glutathione and sulfate depletion	Amino acids, N-Acetylcysteine, Taurine
Pyroglutamate – urine	High	Glutathione depletion	Impaired mercaptan formation	N-Acetylcysteine
Minerals – erythrocyte	Low	Essential element deficiency	Increased heavy metal toxic effects	Mineral supplements and dietary modifications
Antioxidants – serum	Low	Unchecked pro-oxidant effects	Increased oxidative damage	Antioxidant-rich foods and supplements
Metabolic markers – urine	High	B-complex insufficiencies	Loss of energy flow and detoxification enzyme activity	B-complex supplements

Introduction

The ability to monitor some of the major hepatic detoxification reactions adds a valuable clinical tool. Knowing the status of glutathione pathways alone is powerful information for those familiar with the use of strategies for supporting sulfur compound utilization. Detoxification testing can be coupled with measures of heavy metal status and gastrointestinal function to achieve healing by restoring normal biochemical function of cells and tissues.

Although the tests are generally done to detect hidden factors in chronically ill patients, a simple week-long detoxification program has been shown to result in improvements in sense of well-being for disease-free individuals [1]. In this study, a hypoallergic diet supplemented with a broad-spectrum macro- and micro-nutrient drink and intake of at least two quarts of filtered water per day produced significant improvements in objective measures of intestinal permeability and hepatic Phase I and Phase II detoxification capacity. These results serve as reminders of the omnipresent role of toxins in determining disease and wellness.

Terminology and Scope

Terminology in the field of toxicology can be confusing because of the use of duplicate words. For example, the words toxin and toxicant compete for usage while detoxication competes with detoxification. Our preference is to use the shorter, toxin, in the first instance and, in the latter, detoxification, derived from the verb, "detoxify." Historical usage attaches a microbial vector origin to the word, toxin, as in "Cholera toxin." *Toxin* will be used here, however, in a broader context to describe compounds that arise from industrial, microbial, xenobiotic, iatrogenic, or endogenous sources. Examples of each of these are aluminum, aflatoxin, aldrin, acetaminophen, and ammonia, respectively. A complete alphabetical listing of toxins and exposure hazards would require several pages of this text. Tabulations of detailed lists of toxins and levels of release into the environment are available [2-4].

Toxins are chemical agents that produce adverse responses in living things. The last half of the nineteenth century was a time of explosive growth of information detailing the extent of industrial toxin release into the environment and the magnitude of human responses to toxins of any origin [5, 6]. At the same time, our understanding of the mechanisms of human toxin management and removal, known as *detoxification*, multiplied. The scope of this chapter is limited to the description of laboratory evaluations for assessing human toxin exposure and clinical effects related to failure of detoxification.

Human health is impacted by both environmental exposure and endogenous production of toxins. Clinical effects depend on the combination of toxin load and individual capacities to detoxify. Common clinical clues that point to toxin effects are degeneration in hair, skin, nails, mucous membranes, and reproductive function. From these early signs, the list of known toxin-related disorders is very long.

Nutrients vs. Toxins

In the preface to his landmark work on chronic, low level toxic responses sometimes called "chemical sensitivity," Dr. William Rae describes the relationship of nutrients to pollutants: "The final chapter in this volume is on nutrition, which provides fuel for the endocrine, immunologic, and neurological systems and equips the body to respond to pollutant exposure" [7]. Detoxification pathways are metabolically demanding. Nutrients are consumed in numerous toxin-modifying reactions that generally involve the formation of new chemical bonds in energy-requiring steps [8]. Factors that govern the relationship of nutritional status to toxin responses are:

- Biochemical individuality regarding toxin and nutrient responses
- Toxin impairment of energy pathways
- Inefficient nutrient catabolism due to toxins
- Depletion of nutrient pools by detoxification requirements
- Effects of toxins on food quality, digestion, and assimilation
- Direct interference of toxins with nutrient function
- Competition of toxins with nutrients for absorption and utilization
- Toxin impairment of renal retention of nutrients
- Toxin damage to DNA repair mechanisms

The multiple impacts of toxins on nutrient status can create deficiency states. One example of these concepts is the finding that inhibition of energy pathways causes marked reduction in the clearance of acetaminophen [9]. Thus, a patient's response to drugs and environmental toxins depends on their status of nutrients necessary for the operation of central energy pathways. Dietary habits also have direct influence on detoxification pathways. Low protein diets cause increased mortality of laboratory animals from exposure to pesticides, chlorinated hydrocarbons, and organophosphates [10].

As an example of the type of interactions that lead to disease from hepatic detoxification defects, consider the reactions to carbon tetrachloride (CCl_4). Diabetes increases the toxicity of CCl_4 as shown by reduced activities of detoxification pathway enzymes, lowered levels of glutathione, and increased ascorbic acid utilization [11]. Clinical effects will vary depending on the level of exposure to CCl_4, the presence of diabetes, and body status of glutathione and vitamin C.

The laboratory can provide measures of clinical impact of toxins in three ways. Measurements that reflect total body content (body burden) of a toxin or its unique metabolic products can show degrees of exposure. Capacities to process toxins are indicated by challenge tests using relatively harmless substances that are metabolized like toxins by hepatic detoxification pathways. Many of the assessments for nutrient status described in the early chapters of this book have direct bearing on clinical effects of toxins through the protective effects exerted by nutrients. For example, a patient with depleted sulfur amino acids and glutathione status will have diminished capacity for removal of toxic organic compounds through formation of sulfate and mercaptan derivatives. The diverse subjects of this chapter overlap with several others. In order to maintain consistency, some of the information has been included under categories of testing previously presented. Monitoring degrees of exposure to toxic heavy metals is discussed in Chapter 3 – Minerals, and ways of evaluating the gastrointestinal tract as a source of microbial toxins is discussed in Chapter 7 – GI Function.

Sources and Effects of Toxins

Substances that can exert toxic effects in humans abound in nature. Recent activities of humans have engendered thousands of other toxins that would otherwise be very rare. Most people in "advanced" nations no longer scratch out their living from farming and livestock. Countless conveniences have come to be taken for granted. We also are faced with the challenge of living in a world where the "normal" detoxification pathways lack "normal" stimulation and where "abnormal" exposures abound. In this context, "normal" refers to those levels of exposure for pre-industrialized human society. Estimates of routine exposure levels and risks for accidental exposure are under constant revision.

"Xenobiotics" are compounds that are not produced in biological systems. Their occurrence in significant amounts in human tissues is generally due to activities of civilization. The demands of modern society for plastics, alloys, and convenience foods have led to higher levels of exposure and greater variety of toxins. Toxicology, the study of toxic compounds and the reactions that they produce, is a field that has expanded rapidly in the last decade of the twentieth century and now includes specialties of clinical, forensic, and industrial toxicology [12]. Stimulated by public concerns over widespread use of insecticides, especially DDT, the specialty of pesticide toxicology is now providing insight into the range of effects from residues on food [13].

The Federal Clean Water Act currently lists about 300 substances as hazardous industrial discharges and establishes the minimum amount of spillage that must be reported to the National Response Center. The difficulty in controlling small discharges is illustrated by the allowance of companies to release 999 pounds of benzene without reporting [12]. Tens of thousands of pages of regulations are issued annually by the several governmental regulatory agencies that attempt to control population exposures in air, water, and food. Since increased exposure is a given fact of modern life, the clinical impact revolves around abilities to remove toxins from the body.

Humans are protected from the effects of disease-producing substances by physical, chemical, and immunologic barriers. The skin and mucosal surfaces make up the physical barriers. The immunologic barrier is maintained by the secretion of special classes of antibodies, mainly secretory IgA, which bind to large molecules recognized as foreign to the body. The chemical barrier is made up of the gastric secretions of the stomach and the extracellular and intracellular enzyme systems of the tissues.

The gastrointestinal tract is the major source or entry route for toxic substances, and the portal vein delivers all of them (except for lipid soluble compounds) directly to the liver. The ultimate tissue distribution of toxins depends on the portal of entry. The lungs provide a very efficient surface for transfer directly to systemic blood and from there to central nervous tissue or to sites like adipose tissue where accumulation is favored because of slow turnover. Entry through the skin, likewise, is a route that bypasses the liver and favors toxin accumulation in adipose tissue without being modified by detoxification enzymes.

Endogenous toxins and exogenous diet-derived or pharmacological compounds must be converted into non-toxic forms and made water-soluble by enzymatic action (primarily hepatic) in order to have their toxic potential removed. Although detoxification pathways will act upon a wide variety of chemicals, they do not neutralize all foreign compounds. Highly toxic organic mercury and organic phosphorus compounds owe their great potency to the fact that they are both highly reactive and unaffected by the detoxification enzymes of the liver. Cyanide binds so rapidly and strongly to hemoglobin that it also avoids any mechanisms for neutralization. Laboratory evaluations of exposure to compounds in this class of low-molecular weight, volatile toxins are useful, but hepatic detoxification capacity testing may not apply to them.

Testing For Exposures to Toxins

Both toxic metals and small organic molecules may be measured in body fluids and tissues by standard laboratory methods. Various options are available for assessing exposure to toxic heavy metals. Discussion of toxic heavy metals was included in Chapter 3 because elemental screening profiles of blood, urine, or hair frequently include both essential and toxic metals. Measurement of exposure to organic toxins is complicated by migration from body fluids to tissue sites and chemical modifications creating multiple forms of most compounds [14].

Because of the extreme toxicity of methyl mercury and the fact that measurements of concentrations in blood, urine, or hair show total mercury, devices have been developed for the direct measurement of mercury vapors in the oral cavity. Elevated methyl mercury in the mouth is related to the presence of mercury amalgam dental fillings [15]. When high exposures to heavy metals is known or suspected for pregnant women, testing of amniotic fluid may be done to reveal levels of fetal exposure [16].

Testing Detoxification

Detoxification encompasses mechanisms that enhance tolerance, induce mobilization, increase metabolic conversion rates, and induce excretion of toxins. Each of these mechanisms may be addressed by clinical interventions, and each is amenable (to laboratory evaluation) in various degrees. For adverse health consequences from chronic low-level exposures, tolerance is the central issue. Since a toxin-free environment is unrealistic, the enhancement and maintenance of normal detoxification mechanisms through lifestyle modification and use of nutrients is essential. The clinical question is when to use any of the various interventions that are available. Answers to this question for a given patient are provided by laboratory evaluations of detoxification capacities (Table 8-1), total body toxic burden, and toxin exposure (Table 8-2).

TABLE 8-2. Toxin Exposure Testing

Toxic Metals		
	Compound	Test
	Arsenic, Cadmium, Chromium, Cobalt, Lead, Mercury, Nickel	DMSA Challenge 24 hr. Urine, Hair, Whole blood
	Methyl mercury	Oral vapor
Xenobiotics by Direct Concentration Measurement		
	Classes of Compounds Detected	Specimen
	Polyaromatic hydrocarbons (PAHs)	Whole blood or serum
	Benzopyrene	
	Chlorinated hydrocarbons	
	Heptachlor, DDT	
	Polychlorinated biphenyls (PCBs)	
Metabolites Measured in Urine		
	Compound	Origin
	Orotate	Ammonia from impaired urea cycle
	2-Methylhippurate	Xylene exposure
	p-Cresol and other metabolites of amino acids	Products of intestinal bacteria, yeast and fungi
	Tartaric acid and other metabolites of dietary carbohydrates	Products of intestinal yeast and fungi

The question may be broken into one of normal exposure tolerance and another of the ability to handle unusual episodes of high exposure. Answers in both areas are needed for a full picture of a patient's response to toxins. Some toxins accumulate in the body over the course of weeks (or longer) after environmental exposures and the ingestion and processing of food. Others toxins accrue from short-term acute exposures that can vary over intervals of a few seconds to several hours. Episodes of high exposure occur in some occupations such as the life-threatening absorption of methyl mercury solutions that accidentally fall on the skin. Cumulative effects can result from home use of solvents or insecticides [17]. The most outstanding example of endogenous toxin production is the random oxidation effects of free radical oxygen compounds. The damaging compounds are produced as a result of the energy-yielding oxidative pathways in every human cell [18]. Although this may be considered a special form of toxin exposure, the topic of oxidant damage and antioxidant status has been reserved for Chapter 9 – Oxidative Stress.

Another source of chronic exposure requiring detoxification capacity is the flow of compounds from microbial actions on food residues in the gut. This phenomenon is most active in the small intestine as discussed in Chapter 6 – Organic Acids. This source may be considered either normal, chronic background exposure or episodic, acute exposure depending on the extent of microbial overgrowth and the integrity of the physical barrier. Evaluations of total toxin exposure should always include testing for elevated flow of toxic microbial products formed in the gut.

Urinary Markers of Detoxification

Compounds present in unchallenged overnight urine reveal aspects of xenobiotic exposure, endogenous toxin handling, and detoxification functions. Unchallenged urinary markers are summarized in Table 8-1, and they were described in Chapter 6 – Organic Acids. Data from an unchallenged urine specimen cannot show capacities for toxin removal. However, such testing does allow simultaneous examination of the levels of intestinal microbial toxins and markers for hepatic conjugation substrate adequacy.

For example, a patient with elevated pyroglutamate and p-cresol has both a high exposure level (elevated p-cresol) and a reduced detoxification capacity (elevated pyroglutamate). Elevated pyroglutamate indicates specific depletion of glutathione, which takes on special significance when intestinal bacteria are causing increased release of toxic products.

Urinary markers of dysbiosis should be examined carefully in cases of chronic fatigue because signs of small intestinal bacterial overgrowth signal increased transient ammonia loading, which may cause symptoms of lethargy [19]. Intestinal bacteria can contribute significantly to the total load of ammonia [20], and the level of contribution is affected by dietary composition [21, 22]. Exercise also causes a rise in blood ammonia, especially if glucose availability is impaired [23]. This effect helps to explain the association of environmental toxin loading with chronic fatigue [24, 25].

Elevated urinary citrate and orotate are signs of poor ammonia clearance due to arginine insufficiency [26]. Ammonia is extremely toxic to the central nervous system, and the amount of ammonia produced each day by a healthy individual exceeds the CNS toxic limit by several orders of magnitude. Since even transient elevations of blood ammonia can lead to lapses in mental ability and sensation of muscle fatigue [27], evidence of poor ammonia clearance from blood provides valuable insight about patient management. The largest routine source of ammonia is dietary protein because the majority of the amine groups are converted into ammonia. Because it carries two amine groups, excess glutamine is particularly known to challenge the ammonia clearing pathways [28].

Ammonia is normally cleared by the reactions of the urea cycle. This pathway is highly energy (ATP) demanding and dependent on adequate concentrations of arginine and aspartic and glutamic acids [29]. The most critical of these is the essential amino acid arginine.

Testing for Hepatic Detoxification Capacity

The human liver contains six different types of cells that are organized into microscopic arrays called "lobules." Each cell can perform hundreds of different biochemical reactions. In addition, the biochemical activity of each cell depends on its interaction with other cells and with the extracellular matrix. Testing for detoxification capacity can reveal the status of specific functions in these cells.

Any history of hepatic inflammatory disorder is evidence that overall organ reserve is low and that hepatic functions may be globally depressed [30]. Such incidents include infections caused by viruses, bacteria, or parasites (hepatitis, tuberculosis, and amebic liver disease, respectively), drugs (prescription and abused), toxins (DDT or carbon tetrachloride), and excess alcohol intake. Acute toxic exposures can lead to hepatic necrosis and elevated levels of the serum enzymes alanine transaminase (ALT), aspartate transaminase (AST), and creatine phosphokinase (CPK) [31]. Low blood urea nitrogen (BUN) signals hepatic failure and increased risk of critical ammononemic sequelae [32]. Such findings signal general hepatic functional impairment but are not necessarily related to toxin exposure.

For most potentially toxic compounds, the process of preparation for excretion involves two phases. In Phase I (oxidation), substrates are oxidized to introduce functional groups that are more reactive. In Phase II (conjugation), additional polar groups that confer greater water solubility are introduced so that the products can be more easily transported to the kidney and excreted in urine. The common over-the-counter drugs caffeine, acetaminophen, and salicylate have known hepatic detoxification pathways (Figure 8-1). The normal liver possesses multiple enzyme systems to carry out the processes of oxidation and conjugation [33]. Phase I enzymes include cytochrome P450 mixed-function oxidase, peroxidase, hydroxylase, and flavin oxygenase. Phase II reactions include glucuronidation, sulfation, acetylation, and glutathione conjugation. Comprehensive testing for capacities in all areas of hepatic detoxification is not yet available, but tests that show activities in any of the major enzyme systems can give valuable clinical insight regarding how patients respond to toxin loads.

FIGURE 8-1. Use of Challenge Compounds to Evalutate Phases of Detoxification

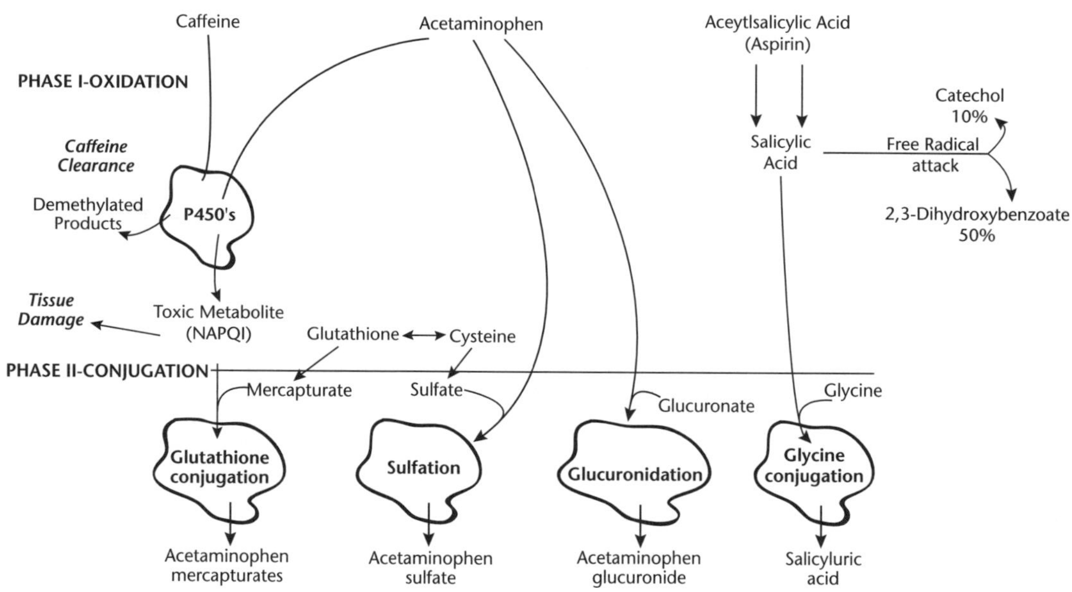

Hepatic detoxification capacity may be evaluated by administering a compound at a sufficiently high dose to challenge the pathway used for detoxification. The challenge compounds must be non-toxic at the levels used and must require the same pathway as the class of toxins under evaluation. Caffeine, acetaminophen, and acetylsalicylic acid are readily available and widely used substances that are suitable for such use. Quantitative analysis of the products resulting from hepatic enzyme actions on these compounds reveals capacities for handling any of the many toxins that use the same enzymes. Although it is not considered a normal part of detoxification, free radical attack on salicylic acid may be measured because these compounds are sometimes reported along with other products from oral challenge testing. Other compounds acted upon by the various pathways are listed in Table 8-3.

TABLE 8-3. Conjugation Pathways Used for Specific Compounds

Pathway	Xenobiotics	Drugs	Natural Compounds
Glutathione Conjugation	Styrene, Acrolein Ethylene Oxide Benzo Pyrenes Methylparathion Chlorobenzene Antracene Toxic Metals Petrolenum Distillates Naphthalene	Acetaminphen Penicillin Ethacrynic Acid Tetracycline	Bacterial toxins, Aflatoxin Lipid Peroxides, Ethyl alcohol Quercitin, N-Aetlcysteine Prostaglandins, Baterial txins Bilirubin, Leukotreine A4
Sulfation	Aniline Pentachbrophenal Terpenes, Amines Hydroxylamines Phenols	Acetaminophen Methyl Dopa Minoxidil Metaraminol Phenylephrine	DHEA, Quercitin Bile acids, Safrole Tyramine, Thyroxine Estrogens, Testosterone Cortisol, Catecholamines Melatonin, 3-Hydroxy coumarin 25-Hydroxy vitamin D, Ethyl alcohol, CCK Cerebrosides
Glycine Conjugation	Napthylacetic acid Alphatic amines	Salicylates Nicotinic Acid Chlorpheniramine Brompheniramine	Bile acids Cinnamic acids, PABA Plant Acids, Benzoic acid Phenylacetic acid
Taurine Conjugation	Propionic acid Caprylic acid		Bile acids, Stearic acid Palmitic acid, Myristic acid Lauric acid, Decanoic acid Buttyric acid
Glucuronidation	Aniline, Carbamates Phenols, Thiophenol Butanol N-Hydroxy-2-napthylamine	Salicylates Acetaminophen Morphine meprobamate Benzodiazephines Clofibric acid Naproxen Digoxin Phenylbutazone Valproic acid, Steroids Lorazepam, Ciramodol Propranolol, Oxazepa	Bilirubin, Estrogens Melatonin, Bile acids Vitamin E, Vitamin A Vitamin K, Vitamin D Steroid hormones
Acetylation	2 Aminofloride, Analine	Clonazepam, Dapsone Mescaline, Isoniazid Hydralazine Procainamide Benzidine, Sulfonamides Promizole	Sertonin, PABA Histamine, Tryptamine Caffeine, Choline Tyramine, Coenzyme A
Methylation	Paraquat Beta carbolines Isoquinolines, Mercury Lead, Arsenic Thallium, Tin Pyridine	Thiouracil, Isoetharine Rimiterol, Doubutamine Butanephine, Elouphed Morphine, Levaphanol Nalorphine	Histamine, Epinephrine Dopamine, Norepinephrine L-dopa, Apomorphine Hydroxyestradiols

Phase I Testing by Caffeine Clearance

The ability to clear a challenge dose of caffeine is an indicator of the detoxification capacity of the hepatic the Phase I pathway. The enzyme system called microsomal P450 mixed-function oxidase is used to oxidize such compounds to prepare them for removal. Its activity towards all such molecules is reflected by how fast caffeine is removed from the body. After an oral challenge dose of 200 mg caffeine, saliva samples are taken at intervals of two and eight hours. The saliva is analyzed for caffeine, the concentration of which closely parallels that in blood [34, 35].

The importance of evaluating this particular part of the enzymes of detoxification is that the cytochrome P450 enzymes handle such a wide variety of toxins and that they have a limited capacity. The rate at which they process substrates is very slow relative to most other enzymes -- only about one molecule per enzyme per minute is transformed [10].

If Clearance Rate Is HIGH:

A high value signifies that the liver is very actively trying to remove substances using the microsomal P450 enzyme. When this activity is elevated it is especially important to minimize exposure to environmental toxins such as car exhausts, food contaminants, and pesticides since the P450 enzyme can convert such substances into procarcinogenic compounds. As exposure decreases, enzyme activity should decline. Smokers and those who maintain a high-protein diet will usually demonstrate a higher caffeine clearance due to continuous induction of this enzyme system.

The gut is a potential source of toxic exposure. Gut microbes are capable of generating many toxins that must be degraded by the P450 enzyme system. If an imbalance exists between the beneficial and pathogenic bacteria, toxic compounds can be absorbed through the gut wall. Routines to reduce such toxic exposure can be helpful (see Intervention Options section).

Because of the mechanism through which the P450 enzyme system causes chemical oxidation, dangerous free radical compounds are formed. The level of protection from the destructive effects of such products depends largely on intake of anti-oxidant nutrients. Therefore, supplementation of the anti-oxidant nutrients such as vitamins C and E, β-carotene, zinc, selenium, and lipoic acid may be helpful (see Chapter 10 – Hormones).

If Clearance Rate Is LOW:

For an individual with average food intake and environmental exposure, a low caffeine clearance rate indicates loss of liver function. Examples of conditions leading to such loss are infectious liver diseases and alcoholic cirrhosis [36].

A low result indicates that the processing and degradation of foreign compounds is slower than normal. This could mean that toxins are accumulating. These substances are generally not soluble in water, and they accumulate in the fatty or membrane-rich tissues. Effects are typically seen in the central nervous system and as rapid aging of neuromuscular function with its degenerative manifestations.

The P450 enzyme is an inducible, on-demand detoxification system. If there has been low exposure to environmental inducers, a low value may not indicate a health problem. In general, however, it is important to decrease toxic exposure and increase the enzyme activity. Supplementation with high-quality protein, iron, riboflavin, and ascorbic acid may be considered, along with routines to stimulate detoxification through the skin and GI tract. Once P450 enzyme production has increased, activity rates depend on the availability of magnesium and niacin (for NADPH synthesis) [10].

Phase II Testing by Acetaminophen Conversion

A 650 mg dose of acetaminophen taken in the evening is converted to various conjugation products overnight, and the levels in the overnight urine reflect the activities of conjugation enzyme activities. Pathways utilizing glutathione, glucuronic acid, or sulfate produce the mercapturate, glucuronate, or sulfate conjugates (of acetaminophen), respectively, (Figure 8-1). The drug is initially oxidized to the reactive metabolite, N-acetyl-p-benzoquinone imine (NABQI), which is highly toxic and must be removed by conjugation with mercapturic acid via glutathione activation [37]. This pathway and others related to detoxification utilizing sulfur compounds are illustrated in Figure 8-2.

Sulfation of acetaminophen can become saturated because of limited availability of the active sulfate donor, 3'-phosphoadenosine 5'-phosphosulfate (PAPS), the supply of which is, in turn, limited by the availability of its precursor, inorganic sulfate. Administration of sodium sulfate and the sulfate precursor N-acetylcysteine can enhance the formation of acetaminophen sulfate [38]. Patients highly susceptible to environmental illness have diminished capacity to form acetaminophen sulfate [39]. A lowered capacity for xenobiotic metabolism, especially of compounds containing sulfur, is a major risk factor for the development of Alzheimer's disease [40]. In ruminants, sulfate metabolism is impaired by excessive dietary molybdenum [41].

When the test is administered and low values are reported for any of the conjugates, that patient has reduced capacity to remove toxic compounds by these pathways. The long-term effects of accumulation of the initial compounds or the intermediate oxidation products can produce widely varying symptoms.

Obese patients may show glucuronide values above normal because obesity is associated with enhanced capacity for biotransformation of drugs via glucuronide conjugation. Conjugating capacity increases in proportion to total body weight and is consistent among drugs biotransformed by this mechanism [42]. Cigarette smoking is more likely to induce drug oxidation rather than

FIGURE 8-2. Sulfur Compounds in Detoxification

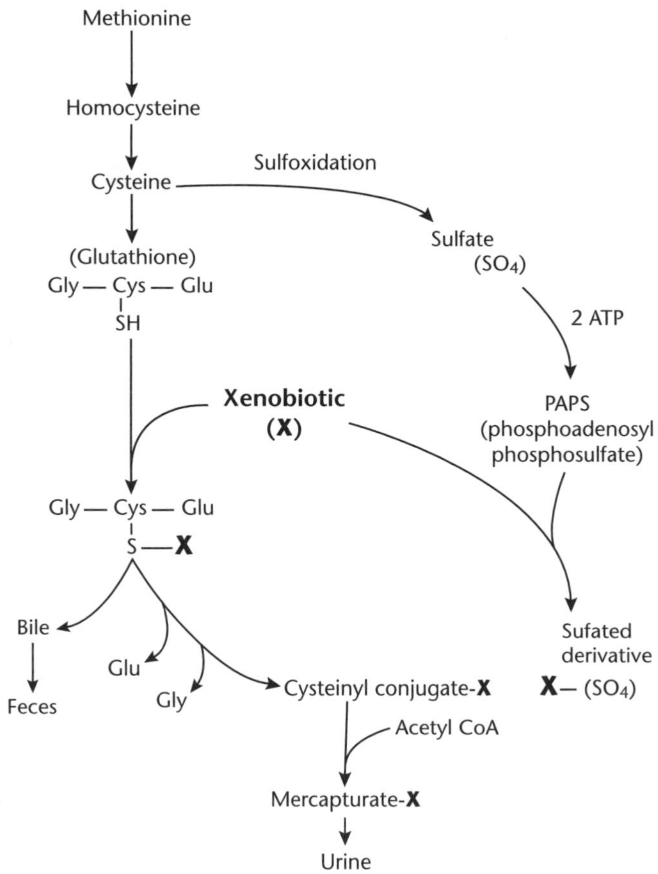

drug conjugation, and thus the habit becomes a factor to consider in the interpretation of results for smokers [43]. They may show a high rate of caffeine clearance but low rates of acetaminophen mercapturate formation. Because oleic acid activates the glucuronidation enzyme system [44], recommendation of olive oil as a dietary fat is one component of intervention when production of acetaminophen glucuronide is low.

When used as an analgesic, recommended doses of acetaminophen should not be exceeded because of hepatotoxic effects. These effects are from the relatively low capacity of the hepatic sulfur conjugation pathway. Acetaminophen, on the other hand, has a high affinity for the enzymes, preventing their activities toward other toxins and leading to hepatic necrosis in the absence of sufficient sulfur compounds. For prevention of such hepatotoxic effects, both inorganic sulfate (as i.v. $MgSO4$) and oral N-acetylcysteine are effective in providing the necessary increase in active sulfate (PAPS) [38].

Phase II Testing by Salicylic Acid Conversion

A 650 mg dose of acetylsalicylic acid (aspirin) taken in the evening is converted to urinary conjugation products that appear in overnight urine (Figure 8-3). Salicylic acid is released in a first step, and the high rate of reactivity of this compound with hydroxyl radicals can be used as a marker of oxidative damage protection [45].

Salicylic acid undergoes simultaneous conjugation with glucuronic acid and glycine, and the proportions of the conjugated products reflect the ability to transform compounds by these pathways [46]. When glycine is adequate, the glycine conjugate, salicyluric acid, is formed more easily than the glucuronate so the relative amounts allow an assessment of these pathways [47].

IGURE 8-3. Salicylic Acid Challenge Test For Detoxification Capacity

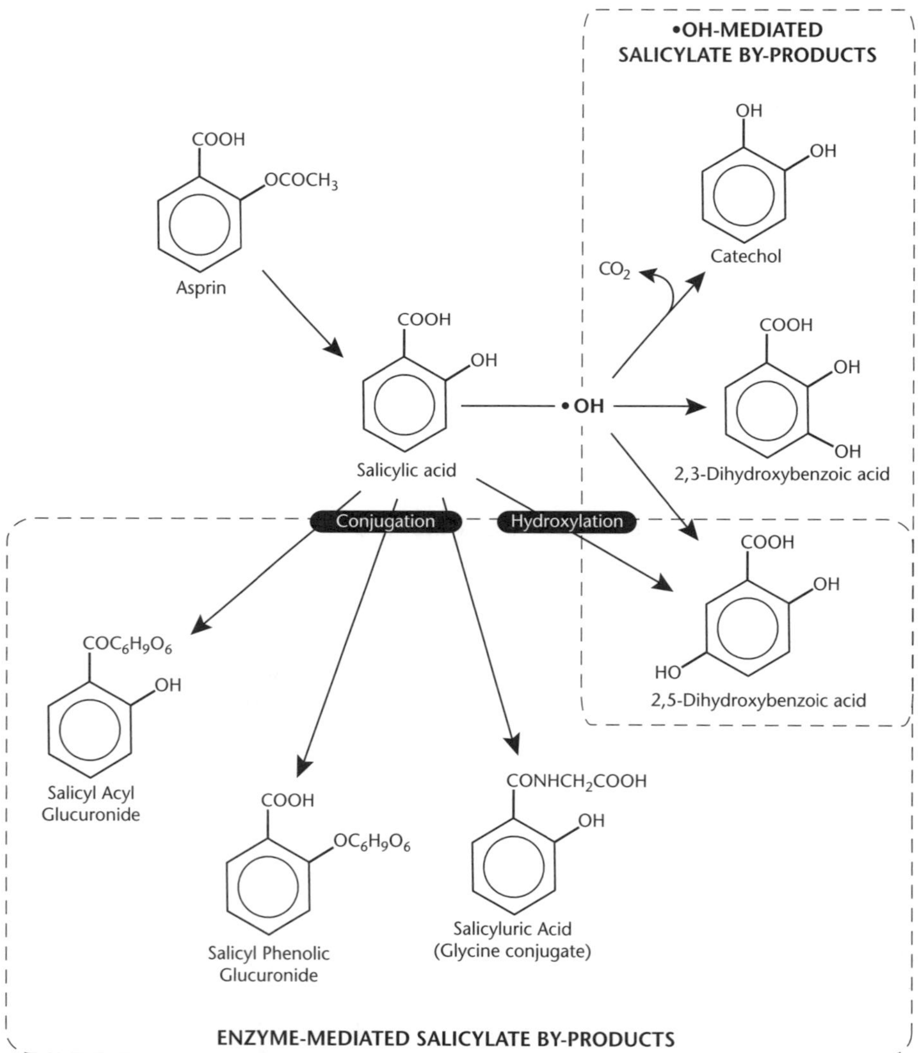

The Phase I/Phase II Ratio

This ratio can warn of early risk for carcinogenic compound accumulation. The oxidation reactions of Phase I frequently produce compounds that are of greater (sometimes extremely greater) carcinogenic potential than the initial xenobiotic. The normal progression would lead to efficient removal via Phase II reactions. Abnormally low Phase II or an increased ratio of Phase I to Phase II are the conditions under which the risk is increased. If either of these conditions is found, detoxification is impaired and the results of toxin exposure are amplified. Patients with these patterns display hypersensitivity to xenobiotics, pharmaceuticals, and other toxin sources.

The Cysteine/Sulfate Ratio

Cysteine is the precursor for sulfate formation (Figure 8-2). Disturbances in this pathway have been observed in Alzheimer's disease with early morning plasma cysteine/sulfate ratios four to five times the mean of controls [48]. This situation leads to concerns about metabolic interference from excess cysteine and reduced xenobiotic detoxification due to sulfate deficiency.

Ethanol Intoxication

Several laboratory evaluations are pertinent to nutrient status critical for alcohol detoxification. The most important nutrient effects for ethanol detoxification relate to a person's ability to maintain adequate hepatic function. Detoxification outcomes may also depend on amino acid supply for maintenance of neurotransmitter levels. Chronic ethanol consumption has negative effects on many aspects of nutrient intake and assimilation.

Ethanol is metabolized through at least two inducible pathways in the liver. Alcohol and aldehyde dehydrogenase enzymes make up the most well known pathway, but ethanol-inducible cytochrome P450 enzymes also can play a role. These inducible forms may be modulated by components in garlic [49]. Variations in the activities of these enzymes and the abilities to induce their synthesis account for the finding that alcohol-dependent individuals show a variation in their rate of clearance of ethanol from blood [50]. Simultaneous exposure to other classes of toxicants adds further variability in responses among alcohol-dependent patients.

The ethanol-inducible P450 is believed to be predominant microsomal P450 cytochrome leading to oxidative damage, including the generation of radical species that contribute to lipid peroxidation. Inhibition of this enzyme by the fatty acid oxidation product trans-4-hydroxy-2-nonenal seems to be a mechanism of negative regulation that can result in self-limiting effects on ethanol oxidation capacity [51] (see Chapter 9 – Oxident Stress). Concurrent ethanol and lead exposure greatly increases hepatic zinc loss [52], and moderate amounts of ethanol decrease the conversion of toluene to hippuric acid by 37% [52]. Formation of hippurates is a glycine conjugation reaction.

Chronic ethanol consumption has a dramatic inhibitory effect on the conversion of 5-hydroxytryptophan to serotonin in the brain [53]. This inhibitory effect may explain the clinically beneficial action of maintaining low concentration of ethanol during alcohol detoxification [54]. The suppression of serotonergic pathways in the brain is a chemical phenomenon relevant to the addictiveness of alcohol. Amino acid testing can be helpful in alcoholic detoxification programs to evaluate the status of neurotransmitter-related amino acids, tryptophan, tyrosine, GABA, and glutamate (see Chapter 4 – Amino Acids).

Intervention Options

Toxin-Related Nutrient Status

Removal of abnormal effects of toxin exposure involves reducing exposure and increasing removal to achieve lowered body burdens. While identification of sources of exposure is of utmost importance, cases of chronic exposure generally present difficulty in locating the multiple specific offending toxin sources. Books dealing with management of the chemically sensitive patient are available [7].

Specific nutrient interventions will be guided by test results. Although all essential nutrients are required to maintain normal metabolic function of hepatic cells, those most closely involved with support of detoxification are shown in Table 8-4. N-Acetylcysteine (NAC) has become a favored oral enhancement of glutathione synthesis, but it must be used with care in patients with high heavy metal body burden because of the potential for increased mobilization from sequestered sites. Selenium is a mineral that is directly involved in glutathione maintenance through activation of glutathione peroxidase.

Because of similarities in chemical bonding, heavy metals show reduced toxic effects when specific inhibitory nutrient elements are increased (Table 8-4). A good example of this effect is the protective action of dietary calcium against the toxic accumulation of lead, both of which compete for the same binding sites. This mechanism even affords protection from transfer of lead to the fetus when mothers who are exposed to lead before pregnancy receive extra calcium during pregnancy [55]. Thus, a test revealing an elevated lead level takes on greater significance when a simultaneous evaluation shows negative calcium balance. The fact that lead deposits in bone where it interacts with the calcium regulatory enzymes [56] clarifies the protective role of calcium in lead toxicity. Recognition of such relationships between nutrients and toxicants leads to clinical insight regarding the large variations in individual responses to toxin exposure. The data presented in Table 8-4 on nutrient-toxic element interactions is a simplification of the total picture of such effects, which is quite complex with multiple cross-element relationships [57].

Sulfur-containing amino acids and essential trace elements are high on the list of priority nutrients in virtually every case of clinical toxicology. Beans, eggs, garlic, and onions are foods that are rich sources for many of these critical nutrients. The cruciferous family of vegetables (cabbage, broccoli, Brussels sprouts) contains substances that stimulate the liver to produce Phase I and II enzymes.

TABLE 8-4. **Detoxification Functions of Specific Nutrients**

Nutrient	Effects
Vitamin C	Increased mobilization, toxic metal binding and antioxidant protection
B-complex vitamins	Hepatic enzyme cofactors
Lipoic acid	Hepatic protection & antioxidant regeneration
N-Acetylcysteine (NAC)	Glutathione formation and direct complexation
Cysteine	Increased glutathione and sulfate production
Methionine	Methyl donor and sulfur supply
S-Adenosylmethionine (SAM)	Active form of methionine
Glycine	Hepatic conjugation
Free-form essential amino acid mixture	Mitochondrial energy production, methyl donors, and organic sulfur
Sulfate	Hepatic conjugation
Calcium	Lead protection
Magnesium	Multiple hepatic and other effects
Selenium	Glutathione regeneration and mercury protection
Manganese	Glutathione regeneration
Copper	Glutathione regeneration

Methods for Increasing Detoxification

When toxin exposure is determined to merit clinical intervention, the first order of business is to reduce the total body burden of toxin by avoiding exposure. For xenobiotics, the source must be identified and measures for avoidance must be devised. The measures sometimes mean extensive modifications of home and work environments [7]. For toxins resulting from intestinal dysbiosis, methods for controlling microbial populations and improving the integrity of the physical barrier may be implemented (see Chapter 7 – GI Function). In addition to the issue of reducing toxin loads, detoxification metabolic rates should be addressed. A summary of such interventions is presented in Table 8-5.

Removal of foods that elicit immune responses can improve gut barrier function and reduce toxin entry. The normal opening of channels for the outward flow of secretory IgA is impaired in chronic

TABLE 8-5. Methods For Increasing Rates of Detoxification

Functional Enhancement	Intervention
Rate of mobilization from soft tissue and bone	Intravenous DMPS Intravenous chelation (EDTA, Penicillamine) Oral DMSA Oral vitamin C
Circulation and sweat flow	Exercise, Sauna, Spa
Hepatic metabolic rates of toxin conversion	Cruciferous vegetables
Urinary excretion	Fluid intake, Diuretic
Fecal excretion	Cholegogues, Dietary fiber

inflammatory states of the gut. Formation of antigen-IgG complexes and loss of antioxidant protection are factors frequently involved in loss of the barrier function against antigen entry. A chemically-defined diet can provide a period of relative absence of antigens to achieve reduction of endogenous toxin exposure. Formulations are widely available containing chemically-defined, balanced proportions of major nutrients supplemented with nutrients that support detoxification. The ultimate elimination of toxin entry via the gut may require total water fasting. Unfortunately, many patients are highly resistant to the concept of fasting. Those with high body burdens of toxins must be managed carefully to successfully employ fasting. The absence of food intake can increase the mobilization of stored toxins, and the heightened detoxification activity demands an extra supply of nutrients.

One effective way to increase the rate of removal of toxins from the body is to stimulate hepaticobiliary flow and maintain a low intestinal transit time. Low transit time is essential because of the high potential for reuptake of toxins released in bile when the transit of intestinal content is slowed. Use of oils (especially olive oil) is sometimes helpful as cholegogues to assure bile flow [58]. Patients who have undergone cholecystectomy will respond more slowly to such interventions since the capacity for bile storage is greatly reduced. High-fiber diets help maintain low intestinal transit to assist toxin removal in feces and avoid re-acquisition via entero-hepatic cycling [59-61]. Unabsorbed dietary fiber binds many toxins and allows them to be removed in stools. The National Cancer Institute recommends between 20-35 grams of fiber daily. The average American eats less than half of this recommended intake. High-fiber cereals are one of the easiest sources to find fiber, as one serving size can be as high as 8-10 grams. Other measures suitable for most patients are listed in Table 8-6.

The skin is a large organ of detoxification via the loss of toxins in sweat [62, 63]. Use of sauna, spa, and massage can accomplish stimulation of toxin losses through increased blood flow and sweating [64].

TABLE 8-6. How to Increase Fiber in Your Diet

1. Eat more fruit. Apples, bananas, apricots, grapes, peaches, and strawberries are rich in fiber.
2. Consume more vegetables. Broccoli, cabbage, cauliflower, corn, green beans, tomatoes, and zucchini all are rich in fiber.
3. Eat legumes. Not only are they rich in fiber, many are high quality protein sources and low in dietary fat.
4. Go whole-grain. Increasing whole-grain bread and wheat bran effectively incorporates fiber.
5. Eat cereals high in fiber. Check the food labels on your breakfast cereal for dietary fiber.
6. Consume more berries. Add berries to pancake batters and muffin mix recipes.
7. Eat brown rice in place of white rice.
8. Get nutty. Many nuts are high in insoluble fiber.
9. Replace processed fruit juice with whole fruit.

Finally, it must be emphasized that the gut is an organ of detoxification at least as important as the liver or skin. The hepatic detoxification reactions discussed in this chapter also occur in the mucosal cells of the gut [65-70]. Laboratory evaluations of gastrointestinal function are discussed in Chapter 7 – GI Function.

References

[1] MacIntosh, A. and K. Ball, The effects of a short program of detoxification in disease-free individuals. *Altern Ther Health Med*, 6(4). 70-6, 2000.

[2] Massaro, E.J., ed. *Handbook of human toxicology.* 1997, CRC: Boca Raton. 1111.

[3] Hayes, W.J. and E.R. Laws, eds. *Handbook of pesticide toxicology*. Classes of pesticides. Vol. 2&3. 1991, Academic Press: San Diego. 1576.

[4] Winter, C.K., J.N. Seiber, and C.F. Nuckton, eds. *Chemicals in the human food chain*. Van Nostrand Reinhold: New York. 276.

[5] Chang, L.W., ed. *Toxicology of metals*. 1996, Lewis: Boca Raton. 1198.

[6] Goldstein, R.S., W.R. Hewitt, and J.B. Hook, eds. *Toxic interactions*. 1990, Academic Press: San Diego. 488.

[7] Rea, W., *Chemical Sensitivity*. 1994, Boca Raton: Lewis. 2015.

[8] Jakoby, W.B., ed. *Enzymatic basis of detoxication - I.* Vol. I. 1980, Academic Press: New York. 415.

[9] Dills, R.L. and C.D. Klaassen, Effect of reduced hepatic energy state on acetaminophen conjugation in rats. *J Pharmacol Exp Ther*, 238(2). 463-72, 1986.

[10] Yu, M., *Environmental toxicology*. 2001, Boca Raton: Lewis. 255.

[11] Mak, D.H. and K.M. Ko, Alterations in susceptibility to carbon tetrachloride toxicity and hepatic antioxidant/detoxification system in streptozotocin-induced short-term diabetic rats: effects of insulin and Schisandrin B treatment. *Mol Cell Biochem*, 175(1-2). 225-32, 1997.

[12] Sacarello, H., *Handbook of hazardous materials*. 1994, Boca Raton: Lewis. 536.

[13] Hayes, W. and E. Laws, eds. *Handbook of pesticide toxicology. General principles. Vol. 1*. 1991, Academic Press: San Diego. 497.

[14] Jakoby, W.B., J.R. Bend, and J. Caldwell, eds. *Metabolic basis of detoxication*. 1982, Academic Press: New York. 375.

[15] Sellars, W.A., M., et al, Methyl Mercury and Dental Amalgams in the Human Mouth. *J Nutr Env Med*, 6. 33-36, 1996.

[16] Lewis, M., *et al.*, Prenatal exposure to heavy metals: effect on childhood cognitive skills and health status. *Pediatrics*, 89(6 Pt 1). 1010-5, 1992.

[17] Kilburn, K.H., Growing up exposed: adult manifestations of protracted childhood insecticide exposure. *Arch Environ Health*, 50(5). 391-2, 1995.

[18] Pryor, W., Free radical involvement in chronic diseases and aging: the toxicity of lipid hydroperoxides and their decomposition products, in *Xenobiotic metabolism: nutritional effects*, J. Finley and D. Schwass, Editors. 1985, American Chemical Society: Washington, DC. p. 77-96.

[19] Prasad, A.S., Zinc in growth and development and spectrum of human zinc deficiency. *J Am Coll Nutr*, 7(5). 377-84, 1988.

[20] Marchese, A., *et al.*, In vitro activity of rifaximin, metronidazole and vancomycin against Clostridium difficile and the rate of selection of spontaneously resistant mutants against representative anaerobic and aerobic bacteria, including ammonia-producing species. *Chemotherapy*, 46(4). 253-66, 2000.

[21] Rychlik, J.L. and J.B. Russell, Mathematical estimations of hyper-ammonia producing ruminal bacteria and evidence for bacterial antagonism that decreases ruminal ammonia production(1). *FEMS Microbiol Ecol*, 32(2). 121-128, 2000.

[22] Sakata, T., *et al.*, Probiotic preparations dose-dependently increase net production rates of organic acids and decrease that of ammonia by pig cecal bacteria in batch culture. *Dig Dis Sci*, 44(7). 1485-93, 1999.

[23] Matthys, D., *et al.*, Carbohydrate availability affects ammonemia during exercise after beta 2-adrenergic blockade. *Med Sci Sports Exerc*, 32(5). 940-5, 2000.

[24] Goubel, F., *et al.*, Endotoxins modify muscle fatigue characteristics. *Fundam Clin Pharmacol*, 9(2). 202-4, 1995.

[25] Pearn, J.H., Chronic fatigue syndrome: chronic ciguatera poisoning as a differential diagnosis. *Med J Aust*, 166(6). 309-10, 1997.

[26] Milner, J.A., Metabolic aberrations associated with arginine deficiency. *J Nutr*, 115(4). 516-23, 1985.

[27] Bogdanovic, M.D., *et al.*, Late onset heterozygous ornithine transcarbamylase deficiency mimicking complex partial status epilepticus. J Neurol *Neurosurg Psychiatry*, 69(6). 813-5, 2000.

[28] Nelson, J., *et al.*, Regulation of orotic acid biosynthesis and excretion induced by oral glutamine administration in mice. *Biochem Med Metab Biol*, 49(3). 338-50, 1993.

[29] Gross, K.L., *et al.*, Arginine-deficient diets alter plasma and tissue amino acids in young and aged rats. *J Nutr*, 121(10). 1591-9, 1991.

[30] Lyoumi, S., *et al.*, Heme and acute inflammation role in vivo of heme in the hepatic expression of positive acute-phase reactants in rats. *Eur J Biochem*, 261(1). 190-6, 1999.

[31] Fodor, O., *et al.*, Further aspects of the protective action of aspartic acid in experimental toxic hepatitis studied by determination of serum and liver tissue enzymes. *Med Interna (Bucur)*, 17(7). 795-806, 1965.

[32] Imler, M. and J.L. Schlienger, The effect of chronic uremia on portal and systemic ammonemia in normal and portal-strictured rats. *J Lab Clin Med*,

94(6). 872-8, 1979.

[33] Gorrod, J.W., H. Oelschlager, and J. Caldwell, eds. *Metabolism of Xenobiotics*. 1988, Taylor & Francis: London. 382.

[34] Zylber-Katz, E., L. Granit, and M. Levy, Relationship between caffeine concentrations in plasma and saliva. *Clin Pharmacol Ther*, 36(1). 133-7, 1984.

[35] Lee, T.C., *et al.*, Saliva as a valid alternative to serum in monitoring intravenous caffeine treatment for apnea of prematurity. *Ther Drug Monit*, 18(3). 288-93, 1996.

[36] Jost, G., *et al.*, Overnight salivary caffeine clearance: a liver function test suitable for routine use. *Hepatology*, 7(2). 338-44, 1987.

[37] Zhou, L., *et al.*, Catalysis of the cysteine conjugation and protein binding of acetaminophen by microsomes from a human lymphoblast line transfected with the cDNAs of various forms of human cytochrome P450. *J Pharmacol Exp Ther*, 281(2). 785-90, 1997.

[38] Liu, L. and C.D. Klaassen, Different mechanism of saturation of acetaminophen sulfate conjugation in mice and rats. *Toxicol Appl Pharmacol*, 139(1). 128-34, 1996.

[39] Steventon, G.B., *et al.*, Metabolism of low-dose paracetamol in patients with chronic neurological disease. *Xenobiotica*, 20(1). 117-22, 1990.

[40] Steventon, G.B., *et al.*, Xenobiotic metabolism in Alzheimer's disease. *Neurology*, 40(7). 1095-8, 1990.

[41] Lamand, M., Influence of molybdenum and sulfur on copper metabolism in sheep: comparison of elemental sulfur and sulfate. *Ann Rech Vet*, 20(1). 103-6, 1989.

[42] Abernethy, D.R., *et al.*, Enhanced glucuronide conjugation of drugs in obesity: studies of lorazepam, oxazepam, and acetaminophen. *J Lab Clin Med*, 101(6). 873-80, 1983.

[43] Scavone, J.M., *et al.*, Differential effect of cigarette smoking on antipyrine oxidation versus acetaminophen conjugation. *Pharmacology*, 40(2). 77-84, 1990.

[44] Ouvina, G., *et al.*, Increase of acetaminophen conjugation ability in experimental cholestasis. *Acta Gastroenterol Latinoam*, 23(2). 71-4, 1993.

[45] Tomita, M., *et al.*, Quantitation of the hydroxyl radical adducts of salicylic acid by micellar electrokinetic capillary chromatography: oxidizing species formed by a Fenton reaction. *Arch Toxicol*, 68(7). 428-33, 1994.

[46] Amsel, L.P. and G. Levy, Drug biotransformation interactions in man. II. A pharmacokinetic study of the simultaneous conjugation of benzoic and salicylic acids with glycine. *J Pharm Sci*, 58(3). 321-6, 1969.

[47] Patel, D.K., *et al.*, Depletion of plasma glycine and effect of glycine by mouth on salicylate metabolism during aspirin overdose. *Hum Exp Toxicol*, 9(6). 389-95, 1990.

[48] Heafield, M.T., *et al.*, Plasma cysteine and sulphate levels in patients with motor neurone, Parkinson's and Alzheimer's disease. *Neurosci Lett*, 110(1-2). 216-20, 1990.

[49] Kishimoto, R., *et al.*, Combined effects of ethanol and garlic on hepatic ethanol metabolism in mice. *J Nutr Sci Vitaminol (Tokyo)*, 45(3). 275-86, 1999.

[50] Jones, A.W. and B. Sternebring, Kinetics of ethanol and methanol in alcoholics during detoxification. *Alcohol Alcohol*, 27(6). 641-7, 1992.

[51] Bestervelt, L.L., A.D. Vaz, and M.J. Coon, Inactivation of ethanol-inducible cytochrome P450 and other microsomal P450 isozymes by trans-4-hydroxy-2-nonenal, a major product of membrane lipid peroxidation. *Proc Natl Acad Sci U S A*, 92(9). 3764-8, 1995.

[52] Santos, J.L., *et al.*, Nonsynergic effect of ethanol and lead on heme metabolism in rats. *Ecotoxicol Environ Saf*, 43(1). 98-102, 1999.

[53] Uzbay, I.T., S.E. Usanmaz, and E.S. Akarsu, Effects of chronic ethanol administration on serotonin metabolism in the various regions of the rat brain. *Neurochem Res*, 25(2). 257-62, 2000.

[54] Keung, W.M., *et al.*, Volitional ethanol consumption affects overall serotonin metabolism in Syrian golden hamsters (Mesocricetus auratus). *Biochem Biophys Res Commun*, 271(3). 823-30, 2000.

[55] Han, S., *et al.*, Effects of Lead Exposure before Pregnancy and Dietary Calcium during Pregnancy on Fetal Development and Lead Accumulation. *Environ Health Perspect*, 108(6). 527-531, 2000.

[56] Hamalainen, M.M. and K.K. Makinen, Alterations in electrolyte and iron metabolism in the rat in relation to peroral administration of galactitol, mannitol and xylitol. *J Nutr*, 116(4). 599-609, 1986.

[57] Miller, G.D. and S.M. Groiak, Essential and nonessential mineral interactions, in *Handbook of human toxicology*, E.J. Massaro, Editor. 1997, CRC: Boca Raton. p. 369-407.

[58] Guarini, S. and W. Ferrari, Olive oil-provoked bile-dependent absorption of heparin from gastro-intestinal tract in rats. *Pharmacol Res Commun*, 17(8). 685-97, 1985.

[59] Brodribb, J., *et al.*, Influence of dietary fiber on transit time, fecal composition, and myoelectrical activity of the primate right colon. *Dig Dis Sci*, 25(4). 260-6, 1980.

[60] Bueno, L., *et al.*, Effect of dietary fiber on gastrointestinal motility and jejunal transit time in dogs. *Gastroenterology*, 80(4). 701-7, 1981.

[61] Kelsay, J.L., K.M. Behall, and E.S. Prather, Effect of fiber from fruits and vegetables on metabolic responses of human subjects I. Bowel transit time, number of defecations, fecal weight, urinary excretions of energy and nitrogen and apparent

digestibilities of energy, nitrogen, and fat. *Am J Clin Nutr*, 31(7). 1149-53, 1978.

[62] Omokhodion, F.O. and G.W. Crockford, Sweat lead levels in persons with high blood lead levels: experimental elevation of blood lead by ingestion of lead chloride. *Sci Total Environ*, 108(3). 235-42, 1991.

[63] Stauber, J.L. and T.M. Florence, A comparative study of copper, lead, cadmium and zinc in human sweat and blood. *Sci Total Environ*, 74. 235-47, 1988.

[64] Sunderman, F.W., Jr., *et al.*, Excretion of copper in sweat of patients with Wilson's disease during sauna bathing. *Ann Clin Lab Sci*, 4(5). 407-12, 1974.

[65] Roediger, W.E. and W. Babidge, Human colonocyte detoxification. *Gut*, 41(6). 731-4, 1997.

[66] Raithel, M., *et al.*, The involvement of the histamine degradation pathway by diamine oxidase in manifest gastrointestinal allergies. *Inflamm Res*, 48 Suppl 1. S75-6, 1999.

[67] Schwelberger, H.G., *et al.*, Expression and cellular localisation of diamine oxidase in the gastrointestinal tract of pigs. *Inflamm Res*, 47(Suppl 1). S62-3, 1998.

[68] Isaacs, C.E., *et al.*, Sulfhydryl oxidase in human milk: stability of milk enzymes in the gastrointestinal tract. *Pediatr Res*, 18(6). 532-5, 1984.

[69] Bennett, P.N., E. Blackwell, and D.S. Davies, Competition for sulphate during detoxification in the gut wall. *Nature*, 258(5532). 247-8, 1975.

[70] Carino, M.A. and A. Horita, The monoamine oxidase activity of the hamster gastrointestinal tract. *Proc West Pharmacol Soc*, 11. 25-6, 1968.

Chapter 9

Oxidant Stress and Aging

TABLE 9-1. Summary of Testing For Oxidant Stress

Category	Test	Abnormal	Intervention
Markers of oxidant damage			
Lipid oxidation	Lipid peroxides – serum or urine	H	Increase antioxidant intake and implement stress management; modulate PUFA and arginine supplementation.
	Isoprostanes – serum	H	
	HNE – serum	H	
	Oxidized LDL – plasma	H	
Protein oxidation	3-Nitroyrosine -plasma	H	
	Methionine sulfoxide – serum	H	
Nucleotide oxidation	Oxoguanosine – serum	H	
	DNA strand breakage – WBC	H	
Antioxidant nutrient testing			
Glutathione	Sulfate – urine	L	N-Acetylcysteine,
	Pyroglutamate – urine	H	Cys, Met (check levels)
Fat-soluble vitamins	Vitamin A, E, β-Carotene, Coenzyme Q_{10} – serum	L	Vitamins as needed
Antioxidant minerals	Selenium, Zinc, Copper – RBC, hair, urine	L	Trace elements, digestive support

Introduction

Figure 9-1. Oxidation Involving Formation of Stable Bonds to Oxygen

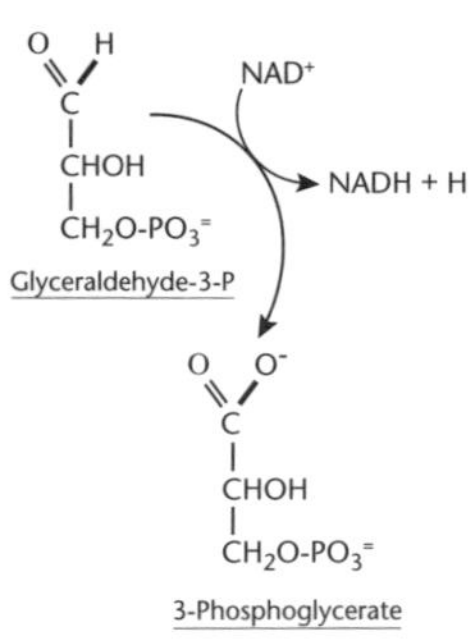

"Oxidation" is the chemical process of electron extraction from an atom. Oxidation reactions in intermediary metabolism usually result in the formation of new bonds to oxygen as electrons are drawn from carbon or hydrogen. Thus, glucose is transformed to carbon dioxide and water. An intermediate step is the oxidation of glyceraldehyde-3-phosphate to 3-phosphoglycerate. In this step, electrons are removed from carbon and transferred to the electron shuttle nicotinamide adenine dinucleotide (NAD) as a new carbon-oxygen bond is formed (Figure 9-1). The chemical bonds in the reactants and products of this reaction are relatively unreactive towards most other molecules within the cell in the absence of enzymes with active sites specific for changing the energy state of the bonds. Note that the electrons that are removed from glyceraldehyde-3-phosphate are transferred to NADH, which carries them to the electron transport system. Other reactions produce compounds that can enter into non-enzymatic chemical bond formation in ways that interfere with normal cell function [1]. Most of the damaging metabolic products are free radical forms of oxygen called "reactive oxygen species" (ROS). Other chemicals, such as carbon tetrachloride, have damaging effects even though they do not contain oxygen. Chemists define free radicals as any atom or molecule containing one or more unpaired electrons (radical) that has escaped (free) the solvent cage or enzyme active site in which it was generated.

The largest source of ROS is mitochondrial respiration. The major energy-yielding step in central energy pathways is the final transfer of the electrons carried by NADH to molecular oxygen via the coenzyme Q_{10} and cytochrome system of the inner mitochondrial membrane. This process, known as oxidative phosphorylation, is carried out by the mitochondrial electron transport system. Several oxidative steps in the electron transport system involve transfer of electrons from one iron atom to another, with no involvement of oxygen until the final transfer that results in formation of H_2O. Whenever molecular oxygen is made available for accepting electrons, the formation of dangerously reactive free radicals is an accompanying phenomenon [2]. Superoxide radicals are formed when molecular oxygen accepts a single electron. Hydrogen peroxide is formed from superoxide by the action of superoxide dismutase; in the presence of free ferrous iron, hydrogen peroxide is converted into the hydroxyl radical (Figure 9-2).

Figure 9-2. Superoxide, Hydrogen Peroxide, and Hydroxyl Radical Generation
The "*" symbol indicates a radical species. Superoxide formation initiates enzymatic and spontaneous reactions that yield hydroxyl radical.

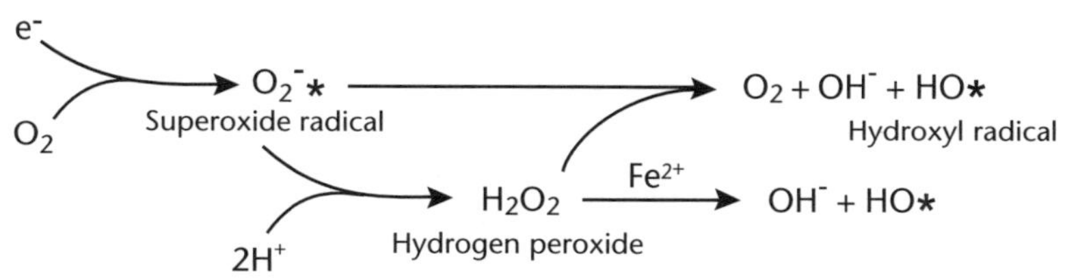

Diabetic neuropathy [3] and Alzheimer's disease [4, 5] are two of the most thoroughly studied disease manifestations of oxidative damage from ROS. In both instances, antioxidant nutrient status has been shown to play a role in protection from the degenerative effects. One way that cells manage the threat of oxygen free radicals is by transferring the electrons to antioxidant vitamins or to the tripeptide, glutathione, via the enzyme, glutathione peroxidase (Figure 9-3). Enzymes necessary for these reactions require the trace elements copper, zinc, manganese (superoxide dismutase), and selenium (glutathione peroxidase). It is widely accepted that unwanted reactions initiated by oxygen free radicals are associated with aging [6]. Nutritional factors afford protection from oxidative damage [7], and laboratory evaluations are available that allow assessment of a patient's current rate of oxidative damage and their adequacy of protection against further damage. Markers of oxidant damage to fatty acids (membranes), proteins, and DNA are available as well as various indicators of overall or specific antioxidant protection.

Oxidative damage caused by free radicals is thought to be a basic mechanism underlying aging and many diverse disorders, including atherosclerosis, cancer, rheumatic diseases, allergic inflammation, cardiac and cerebral ischemia, respiratory distress syndrome, various liver disorders, irradiation, thermal injury, and toxicity induced by certain metals, solvents, pesticides, and drugs [8-11]. The increasing evidence for disease protection from a single antioxidant, vitamin E, has led to the launching of large-scale studies of vitamin E supplementation of patients with early Parkinson's disease [12], and as a suppressant of free radical damage during cardiopulmonary bypass operations [13].

Figure 9-3. Generation and Removal of Oxygen Free Radicals

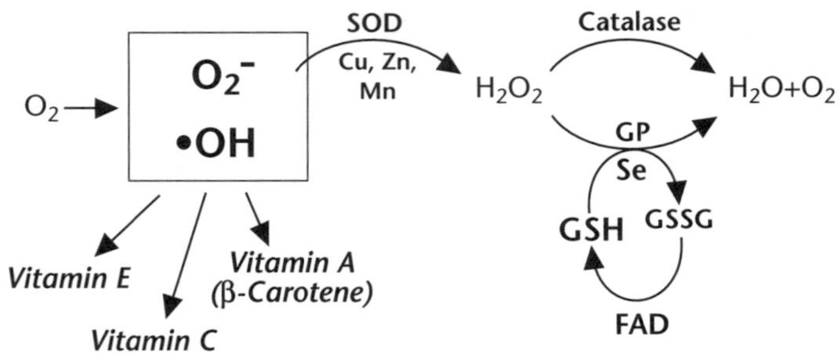

O_2^{-} = superoxide radical, **•OH** = hydroxyl radical, **SOD** = superoxide dismutase, **GSH** = reduced glutathione, **GSSG** = oxidized glutathione, **GP** = glutathione peroxidase, **GR** = glutathione reductase, **ROOH** = hydroperoxide of fatty acids

Markers of Oxidant Damage

Oxidative effects in Down's syndrome illustrate both the origins of oxidative stress and the value of testing that allows monitoring the adequacy of antioxidant controls. The extra copy of chromosome 21 in Down's syndrome contains a gene for the superoxide dismutase enzyme. This is an enzyme critical for oxidant protection, but it is only one of the links in the chain of normal electron transfer. Tissue levels of superoxide dismutase can be 40% higher in Down's patients compared to individuals with the normal chromosome count. The increased conversion of superoxide to hydrogen peroxide leads to increased rates of oxidant damage, and the damage to membrane lipids and DNA can be reflected in the markers described below [14].

Concentrations of specific antioxidants in body fluids are useful parts of the free radical pathology puzzle. However, they do not indicate to what degree the body is experiencing oxidant damage. Procedures that challenge white cells with a pro-oxidant and then measure proliferation are of theoretical interest, but no consensus exists in the scientific community that this measure accurately reflects capacity for free radical protection.

It is extremely difficult to measure free radicals directly since they exist for only fractions of a second. Tests that measure the by-products of free radical pathology are generally considered good indicators of this process [15].

Such assays include measurements of cellular deposition of lipofuscin and ceroid, erythrocyte oxidative hemolysis, plasma conjugated dienes, breath pentane, and red cell malonaldehyde. Except for the latter two assays, problems of tissue availability (brain), sample stability, and complex sample handling requirements make these impractical for routine clinical use.

Endogenous Auto-Oxidation Rate Marker

p-Hydroxyphenyllactate

p-Hydroxyphenyllactate (HPLA) is reported in urinary organic acid profiles. A discussion of its role as a marker of increased oxidative status is presented in Chapter 6 – Organic Acids. The compound offers a unique view of the regulation of endogenous auto-oxidation processes. Unlike other markers that show how much challenge can be tolerated or how much damage has occurred, HPLA reveals how much challenge is being generated within the tissues. Elevated HPLA, a carcinogenic tyrosine metabolite, decreases ascorbic acid (vitamin C) in liver, adrenal glands, and blood [16].

Lipid Peroxidation

Membrane components are especially susceptible to oxidative damage because they are the sites of greatest exposure to oxygen radicals and they possess high concentrations of molecules that are easily oxidized (unsaturated fatty acids). Cell membranes are the first subcellular domain to encounter molecular oxygen when it is released from hemoglobin, and mitochondrial membranes are the site where molecular oxygen is utilized constantly for ATP production and where single electron transfer reactions occur.

Malondialdehyde (Lipid Peroxides)

Lipid peroxides are the products of the chemical damage done by oxygen free radicals to the lipid component of cell membranes [17, 18]. A serum lipid peroxide level measures the overall effect of oxygen free radical pathology, the risk for degenerative processes, and the need for compensatory antioxidant supplementation and/or lifestyle modification. High serum lipid peroxide levels indicate excessive lipid peroxidation.

Evaluation of total lipid peroxides has proven to be a simple, inexpensive, and accurate means of reflecting whole body free radical activity. The assay measures malondialdehyde, a major product of lipid peroxidation, by reaction with thiobarbituric acid. Thus, it is also known as the thiobarbituric acid reactive substance (TBARS) test. Whenever total antioxidant capacity is inadequate to meet the oxidative challenge, cell membrane oxidation increases, releasing lipid peroxides.

Concentrations of lipid peroxides may be measured in urine or serum, but serum has superior sensitivity to slight increases in lipid peroxidation. It has gained general acceptance in research and clinical laboratories as a standard means of assessing the body's antioxidant capability and overall oxidative stress [19, 20].

Isoprostanes

Isoprostanes are prostaglandin-like compounds produced by nonenzymatic peroxidation of arachidonic acid. Elevated urinary excretion of isoprostane F2 reflects inadequate antioxidant status. Isoprostane F2 is thought to be a factor that contributes to the genesis of arteriosclerosis and to the activation of the clotting system [21]. Accumulation of isoprostanes may mediate the atherosclerotic effects of homocysteine because elevated fasting plasma homocysteine is associated with enhanced *in vivo* lipid peroxidation of LDL and high levels of isoprostane [22]. Patients with hypercholesterolemia show increased levels of isoprostanes [23].

Recent *in vitro* observations show the potential for formation of extremely reactive products of the isoprostane pathway. They rapidly form covalent adducts with lysine residues on proteins, which may cause enzyme inactivation or formation of antigenic macromolecules. Such reactions provide one mechanism of loss of tissue function from oxidative damage [24]. Isoprostane excretion has been used to monitor disease progression or improvement with abstinence in patients with alcoholic liver disease [25]. The marker is also an indicator of amyloid beta-peptide-induced oxidative damage in Alzheimer's disease [26]. Zinc protects against the toxicity of beta-amyloid peptide [27]. Both L-arginine and alpha tocopherol reduce urinary isoprostane excretion and improve endothelium-dependent vasodilatation, although by different mechanisms [28].

4-Hydroxy-2-nonenal

The aldehyde product of lipid peroxidation 4-hydroxy-2-nonenal (HNE) produces a range of powerful biological effects. It is a highly mutagenic and genotoxic product generated during oxidative stress. HNE acts as a part of the signaling pathway in the cellular response to oxidative stress by inducing gene expression of cyclooxygenase 2, which causes increased synthesis of prostaglandins and thromboxanes [29]. Congestive heart failure is one state of oxidative stress in which HNE is consistently elevated [30]. Hyperglycemic conditions, such as diabetes mellitus, present another state of HNE elevation. The insulin-releasing response of pancreatic islet cells is inhibited by HNE. The oxidative state stimulated by hyperglycemia causes further inhibition of the normal glucose-lowering response with a "feed-forward" (self-propagating) effect. This linkage explains the action of lipoic acid as an antioxidant in improving insulin response [31, 32]. A similar antioxidant role of phenolic compounds from olive oil is associated with lowered HNE production in cancer patients and patients with inflammatory bowel conditions such as ulcerative colitis and Crohn's disease [33]. The fecal matrix supports abundant generation of ROS, and high HNE levels are generated in the colonic mucosa of patients with Crohn's disease. In neurodegenerative disease, elevated HNE reflects the destructive cycle of oxidative stress and mitochondrial dysfunction [34]. High HNE in Alzheimer's disease provides evidence of oxidative stress-induced pathology in the formation of neurofibrillary tangles [35].

HNE reacts rapidly with DNA to form adducts [36]. Such DNA adducts to HNE are elevated in patients with Wilson's disease. The accumulation of copper and iron in this disease leads to increased oxidative damage and high risk of primary liver cancers. Such DNA adducts are also greatly increased in humans placed on diets high in omega-6 polyunsaturated fatty acids without adequate antioxidant protection [37]. Thus, HNE and its DNA adducts are promising biomarkers for monitoring effects of imbalanced dietary fat intake, oxidative stress, and antioxidant protection. Lowered HNE also reveals an important aspect of positive health benefits from calorie-restricted diets [38].

Plasma HNE, thus, is a marker of the early stages of degenerative effects in many diseases and, conversely, can be used to forecast wellness. The clearance of HNE requires biotransformation via omega oxidation of the aldehyde [39] and conjugation with glutathione to allow excretion of a mercaptan derivative [40] (See Chapter 8 – Detoxification).

Oxidized Low-Density Lipoprotein

Evidence implicating oxidation products in lipoprotein particles with atherogenesis is increasing [41, 42]. Such a mechanism may explain the anti-atherogenic effects of antioxidants, which can reduce the oxidation rate of LDL [43]. The ensuing immune response against oxidized low-density lipoprotein (LDL) may have a protective role in the early stages of atherogenesis [44], but this same immune response may be a trigger for setting in motion the cascade of events that culminates in advanced atherosclerotic lesions [45]. Changes in age-related immune parameters are also mediated by oxidized cholesterol [46].

The endothelial cells are constantly challenged to resist oxidative damage. The largest source of this challenge is oxidized plasma lipoprotein components containing oxidized fatty acids and cholesterol [47]. Cholesterol is easily oxidized to cholest-4-en-3-one. This and other partial oxidation products accumulate in lipoprotein particles where they are presented to the luminal surface of endothelial cells. Lipid hydroperoxides in plasma, for example, are carried almost exclusively (89%) in LDL. In 70 patients tested three months after coronary artery bypass graft surgery, plasma lipid hydroperoxide levels were significantly increased when compared with matched healthy controls [48], providing further evidence linking oxidative damage and heart disease.

Generally, the assay of oxidized LDL is performed on lipids

extracted from plasma on Liposorb gel. Results should be less than 2.5 nmol/ml. A related assay measures a sub-population of LDL particles with relatively high density. Individuals with elevated levels of this variety of LDL are at high risk of heart disease even if their total LDL concentration is not high [49]. The two phenomena discussed here may be linked because oxidized cholesterol may accumulate in "small, dense" LDL. The LDL subfractions are measured by analytical ultra-centrifugation or electrophoresis of serum lipoproteins.

Protein Oxidation

3-Nitrotyrosine

The formation of peroxynitrite by reaction of nitric oxide with superoxide radical is discussed in Chapter 4 under the heading "3-Nitrotyrosine," which is an amino acid derivative that serves as a marker for cellular levels of nitrating species. Although the nitration of a tyrosine residue in a protein does not necessarily mean inactivation of function, such protein modifications are one reason for loss of function due to increased oxidant exposure. Antioxidant nutrients, including the polyphenolic compounds abundant in tea [50], offer protection against the formation of peroxynitrite.

The retina has special need for oxidative damage protection because of the free radical-generating effects of ultraviolet radiation. Soluble retinal proteins induce nitric oxide and superoxide production by macrophages. Increased production of ROS by macrophages in the presence of these soluble retinal proteins in vivo may accelerate photoreceptor degeneration [51, 52]. Vitamin E pretreatment protects against the loss of free radical defense in kittens exposed to high oxygen atmosphere [53].

Methionine Sulfoxide

Methionine sulfoxide is a marker of cellular oxidative stress. This marker reveals stress specifically involving enzymes and structural proteins because it is formed from exposed methionine residues. Elevated levels indicate inadequate overall antioxidant protection. The enzyme required to regenerate methionine, peptide methionine sulfoxide reductase (PMSR), is involved in the resistance of mammalian cells to oxidative stress [54]. Methionine sulfoxide has been shown to cause enzyme inactivation, as it accumulates in the ionized calcium regulating protein, calmodulin, in senescent brain under conditions of oxidative stress [55], including Alzheimer's disease [56]. Dihydrolipoic acid has specific antioxidant properties and may exert a curative effect in diseases accompanied by oxidative stress by stimulating the methionine sulfoxide reductase enzyme [57]. In blood, methionine sulfoxide appears along with oxidized cholesterol as the process of oxidative damage acts on blood lipoproteins [58].

Nutrient effects may extend to reversing, as well as preventing, oxidative damage. Oxidative damage to the protein alpha-1-antiprotease (alpha 1-AP) occurs with formation of 3-nitrotyrosine residues [59]. Oxidized alpha 1-AP has been implicated in the etiology of lung diseases. Repair of oxidized alpha 1-AP is catalysed by PMSR, which is activated by dihydrolipoic acid to reverse oxidative damage to the damaged alpha 1-AP [57]. Thus, there is the potential to use methionine sulfoxide to monitor not only oxidative stress at the biochemical level but to demonstrate reversal of oxidative damage by increasing dietary lipoic acid. If the damage is detected early enough, reversing oxidative damage to proteins can reverse chronic, degenerative diseases.

Nucleotide Oxidation

8-Oxo-deoxyguanosine

The rate of oxidative DNA damage may be estimated by measuring 8-oxo-2'-deoxyguanosine (8-oxodG), a repair product of the highly mutagenic oxidation of guanine in DNA or the cellular pool of GTP, from a 24-hour urine collection. Bus drivers from Copenhagen showed increased excretion of 8-oxodG compared with drivers from rural/suburban greater Copenhagen, suggesting that exposure to ambient air pollution can cause oxidative damage to DNA [60].

The intracellular antioxidants glutathione and ascorbate, can protect human lymphocytes against oxidative DNA damage. The antioxidant levels are negatively correlated with the levels of oxidative lymphocyte DNA damage, and the strongest correlation observed is with glutathione and ascorbate [61]. Even among populations that might be considered normal, increasing intake of antioxidant vitamins with a nutrient-rich supplement reduces 8-oxodG by 22% compared to unsupplemented adults [62]. 8-OxodG also increases with high fat diets, showing the hyperoxidative response associated with excessive fat intake [60].

Age-associated decline in mitochondrial function due to mitochondrial DNA damage is indicated by the finding of increased 8-oxodG with age in laboratory animals [63].

DNA Strand Breakage (Comet assay)

One of the ultimate damaging consequences of oxidative challenge is damage to DNA. Such damage may be measured directly by looking for breaks in DNA strands as revealed on single cell (WBC) gel electrophoresis in the Comet assay [64]. This assay measures DNA damage, particularly DNA strand breakage. It has been used to measure the oxidative damage resulting from iron overload in thalassemia [65]. The test provides a rapid measure for biomonitoring of occupational exposure to DNA-damaging agents [66].

Antioxidant Nutrients

Human cells possess intricate protective systems against oxidative damage. An important concept in antioxidant interventions is the potential for imbalance in antioxidants because they function in concert. Oxygen free radicals are normal components of mitochondrial membranes and are a problem only when they are formed in an uncontrolled fashion. Similarly, if antioxidants like vitamins A and E are consumed in amounts out of proportion with the total spectrum of electron acceptors, the antioxidants can become a part of the problem.

Each type of antioxidant accepts electrons from free radicals and then passes them to another in the chain of electron acceptors until a stable compound (ultimately water) is formed. The electrons must be passed down in a controlled process from one acceptor to another. Vitamin E is one of the major protective compounds in cell membranes because it can accept high-energy electrons from compounds like hydroxyl radicals. When a free-radical electron is transferred to a molecule of vitamin E, it has lower reactivity. However, the vitamin E radical is still an unwanted oxidant that should pass the electron to a carrier with intermediate reactivity like that of β-carotene or cytochromes. Increased rates of lung cancer in smokers who supplemented β-carotene may be due to the phenomenon of increased free radical effects when single antioxidant nutrients are supplemented.

Glutathione

The disulfide chemical group in glutathione makes a particularly suitable, soft landing point for electrons that cause damage if left associated with reactive oxygen molecules. This chemical uniqueness allows understanding of why glutathione is a universal cell reduction-oxidation balancing molecule. It is ubiquitous in human tissues, and deficiencies are related to many and varied causes of disease. Markers of glutathione adequacy (pyroglutamate and sulfate) are discussed in Chapter 6 – Organic Acids.

Vitamins A, C, and E, and β-Carotene

Laboratory evaluations for these nutrients are discussed in Chapter 2 – Vitamins. They all play critical roles in the chain of electron acceptors that allows removal of oxygen radicals. Vitamin A and its precursor, β-carotene, have independent actions in this process and the various isomers of tocopherol likewise operate with redox potentials uniquely beneficial to specific tissues.

Vitamins C and E are major players in antioxidant protection. The clinical value of β-carotene testing may be as a marker of good dietary habits rather than as a direct indicator of antioxidant status since it may be an insignificant factor in free radical protection [67]. Profiling fat-soluble vitamins in serum has been reviewed in Chapter 2 – Vitamins. Such testing provides direct concentration measures for vitamins A and E, along with β-carotene and coenzyme Q_{10}. These are the principal molecules examined for deficiencies to reveal need for augmenting dietary intake. More comprehensive profiles are available that include the various minor isomers of vitamin A (carotenoids) and vitamin E (tocopherols). Typical components are shown in Table 9-2.

Table 9-2. Some Antioxidants Measured in Human Serum

Carotenoids
• Retinol
• Lutein
• Zeaxanthin
• β-Cryptoxanthin
• *trans*-Lycopene
• Total Lycopene
• α-Carotene
• trans-β-Carotene
• Total β-Carotene
Tocopherols
• α-Tocopherol
• γ-Tocopherol
• δ-Tocopherol

Copper, Manganese, Selenium, Zinc, and Riboflavin

Copper, manganese, selenium, zinc, and riboflavin are considered antioxidant nutrients because they play specific roles as cofactors for the enzymes that catalyze reactions which remove oxygen radicals (Figure 9-3). These nutrients are cofactors and precursor vitamins for the enzymes glutathione reductase [68], glutathione peroxidase [69], and superoxide dismutase [70]. Total body selenium is so largely dedicated to this role that some studies have evaluated overall oxidative protection by measuring serum and urinary selenium along with red cell enzymes [71]. Laboratory evaluations and specific metabolic roles for the antioxidant trace elements are discussed in Chapter 3 – Minerals.

Total Antioxidant Capacity

Because of the interactions between antioxidants in cells, the critical issue is a patient's overall capacity to resist oxidation. Studies focused on single antioxidants have reported that they have either no effect or slight negative correlation with incidence of cancers of the lung and breast [72-74]. As discussed above, it may be that the various electron acceptor/donor roles of the antioxidants must be present in a narrow range of proportional concentrations. If any one component, like β-carotene, is increased dramatically, then it can become a source of electrons for further damaging radical formation. A test to evaluate the overall ability to meet the challenge of oxidant stress would be quite helpful.

Oxygen Radical Absorption Capacity

One promising antioxidant capacity test is the oxygen radical absorbent capacity (ORAC) test. When serum is challenged with a chemical oxidant, the time course of resistance to protein destruction is a measure of the overall ability to resist the oxidative challenge [75, 76]. The test is useful for assessing antioxidant capacities of foods and for comparative antioxidant capacities among species of animals [77]. A difficulty of the test arises from the fact that serum components like albumin and uric acid that are present at high concentrations make major contributions to the resistance against oxidation. Their contributions are so great that they obscure the more important factors (such as ascorbate and vitamin E) that are major factors for protection in sub-cellular compartments. The use of ORAC to detect susceptibility for human disease has been disappointing because of the lack of discrimination among healthy and diseased individuals. Due to the lack of sensitivity, serum from a range of normal individuals has ORAC values very similar to those from diabetics, cancer patients, and HIV-infected individuals [78].

Treatment Options

In the case of markers of oxidant damage, the lower the test result level, the better. However, since free radicals are produced in the normal processes of metabolism, the body can tolerate a certain level without significant health risk.

Treatment of patients with elevated lipid peroxide levels involves the two-fold approach of reducing exposure to free radical generating agents and increasing antioxidant levels in the body.

Reduce Pro-Oxidants

It is clear that environmental agents increase free radical activity. The toxicity of heavy metals, pesticides, ionizing radiation, and cigarette smoke may be due to their free radical initiating ability. Lifestyle factors such as stress, heavy exercise, smoking, excess alcohol, and infections tend to raise lipid peroxide levels.

The body burden of pro-oxidant compounds may be lowered by reducing exposure and increasing removal or detoxification capacity. Individuals who may have greatly increased pro-oxidant responses to even slight xenobiotic exposures must take extra measures to lower environmental levels. Other sources provide more details about how to manage environmentally sensitive patients [79].

Increase Antioxidants

If lipid peroxide levels are high, supplementation of natural antioxidants is indicated. These include vitamins A, B_2, B_3, C, E, and β-carotene, selenium, coenzyme Q_{10}, taurine, and certain bioflavonoids [7]. Balanced intake of all of the antioxidants is superior to individual supplementation because of the demonstrated co-dependence of antioxidant reaction sequences. In this regard, it is worth considering the use of botanical sources of antioxidants such as Ginkgo biloba, Panax ginseng, or other rich sources of complex mixtures of polyphenols with antioxidant vitamins [80].

Magnesium deficiency can increase oxidant loads because of its effect of increasing stress responses [81]. Magnesium was not included above as an antioxidant nutrient because it has no specific free-radical scavenging cofactor role. However, when implementing overall oxidant reduction strategies, magnesium status must be assured.

Other Lifestyle Factors

Sleep disorders interfere with normal tissue restoration and hormonal control patterns. The pineal hormone melatonin, in addition to its role as functional regulator of the hypothalamo-pituitary-gonadal system, is a highly efficient free radical scavenger and general antioxidant [82]. It is roughly twice as effective as vitamin E for protection against the peroxyl radical. Regular adequate sleep is the major controlling factor in maintaining adequate levels of melatonin.

References

[1] Smith, C., *Free radical mechanisms of tissue injury, in Free radical mechanisms of tissue injury,* M. Moslen and C. Smith, Editors. 1992, CRC Press: Boca Raton. p. 224.

[2] Levine, S. and P. Kidd, *Antioxicant adaptation; Its role in free radical pathology.* 1985, SanLeandro: Biocurrents-Allergy Research Group. 367.

[3] Lelkes, E., B. Unsworth, and P. Lelkes, Reactive Oxygen Species, Apoptosis and Altered NGF-Signaling in PC12 Pheochromocytoma Cells Cultured in Elevated Glucose: An in vitro Cellular Model for Diabetic Neuropathy. 2000, *6th Internet World Congress for Biomedical Sciences Invited Symposium: Reactive Oxygen Species and Neurodegenerative Diseases.*

[4] Smythies, J., A note on the role of endocytosis in the etiology of Alzheimer´s disease: a new hypothesis. 2000, *6th Internet World Congress for Biomedical Sciences Index Invited Symposium: Reactive Oxygen Species and Neurodegenerative Diseases.*

[5] Sun, A., B. Draczynska-Lusiak, and G. Sun, Oxidized lipoproteins, beta amyloid peptides and Alzheimer's disease. 2000, *6th Internet World Congress for Biomedical Sciences Index Invited Symposium: Reactive Oxygen Species and Neurodegenerative Diseases.*

[6] Knight, J.A., Free radicals: their history and current status in aging and disease. *Ann Clin Lab Sci,* 28(6). 331-46, 1998.

[7] Bland, J., The nutritional effects of free radical pathology, in *1986 A year in nutritional medicine,* B. JS, Editor. 1986, Keats: New Canaan. p. 293-322.

[8] Comporti, M., Lipid peroxidation and cellular damage in toxic liver injury. *Lab Invest,* 53(6). 599-623, 1985.

[9] Pryor, W., Free radical involvement in chronic diseases and aging: the toxicity of lipid hydroperoxides and their decomposition products, in *Xenobiotic metabolism: nutritional effects,* J. Finley and D. Schwass, Editors. 1985, American Chemical Society: Washington, DC. p. 77-96.

[10] Ledwozyw, A., *et al.*, The relationship between plasma triglycerides, cholesterol, total lipids and lipid peroxidation products during human atherosclerosis. *Clin Chim Acta,* 155(3). 275-83, 1986.

[11] Francesco, V., *et al.*, Malondialdehyde-like material and beta-thromboglobulin plasma levels in patients suffering from transient ischemic attacks. *Stroke,* 16(1). 14-6, 1985.

[12] Lewin, R., Drug trial for Parkinson's [news]. *Science,* 236(4807). 1420, 1987.

[13] Cavarocchi, N.C., *et al.*, Superoxide generation during cardiopulmonary bypass: is there a role for vitamin E? *J Surg Res,* 40(6). 519-27, 1986.

[14] Jovanovic, S.V., D. Clements, and K. MacLeod, Biomarkers of oxidative stress are significantly elevated in Down syndrome. *Free Radic Biol Med,* 25(9). 1044-8, 1998.

[15] Lonsdale, D., Free oxygen radicals and disease., in *1986 A year in nutritional medicine,* J. Bland, Editor. 1986, Keats: New Canaan. p. 85-114.

[16] Levchuk, A.A., *et al.*, Effect of the carcinogenic tyrosine metabolite p-hydroxyphenyllactic acid on the ascorbic acid concentration in the organs and blood of mice. *Biull Eksp Biol Med,* 102(10). 462-3, 1986.

[17] Slater, T.F., *et al.*, Free radical mechanisms in relation to tissue injury. *Proc Nutr Soc,* 46(1). 1-12, 1987.

[18] Sevanian, A. and P. Hochstein, Mechanisms and consequences of lipid peroxidation in biological systems. *Annu Rev Nutr,* 5. 365-90, 1985.

[19] Satoh, K., Serum lipid peroxide in cerebrovascular disorders determined by a new colorimetric method. *Clin Chim Acta,* 90(1). 37-43, 1978.

[20] Wartanowicz, M., *et al.*, The effect of alpha-tocopherol and ascorbic acid on the serum lipid peroxide level in elderly people. *Ann Nutr Metab,* 28(3). 186-91, 1984.

[21] Pratic, D., *et al.*, Ongoing Prothrombotic State in Patients With Antiphospholipid Antibodies: A Role for Increased Lipid Peroxidation. *Blood,* 93(10). 3401-3407, 1999.

[22] Voutilainen, S., *et al.*, Enhanced In vivo lipid peroxidation at elevated plasma total homocysteine levels. *Arterioscler Thromb Vasc Biol,* 19(5). 1263-6, 1999.

[23] Oguogho, A., M. Mehrabi, and H. Sinzinger, Increased plasma, serum and urinary 8-epi-prostaglandin F2 alpha in heterozygous hypercholesterolemia. *Wien Klin Wochenschr,* 111(3). 113-8, 1999.

[24] Brame, C.J., *et al.*, Identification of extremely reactive gamma-ketoaldehydes (Isolevuglandins) as products of the isoprostane pathway and characterization of their lysyl protein adducts. *J Biol Chem,* 274(19). 13139-46, 1999.

[25] Hill, D. and J. Awad, Increased urinary F2-isoprostane excretion in alcoholic liver disease. *Free Radic Biol Med,* 26(656-60)1999.

[26] Aksenov, M.Y., *et al.*, Amyloid beta-peptide (1-40)-mediated oxidative stress in cultured hippocampal neurons. Protein carbonyl formation, CK BB expression, and the level of Cu, Zn, and Mn SOD mRNA. *J Mol Neurosci,* 10(3). 181-92, 1998.

[27] Lovell, M.A., C. Xie, and W.R. Markesbery, Protection against amyloid beta peptide toxicity

by zinc. *Brain Res*, 823(1-2). 88-95, 1999.
[28] Boger, R.H., *et al.*, Dietary L-arginine and alpha-tocopherol reduce vascular oxidative stress and preserve endothelial function in hyper-cholesterolemic rabbits via different mechanisms. *Atherosclerosis*, 141(1). 31-43, 1998.
[29] Kumagai, T., *et al.*, 4-hydroxy-2-nonenal, the end product of lipid peroxidation, is a specific inducer of cyclooxygenase-2 gene expression. *Biochem Biophys Res Commun*, 273(2). 437-41, 2000.
[30] Mak, S., *et al.*, Unsaturated aldehydes including 4-OH-nonenal are elevated in patients with congestive heart failure. *J Card Fail*, 6(2). 108-14, 2000.
[31] Jacob, S., *et al.*, Oral administration of RAC-alpha-lipoic acid modulates insulin sensitivity in patients with type-2 diabetes mellitus: a placebo-controlled pilot trial. *Free Radic Biol Med*, 27(3-4). 309-14, 1999.
[32] Yaworsky, K., *et al.*, Engagement of the insulin-sensitive pathway in the stimulation of glucose transport by alpha-lipoic acid in 3T3-L1 adipocytes. *Diabetologia*, 43(3). 294-303, 2000.
[33] Owen, R.W., *et al.*, The antioxidant/anticancer potential of phenolic compounds isolated from olive oil. *Eur J Cancer*, 36(10). 1235-47, 2000.
[34] Picklo, M.J., *et al.*, 4-Hydroxy-2(E)-nonenal inhibits CNS mitochondrial respiration at multiple sites. *J Neurochem*, 72(4). 1617-24, 1999.
[35] Takeda, A., *et al.*, In Alzheimer's disease, heme oxygenase is coincident with alz50, an epitope of tau induced by 4-hydroxy-2-nonenal modification. *J Neurochem*, 75(3). 1234-41, 2000.
[36] Lee, S.H., *et al.*, Characterization of 2'-deoxyadenosine adducts derived from 4-oxo-2-nonenal, a novel product of lipid peroxidation. *Chem Res Toxicol*, 13(7). 565-74, 2000.
[37] Bartsch, H., J. Nair, and I. Velic, Etheno-DNA base adducts as tools in human cancer aetiology and chemoprevention. *Eur J Cancer Prev*, 6(6). 529-34, 1997.
[38] Zainal, T.A., *et al.*, Caloric restriction of rhesus monkeys lowers oxidative damage in skeletal muscle. *Faseb J*, 14(12). 1825-1836, 2000.
[39] Laurent, A., *et al.*, Metabolism of 4-hydroxy-nonenal, a cytotoxic product of lipid peroxidation, in rat precision-cut liver slices. *Toxicol Lett*, 114(1-3). 203-14, 2000.
[40] Alary, J., *et al.*, Mercapturic acid conjugates as urinary end metabolites of the lipid peroxidation product 4-hydroxy-2-nonenal in the rat. *Chem Res Toxicol*, 8(1). 34-9, 1995.
[41] Carew, T.E., Role of biologically modified low-density lipoprotein in atherosclerosis. *Am J Cardiol*, 64(13). 18G-22G, 1989.
[42] Witztum, J.L. and D. Steinberg, Role of oxidized low density lipoprotein in atherogenesis. *J Clin Invest*, 88(6). 1785-92, 1991.
[43] Crawford, R.S., *et al.*, Dietary antioxidants inhibit development of fatty streak lesions in the LDL receptor-deficient mouse. *Arterioscler Thromb Vasc Biol*, 18(9). 1506-13, 1998.
[44] Fukumoto, M., *et al.*, Antibodies against oxidized LDL and carotid artery intima-media thickness in a healthy population. *Arterioscler Thromb Vasc Biol*, 20(3). 703-7, 2000.
[45] Nicoletti, A., *et al.*, Induction of neonatal tolerance to oxidized lipoprotein reduces atherosclerosis in ApoE knockout mice. *Mol Med*, 6(4). 283-90, 2000.
[46] Osada, K., *et al.*, Effect of oxidized cholesterol on age-associated changes to immune parameters in spleen lymphocytes and peritoneal exudate cells derived from rats. *Biosci Biotechnol Biochem*, 64(5). 1047-51, 2000.
[47] Liu, P., *et al.*, Presence of oxidized cholesterol in caveolae uncouples active platelet- derived growth factor receptors from tyrosine kinase substrates. *J Biol Chem*, 275(41). 31648-54, 2000.
[48] Kovacs, I.B., *et al.*, Elevated plasma lipid hydroperoxides in patients with coronary artery disease. *Am Heart J*, 134(3). 572-6, 1997.
[49] Packard, C., M. Caslake, and J. Shepherd, The role of small, dense low density lipoprotein (LDL): a new look. *Int J Cardiol*, 74 Suppl 1. S17-22, 2000.
[50] Fiala, E.S., *et al.*, (-)-Epigallocatechin gallate, a polyphenolic tea antioxidant, inhibits peroxynitrite-mediated formation of 8-oxodeoxyguanosine and 3- nitrotyrosine. *Experientia*, 52(9). 922-6, 1996.
[51] Shimizu, K., *et al.*, Stimulation of macrophages by retinal proteins: production of reactive nitrogen and oxygen metabolites. *Invest Ophthalmol Vis Sci*, 40(13). 3215-23, 1999.
[52] Wu, G.S., J. Zhang, and N.A. Rao, Peroxynitrite and oxidative damage in experimental autoimmune uveitis. *Invest Ophthalmol Vis Sci*, 38(7). 1333-9, 1997.
[53] Bougle, D., *et al.*, Retinal superoxide dismutase activity in newborn kittens exposed to normobaric hyperoxia: effect of vitamin E. *Pediatr Res*, 16(5). 400-2, 1982.
[54] Moskovitz, J., *et al.*, Overexpression of peptide-methionine sulfoxide reductase in Saccharomyces cerevisiae and human T cells provides them with high resistance to oxidative stress. *Proc Natl Acad Sci U S A*, 95(24). 14071-5, 1998.
[55] Sun, H., *et al.*, Repair of oxidized calmodulin by methionine sulfoxide reductase restores ability To activate the plasma membrane Ca-ATPase. *Biochemistry*, 38(1). 105-12, 1999.
[56] Haghighi, A.Z. and K.R. Maples, On the mechanism of the inhibition of glutamine synthetase and creatine phosphokinase by methionine sulfoxide. *J Neurosci Res*, 43(1). 107-

11, 1996.

[57] Biewenga, G.P., *et al.*, Effects of dihydrolipoic acid on peptide methionine sulfoxide reductase. Implications for antioxidant drugs. *Arzneimittelforschung*, 48(2). 144-8, 1998.

[58] Garner, B., *et al.*, Oxidation of high density lipoproteins. I. Formation of methionine sulfoxide in apolipoproteins AI and AII is an early event that accompanies lipid peroxidation and can be enhanced by alpha-tocopherol. *J Biol Chem*, 273(11). 6080-7, 1998.

[59] Khan, J., *et al.*, 3-Nitrotyrosine in the proteins of human plasma determined by an ELISA method [published erratum appears in Biochem J 1998 Jun 15;332 (Pt 3):808]. *Biochem J*, 330(Pt 2). 795-801, 1998.

[60] Loft, S., *et al.*, Increased urinary excretion of 8-oxo-2'-deoxyguanosine, a biomarker of oxidative DNA damage, in urban bus drivers. *Mutat Res*, 441(1). 11-9, 1999.

[61] Lenton, K.J., *et al.*, Glutathione and ascorbate are negatively correlated with oxidative DNA damage in human lymphocytes. *Carcinogenesis*, 20(4). 607-13, 1999.

[62] Chen, L., *et al.*, Diet modification affects DNA oxidative damage in healthy humans. *Free Radic Biol Med*, 26(5-6). 695-703, 1999.

[63] Souza-Pinto, N.C., *et al.*, Age-associated increase in 8-oxo-deoxyguanosine glycosylase/AP lyase activity in rat mitochondria. *Nucleic Acids Res*, 27(8). 1935-42, 1999.

[64] Thomas, S., *et al.*, Measurement of DNA damage using the comet assay. *Methods Mol Biol*, 100. 301-9, 1998.

[65] Anderson, D., *et al.*, Effects of iron salts and haemosiderin from a thalassaemia patient on oxygen radical damage as measured in the comet assay. *Teratog Carcinog Mutagen*, 20(1). 11-26, 2000.

[66] Moller, P., *et al.*, The comet assay as a rapid test in biomonitoring occupational exposure to DNA-damaging agents and effect of confounding factors. *Cancer Epidemiol Biomarkers Prev*, 9(10). 1005-15, 2000.

[67] Cooper, D.A., A.L. Eldridge, and J.C. Peters, Dietary carotenoids and lung cancer: a review of recent research. *Nutr Rev*, 57(5 Pt 1). 133-45, 1999.

[68] Nuttall, K.L., Elemental selenium and glutathione reductase. *Med Hypotheses*, 16(2). 155-8, 1985.

[69] Hatanaka, N., *et al.*, Selenium kinetics and changes in glutathione peroxidase activities in patients receiving long-term parenteral nutrition and effects of supplementation with selenite. *Nutrition*, 16(1). 22-6, 2000.

[70] Neve, J., *et al.*, Selenium, zinc and copper in Down's syndrome (trisomy 21): blood levels and relations with glutathione peroxidase and superoxide dismutase. *Clin Chim Acta*, 133(2). 209-14, 1983.

[71] Fenech, A.G. and R. Ellul-Micallef, Selenium, glutathione peroxidase and superoxide dismutase in maltese asthmatic patients: effect of glucocorticoid administration. *Pulm Pharmacol Ther*, 11(4). 301-8, 1998.

[72] Ocke, M.C., *et al.*, Repeated measurements of vegetables, fruits, beta-carotene, and vitamins C and E in relation to lung cancer. The Zutphen Study. *Am J Epidemiol*, 145(4). 358-65, 1997.

[73] Verhoeven, D.T., *et al.*, Vitamins C and E, retinol, beta-carotene and dietary fibre in relation to breast cancer risk: a prospective cohort study. *Br J Cancer*, 75(1). 149-55, 1997.

[74] Omenn, G.S., *et al.*, Risk factors for lung cancer and for intervention effects in CARET, the Beta-Carotene and Retinol Efficacy Trial [see comments]. *J Natl Cancer Inst*, 88(21). 1550-9, 1996.

[75] Cao, G., H.M. Alessio, and R.G. Cutler, Oxygen-radical absorbance capacity assay for antioxidants [see comments]. *Free Radic Biol Med*, 14(3). 303-11, 1993.

[76] Cao, G., *et al.*, Automated assay of oxygen radical absorbance capacity with the COBAS FARA II. *Clin Chem*, 41(12 Pt 1). 1738-44, 1995.

[77] Ninfali, P. and G. Aluigi, Variability of oxygen radical absorbance capacity (ORAC) in different animal species. *Free Radic Res*, 29(5). 399-408, 1998.

[78] Lord, R.S. and J.A. Bralley, *Notes on the development of ORAC assay*. Private communication, 1996.

[79] Rea, W., *Chemical Sensitivity*. 1994, Boca Raton: Lewis. 2015.

[80] Mantle, D., F. Eddeb, and A.T. Pickering, Comparison of relative antioxidant activities of British medicinal plant species in vitro. *J Ethnopharmacol*, 72(1-2). 47-51, 2000.

[81] Seelig, M.S., Consequences of magnesium deficiency on the enhancement of stress reactions; preventive and therapeutic implications (a review). *J Am Coll Nutr*, 13(5). 429-46, 1994.

[82] Reiter, R.J., Functional pleiotropy of the neurohormone melatonin: antioxidant protection and neuroendocrine regulation. *Front Neuroendocrinol*, 16(4). 383-415, 1995.

Chapter 10

Hormones

Introduction

The great fluctuations of food intake from day to day are in contrast to the great stability of body processes. Maintenance of constant energy flow, end product removal, and neuromuscular readiness in states of food overabundance or deprivation is accomplished by the elaborate network of hormone responses. For example, when high carbohydrate foods are consumed, insulin and insulin-like growth factor 1 (IGF-1) stimulate the lowering of blood sugar. If blood sugar drops too low or too fast, the adrenal glands start to produce greater amounts of adrenaline and cortisol. Circulating levels of thyroid hormone, estrogen, and testosterone modulate tissue responses to adrenaline and cortisol. Thus, the web-like interplay of hormones works as part of the total neuroimmuno-hormonal regulation of the body.

Hormones are a class of chemical messengers that control cell function. The full spectrum of cellular controls includes hormones and newer findings of short-range interactions. Classical endocrinology is concerned with long-range interactions in which chemical messengers produced in one organ exert control on specific target tissues outside of the site of synthesis. Hormones may be defined as those compounds that display such properties. We now know that many other cell controls are present that operate through more short-range effects.

The catalog of one specialty laboratory lists over 350 separate single hormone or multiple hormone profiles [1]. Well over 100 direct hormone and challenge response tests are available for GI hormones alone. Endocrinology is a burgeoning field where the elucidation of cell differentiation and regulation has led to a virtual explosion of knowledge about the rapidly growing list of compounds used for cellular communication. We will confine our discussion to a few of the tests that are most directly related to nutrient status.

The most-well known example of nutrient dependency on hormone activity is increased urinary losses of B-vitamins due to the stress response. Chronic stress causes glucocorticoid elevation, which increases urinary losses of water-soluble vitamins. Chronic stress also increases amino acid utilization, while high carbohydrate diets demand increased insulin output. Understanding the hormone balance in each case allows accurate assessment of where the focus of attention should be placed. Laboratory evaluations reflect the sum total of nutrient intake and hormone controls on the current body pools of nutrients. There are varied opinions about which is the most important hormone to investigate in routine clinical evaluations. The list of high priority hormones to evaluate includes insulin, estrogen, testosterone, thyroxin, and the stress hormones cortisol and DHEA. The related releasing factors TSH, FSH, LH, and ACTH may also be considered. Gastrointestinal hormones have major impact on digestion and assimilation of nutrients. Tests for these hormones are used in nutritional status assessments and clinical investigations of childhood growth problems. The pancreatic polypeptide hormone secretin has come to be recognized as an agent capable of far-reaching consequences due to the discovery that it may be greatly reduced in autistic spectrum disorders [1].

Studies of societies that have remarkably long life spans universally describe people with lifestyles of low stress. Factors like anxiety, competition, and poor personal relationships contribute to degenerative diseases and aging. Laboratory evaluation can give direct clinical insight about the level of stress and other adverse lifestyle factors that bear on outcomes from nutritional interventions. The physiological responses to stressors of various origins are initiated by modulating the output of cellular controls such as hormones, neurotransmitters, nitric oxide, and eicosanoids. Other chapters have covered testing for amino acid neurotransmitter precursors, organic acid excretory products, and fatty acid flow into eicosanoid pathways. Measuring polypeptides, catecholamines, thyroid hormones, and sex hormones adds valuable clinical information that can indicate specific nutrient interventions.

The relationships between nutrients and hormones may be viewed from two perspectives. The output of hormones in response to physiological stresses can impact the demands for specific nutrients (Table 10-1). For example, hypothyroidism causes increased utilization of the antioxidant vitamins C and E, and the abnormality is accompanied by decreased conversion of β-carotene to vitamin A [3]. The other nutrient effects listed in Table 10-1 illustrate the range of hormonal impacts on nutrient status.

TABLE 10-1. Hormonal Effects on Nutrient Status

Hormone	Abnormal	Nutrient	Hormone Effect on Nutrient(s)
Thyroid	High	Vitamins C, E	Increased utilization
	Low	Vitamin A	Decreased carotene conversion
	Low	Iodine	Indication of deficiency
Testosterone	Low (M)	Zinc, Amino Acids	Increased utilization
Estrogen	Low (F)	Calcium, Magnesium, Vit. D	Decreased mineralization of bone
ACTH	High	Amino Acids	Increased utilization
Cortisol	High	B-Complex Vitamins, Amino Acids, Minerals, Iron	Increased losses
		Fatty Acids	Lowered omega-6 levels
DHEA	High	Vitamin E	Increased oxidant stress
Growth Hormone	Low	Fatty Acids, Amino Acids	Decreased fat oxidation
		Vitamin D	Increased catabolism
Insulin	High	Minerals, Amino Acids, Fatty Acids	Increased losses
	Low	Glucose	Decreased metabolic utilization and increased urinary losses

TABLE 10-2. Nutrient Influences on Cellular Control

Nutrient or Dietary Component	Cellular Control Factor(s)	Effect of Low Nutrient Supply on Hormone
Iodine	Thyroxin	Reduced biosynthesis
Amino Acids	Tryptophan – Serotonin Tyrosine – Thyroxin Arginine – Nitric oxide	Reduced biosynthesis
Essential Fatty Acids	Eicosanoids $T_4 \rightarrow T_3$	Reduced biosynthesis Reduced conversion rate
Zinc	Growth hormone Testosterone Thyroxin Corticosteroids	Reduced concentrations Primary failure of Leydig cells Reduced synthesis and reduced tissue conversion to T_3 Impaired ACTH response
Selenium	T_4	Reduced conversion to T_3
Molybdenum & Boron	Estrogen	Decreased metabolic conversion to 2-methoxyestrone
Dietary Fiber	Butyrate	Unregulated colonic epithelial growth
Soy Products	Isoflavone derivatives	Altered estrogen activity

On the other hand, nutrient intake can have effects on the circulating levels of hormones (Table 10-2). Examples are the restriction of thyroxin synthesis by iodine deficiency [4] and the lowered serotonin levels in brain when tryptophan is depleted [5]. Nutrients play important roles in regulation of cell function by their actions on enzyme activities and gene expression [6]. Hormone actions are also mediated by changes in nutrient intake. Marked increases in glucagon concentrations, for example, are observed after high protein feeding [7]. Diets high in carbohydrate induce insulin secretion with consequently lowered concentrations of amino acids in plasma and increased deposition of body fat.

Deficiency of iodine has long been recognized as the cause of goiter. Iodine is required for the synthesis of thyroxin, which regulates cell activities rather than directly affecting enzyme activities like most other nutrients discussed in this book. Lack of the precursor iodine causes low thyroid hormone output from the thyroid gland. Iodine biochemistry is unique, but it is not alone in the class of nutrient requirements for maintenance of cell controls. Nutrient insufficiencies and toxic burdens are best evaluated with knowledge of cell regulation signals that govern normal function. Not only are molecules that control the functions of cells derived from nutrients, but also many natural food ingredients have direct effects on regulation of metabolism and cell communication. At another level, changes in dietary composition of carbohydrate, fat, and protein exert changes in cell controls that effect micronutrients. Laboratory assessment of hormones is pertinent to the practice of nutritional medicine because their impact on metabolism causes shifts in nutrient demands. The stress response hormones increase the rate of loss of the B-complex vitamins while the mineralocorticoids increase the rate of bone loss, thus raising need for bone-forming minerals. Steroids, especially the sex hormones, exert controls on body fat distribution and lean body mass, influencing the rates of fatty acid and amino acid metabolism. It is because of these controls that hormonal analysis is a powerful component of comprehensive laboratory evaluation of nutritional status. Test profiles commonly used to evaluate hormone effects are thyroid, insulin and insulin-like growth factor, the stress hormones (primarily DHEA and cortisol), and the sex hormones (testosterone and the estrogens). The following is a brief discussion of tests that have been useful in evaluating these hormones and their interaction with nutritional status.

Thyroid Hormone

Thyroid hormone evaluation was one of the first to be routinely done in the clinical laboratory. Protein-bound iodine (PBI) analysis was replaced by tetraiodothyronine (T4) and, more recently, by thyroid profiles of free and bound T4 along with the metabolite triiodothyronine (T3) and thyroid-stimulating hormone (TSH). Calculated values of T3 uptake and free thyroxine index (T7) have been added to improve discrimination of primary thyroid disorders and conditions that stimulate the synthesis of serum thyroid-binding proteins. Still, some physicians prefer to rely on patient history and physical exam for final evaluation of thyroid status. Elevated TSH with normal (or low) T3 and T4 is a pattern indicating potential essential nutrient limitation of thyroid gland hormone T4 synthesis (Table 10-3). Since iodine and tyrosine are the direct precursors for thyroid hormone synthesis, supplementation may help to maintain normal thyroid status in patients with this pattern. When plasma

TABLE 10-3. Thyroid Profile Data in Nutrient-Related Hormone Response Failure. Thyroid profile for a patient with impaired TSH response, which may be due to insufficiency of iodine or tyrosine.

Analyte	Result	Range
Total T3	0.86	0.45 - 1,37 mcg/mL
T3 Uptake	34	23 - 40%
T4	6.3	4.5 - 12.0 ug/dL
T7 (FTI)	2.1	1.4 - 3.8
TSH	5.3 (H)	0.3 - 4.7 UIU/mL

amino acid profile data is available, low phenylalanine and tyrosine are indicators that corroborate substrate limitation for T4 synthesis.

The activity of the primary circulating thyroid hormone, T4, depends on its conversion to T3 by enzymes requiring selenium [8]. Patients with normal circulating hormone levels who display clinical hypothyroid symptoms may be selenium depleted. Thyroid function is particularly mineral dependent. Single and multiple deficiencies of selenium, zinc, and iodine have distinct effects on thyroid metabolism and structure in rats [9]. Human populations have developed Keshan disease, Keshan-Beck disease, and hypothyroid cretinism when they become deficient in iodine and selenium simultaneously [10]. Thyroidectomized animals have decreased skeletal muscle breakdown (low 3-methylhistidine), low insulin, and increased TSH levels [11]. This effect of thyroid hormone on protein turnover suggests that amino acid evaluation is indicated for patients with abnormal thyroid status. Vitamin A status also should be evaluated for hypothyroid patients who have impaired conversion of β-carotene to vitamin A as an indirect effect of the hypothyroid state [3].

High blood levels of fatty acids can inhibit the conversion of T4 to T3. Extrathyroidal conversion of T4 to T3 is directly correlated with plasma free fatty acid concentration in patients with various non-thyroidal illnesses including diabetes mellitus, liver cirrhosis, chronic obstructive pulmonary disease, and chronic heart failure [12]. The effect is enhanced by adding oleic acid, especially in patients with lower serum albumin. This effect indicates that the composition of fatty acids, not just their total concentrations, determines the magnitude of inhibition [13]. On the other hand, fatty acid composition is influenced by the action of thyroid hormone. Hypothyroid patients have inhibited delta-6- and delta-5-desaturase pathways that cause alterations of membrane fatty acid composition [14].

In hyperthyroid patients, risk of oxidative damage is increased due to the higher metabolic rate. Lipid peroxidation is increased compared to controls, and activities of antioxidant enzymes are altered. These effects are significantly reduced when ascorbic acid is used at 1000 mg/day [15]. Other markers of oxidative damage provide insight about how well a hyperthyroid patient is controlling oxidative damage (see Chapter 9 – Oxidant Stress).

Growth Hormone & Insulin-like Growth Factor 1

The principal effect of growth hormone (GH), a small peptide hormone secreted by the pituitary gland, is to stimulate the liver and other tissues to synthesize insulin-like growth factor 1 (IGF-1) [16]. Thus, IGF-1 mediates the action of growth hormone. The levels of IGF-1, unlike growth hormone, are relatively constant during the day making its measurement more reliable. The greater stability of serum

IGF-1 is due in part to the buffering action of IGF binding protein that is also present in serum. The presence of a binding protein raises the question of whether bound, free, or total forms should be measured. Salivary IGF-1 levels measure the active, free form, and they reflect the GH status of the patient [17]. The concentrations of serum IGF-1 are low at birth (< 100 ng/ml) and rise steeply during childhood and puberty, reaching values as high as 1000 ng/ml by age 15. Levels generally begin to fall by the mid-twenties, and they continue to decline thereafter [16 , 18]. Although the absence of GH and IGF-1 may contribute to the phenotype of aging, increased IGF-1 is an important risk factor in age-related pathologies including lung, breast, and prostate cancer. This apparent contradiction raises the question of whether levels of these hormones change in response to other factors as a way of normalizing physiological stressors.

Nutritional status affects serum IGF-1 concentrations, and nutrient responses are dependent on IGF-1. Fasting results in resistance to growth hormone. Caloric restriction induces moderate endocrine compensatory mechanisms including increased GH and decreased IGF-1. It has been proposed that the decline in pathologies and increased life span observed in caloric restriction result from these changes causing decreased stimulus for cellular replication [19]. Plasma amino acid concentrations reflect changes in amino acid metabolism associated with changes in IGF-1 and IGF binding protein levels. In rats, lower dietary protein intake and lower plasma EAA/NEAA ratios are associated with lower IGF-1 [20]. GH treatment can reduce post-menopausal osteoporosis by up-regulation of intestinal vitamin D receptors similar to the action of estrogen [21]. Both GH and IGF-1 have been shown to modulate renal vitamin D through regulation of inorganic phosphate balance [22].

Fat-soluble nutrients affect the GH and IGF-1 activities also. A direct correlation between deficient levels of vitamin A and both 24 hour GH output and IGF binding protein was found in short, slowly growing children [23]. Dietary conjugated linoleic acid (CLA) increases serum IGF-1 and IGF binding protein concentrations and reduces bone formation in rats fed n-6 or n-3 fatty acids [24]. This observation may be important to consider in cases where CLA is used for reduction of body fat [25].

Several factors have been identified to explain improvements of tissue function in aged animals resulting from administration of GH. GH has been successfully used in reducing cholesterol [26, 27], body fat [28], blood pressure, anxiety, and increasing lean muscle [29], bone density, HDL, and energy levels [30]. Individual differences in IGF-1 synthesis are significant. Dose titration to the upper normal mid-life range reduces the incidence of side effects and GH insensitivity. Women have lower sensitivities than age-matched men and are likely to require slightly higher doses of GH. Starvation, cachexia, hyperalimentation, and insulin-dependent diabetes mellitus are associated with functional GH resistance in which, despite normal or high GH levels, circulatory IGF-1 levels are low. In such conditions, human recombinant IGF-1 is available for direct intravenous or subcutaneous administration [31], but doses must be adjusted carefully to avoid side effects.

Insulin

Diagnosis, monitoring, and patient education in Type I and Type II diabetes have been extensively reviewed [32-34]. Recent attention also has been focused on the occurrence of another aspect of glycemic control, insulin resistance, sometimes called "Syndrome X" [35]. As resistance to insulin increases, the magnitude of the insulin response to a carbohydrate load increases. The higher circulating insulin concentration affects energy metabolism by increasing synthesis of triglycerides and cholesterol. In other words, hypercholesterolemia and hypertriglycreridemia can be the result of hyperinsulinemia.

Fasting serum insulin concentrations above 10 IU/mL indicate insulinemia associated with insulin insensitivity. Follow up testing with a two-hour glucose tolerance test with insulin readings aids in the differential diagnosis of dysglycemia and dysinsulinemia. Low carbohydrate diets help to lower insulin output and reduce the impact of hyperinsulinism.

Macronutrient control is a major focus of diabetic patient management. Restricted carbohydrate and simple sugar intake with adequate protein supplied in several small meals can help stabilize blood sugar in both Type I and II diabetes. Interventions with other essential nutrient supplements have also been shown to be helpful. Nutritional interventions with fish oils [36], chromium [37, 38], and vanadium [39] are effective for aiding the insulin responsiveness of Type II diabetics. Profiles of fatty acids and trace elements frequently reveal need for these nutrients in diabetic patients. Omega-3 fatty acids improve insulin sensitivity [40, 41].

Testosterone

Testosterone is the major anabolic hormone in males. Levels reach a peak at around 20 years of age and decline steadily thereafter. The negative impact of zinc deficiency on testosterone output is the strongest nutrient effect on this hormone [42]. The negative effect of testosterone deficiency on nitrogen balance and muscle development may be the major impact of testosterone on nutrient status. Myopathy has been reported in patients with low testosterone [43]. Muscle loss may be reversed by treatment with testosterone and by assuring amino acid adequacy (see Chapter 4 – Amino Acids). Other effects of low testosterone in adults are gaining recognition. Testosterone hormone replacement therapy is becoming more common [44].

Testosterone testing may be done in saliva, in which case the value reflects the free testosterone in serum. Total serum testosterone is also useful to determine whether there is normal response of the Leydig cells to FSH and LH. In blood, testosterone is tightly bound to a globulin called sex hormone-binding globulin (SHBG) and weakly bound to albumin. Simultaneous determination of total testosterone and SHBG allows the calculation of free testosterone concentration, sometimes called the free androgen index (FAI). Bone mineral density is positively correlated with FAI in elderly males [45]. Measurement of FAI in infertile women reveals an association of ovarian dysfunction with hyperandrogenemia [46].

Estrogens and Estrogen Metabolites

Estrogens are the major steroid anabolic hormones in females before menopause. Estrogen hormone replacement therapy (HRT) is widely used to prevent the post-menopausal, accelerated loss of bone and other aging factors. Testing for hormone levels must account for menstrual cycle variations in pre-menopausal women. Estradiol and progesterone measurements allow the determination of peak values within broad ranges of cycle phases (Table 10-4) and, when 15-20 points are tested over a month interval, hormone output may be assessed by observing the degree of rise at ovulation (Figure 10-1). In order to evaluate the full physiological impact of estrogen hormones, a more complete profile that includes estrone and estriol may be done so that total estrogen can be calculated. Direct assay of total estrogens is also available.

TABLE 10-4. Female Cycle Mapping Reference Values

	Follicular	Midcycle	Luteal	Post Menopause ng/ml
Progesterone	< 0.5	0.05 - 0.1 ng/ml	0.06 - 0.5 ng/ml	< 0.05
Estrone (E1)	2.0 - 39	11 - 45 pg/ml	3 - 52 pg/ml	1 - 7
Estradiol (E2)	0.5 - 5.0	2.5 - 7.0 pg/ml	0.2 - 4.0 pg/ml	< 0.05
Estriol (E3)	3 - 48	20 - 130 pg/ml	9 - 60 pg/ml	0 - 30
Total Estrogen	7 - 110	38 - 221 pg/ml	13 - 139 pg/ml	0 - 60

The potential pro-carcinogenic effects of estrogen have dampened the enthusiasm about HRT. In fact, studies focused on problems with estrogen replacement have led to concerns about the risk of aberrant estrogen metabolism in premenopausal women [47]. The role played by estrogen oxidative metabolism in hormone-sensitive diseases has come from studies of estrogen-dependent neoplasms of reproductive organs. With respect to breast cell malignancies, estradiol is believed to act primarily as a promotional

FIGURE 10-1. Female Cycle Mapping

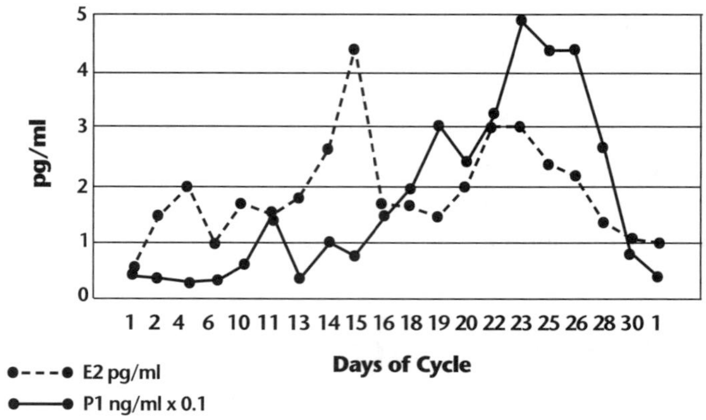

E_2: estradiol; P_1: progesterone

factor, causing increased growth rates in breast cells already transformed to a cancerous state. However, estradiol is not the only active estrogen in the human body – other metabolites formed from estradiol have the capacity to act as estrogens, and in some cases, as anti-estrogens.

Considerable work has shown that the major metabolites of estradiol and estrone are those hydroxylated at either the C-2 or the C-16α positions (Figure 10-2). 2-Hydroxyestrone ($2OHE_1$) is essentially devoid of peripheral biological activity, as shown in studies on uterine weight, gonadotrophin secretion, and cell

FIGURE 10-2. Pathways That Compete for Estrone

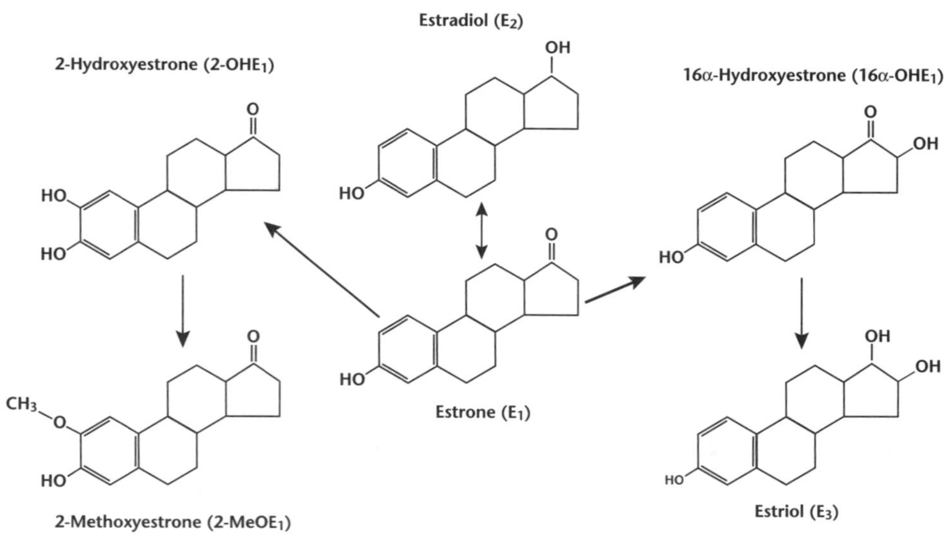

proliferation [48]. 2OHE1 has even been found to exert a modest anti-estrogenic effect [49, 50], and has been called "the good estrogen" [51]. The 16α-hydroxylated metabolites, 16α-hydroxyestrone (C) and estriol (E3), are estrogen agonists [52]. Women with breast and endometrial cancers have marked elevation of 16α-hydroxylation, which is a significant risk factor for such estrogen-dependent tumors [53]. Tumors in other estrogen-sensitive tissues are also promoted by 16OHE1. As the ratio of 2OHE1 to 16OHE1 decreases, the severity of recurrent respiratory papillomatosis increases [54]. Testing for the urinary levels of 2-hydroxyestrone and 16α–hydroxyestrone provides valuable insight regarding risk and dietary patient counseling [48, 50, 52]. The ratio of 2OHE1 to 16OHE1 (Estrogen Metabolite Index or EMI) should be greater than 2.0 and values in the upper normal range are advisable (Table 10-5). Any woman using HRT who has a low EMI should be monitored closely for improvements in the urinary metabolites.

TABLE 10-5. Expected Values for Urinary Estrogen Metabolites (1st Morning Urine)

Analyte	Premenopausal	Postmenopausal	Postmenopausal with Hormone Therapy
16OHE1	3 - 30 ng/mg creatinine	2 - 8 ng/mg creatinine	5 - 25 ng/mg creatinine
2OHE	3 - 40 ng/mg creatinine	2 - 10 ng/mg creatinine	10 - 75 ng/mg creatinine
Urinary Estrogen Metabolites (UEM) Index	4 - 110 ng/mg creatinine	5 - 18 ng/mg creatinine	6 - 158 ng/mg creatinine
Estrogen Metabolite Ratio (2OHE/16OHE1) (EMI)	> 2.0 ("healthy" pre- and postmenopausal women)		

The increased risk from estrogen and estrogen metabolites have led to a search for compounds that produce estrogen-like effects safely and those that decrease the production of 16OHE1. Soy isoflavones (*e.g.*, daidzein and genistein) are natural compounds that have come closest to meeting these criteria. They are increased by eating a diet rich in soy products [55]. Soy isoflavones have little or no affect on induction of tumors, and they have many estrogen-like properties even though they operate through different receptor sites on target tissues [56-58]. Dietary intake of soy products [59] and flax oil [60] has been shown to favorably modulate the rates of 2- vs. 16-hydroxyestrone production.

Diet also can give chemopreventive activity against breast cancer by increasing the removal of circulating estrogen. Estrogens are metabolized by cytochrome P-450 enzymes that are inducible by compounds found in vegetables of the brassica family like cabbage, Brussels sprouts, and broccoli [61]. Two phytochemicals contained in these foods, indole-3-carbinol (I3C) and diindolylmethane (DIM), have been identified as active inducers of certain P450-isozymes [62].

The reaction catalyzed by these P450-isozymes produces 2-hydroxylation of estradiol. Induction of P450 by I3C results in decreased concentrations of several metabolites known to activate the estrogen receptor [63]. If the P450 enzyme is not sufficiently active, a competing pathway causes 16-hydroxylation. Flaxseed supplementation at 10 g/d significantly increases the urinary 2/16-hydroxyestrone ratio, suggesting a breast cancer chemo-protective effect in premenopausal women [60]. Women who have a low EMI may be treated with I3C and DIM supplements, as well as with ground flaxseed.

Laboratory Evaluations of the Stress Response

Adrenal Medullary Response to Stress

Epinephrine and Norepinephrine

Increased output of catecholamines to modulate the autonomic nervous system is the most rapid chemical response to psychological stressors. In Chapter 6 – Organic Acids we described the association of low CNS levels of catecholamine neurotransmitters with low urinary levels of vanilmandelate (VMA) and homovanillate (HVA) [64]. An example of the relationship of these hormones to mental stress is the 42% increase in urinary VMA in cows in response to the stress of exposure to flies [65]. Sympathetic nerve endings in the adrenal gland release epinephrine and those in all other tissues release norepinephrine. Treatments aimed at improving protein digestion and supplementation of the amino acid precursor tyrosine can normalize neurotransmitter levels in the CNS [66]. The precursor relationship of tyrosine to epinephrine also gives rise to the antihypertensive role of tyrosine [67]. Patients with hypertension show lower VMA and 5-HIA levels when treated with a selective serotonergic antagonist [68]. Thus the levels of VMA in urine are shown to be responsive to agents designed to block hormone action as well as to externally imposed stress.

Stress-Induced Symptoms

Acute

- Dry mouth
- Elevated heart rate
- Inhibited digestion
- Increased blood glucose
- Hypertension

Chronic

- Hypotension
- General Effects
  - -Depression
  - -Sleep disturbances
  - -Inability to deal with stress
  - -Fatigue
- Insulin effects
  - -Insulin insensitivity
  - -High cholesterol
  - -High triglycerides
  - -High plasma sat. fatty acids
- Thyroid hormone effects
  - -Cold extremities
  - -Hypotension

Adrenal Cortex Response to Stress

Adrenal Steroids

The adrenal steroids are the family of hormones that permit us to make long-term adjustments to stress. Stress activates the hypothalamic-pituitary-adrenal (HPA) axis (Figure 10-3). The sequence of hormone actions begins in the brain and ultimately stimulates the adrenal cortex to produce hormones that affect many body functions. In health, when stressors are removed, the hormone levels fall back into the normal range. Patients who maintain elevated levels of cortisol even under psychologically non-stressful situations have a habituated endocrine stress response. They may have greater urinary losses of B-vitamins [69] and reduced absorption of calcium [70].

FIGURE 10-3. Metabolic Responses to Stress In The Hypothalamic-Pituitary-Adrenal Axis

The primary, immediate response to stress perceived in the brain is translated into chemical signals via releasing hormones that flow from the hypothalamus to the pituitary gland. Of the many pituitary polypeptide hormones, adrenocorticotropic hormone (ACTH) and growth hormone (GH) are released in response to hypothalamic corticotropin releasing factor. The chemical signal is further amplified and modulated by release of somatomedin from the liver and the stimulation of the major adrenal corticosteroids, dehydroepiandosterone (DHEA) and cortisol. Activities of the immune system are effected by both central nervous system factors and adrenal hormones.

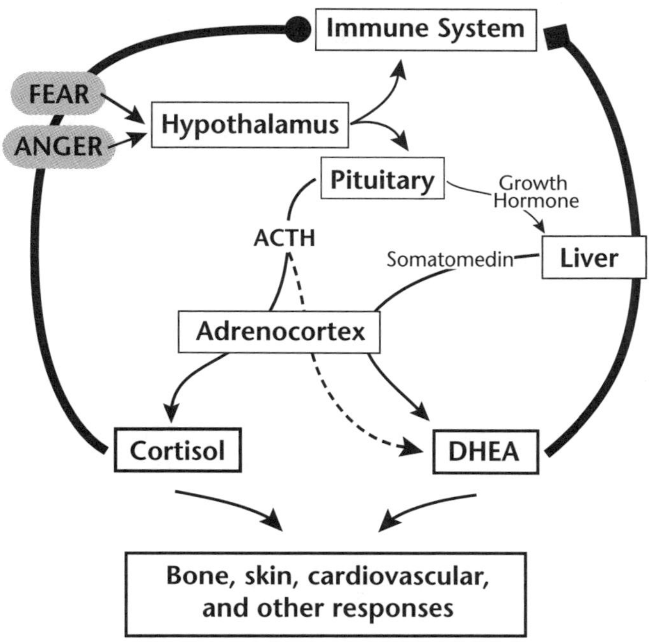

Dehydroepiandosterone (DHEA), the most abundant steroid in the body, is a precursor produced by the adrenal gland and converted to testosterone or the estrogens by the body's tissues (Figure 10-4). Levels of DHEA decrease with age and are inversely associated with coronary artery disease. Cortisol is the primary glucocorticoid that regulates glucose metabolism and the body's response to stress. During times of stress, cortisol levels increase and accelerate the breakdown of proteins to provide the fuel to maintain body functions. This catabolic activity needs to be balanced with periods of rebuilding to maintain good health. Under the influence of cortical steroids, humans can adapt to new stresses and survive until the physiological defense mechanisms are exhausted or until the stress subsides. Selye, a pioneer in the study of the effects of stress on the body, termed this phenomenon the "General Adaptation Syndrome" [71]. The ability to avoid multiple stress effects from mineral and water imbalances is due to adrenal steroid hormones. Animals can survive without adrenal glands if they are given small doses of adrenal hormones, but they are unable to adapt to any major chemical or physical stress, and cannot tolerate temperature extremes, starvation, infection, sensitizing agents, or noxious chemicals [72].

FIGURE 10-4. Biosynthesis of Steroid Hormones
Four classes of steroid hormones are produced from cholesterol. Key intermediates are pregnenolone, progesterone, and DHEA. The boxed numbers indicate the relative concentrations of circulating estrone, estradiol, and estriol.

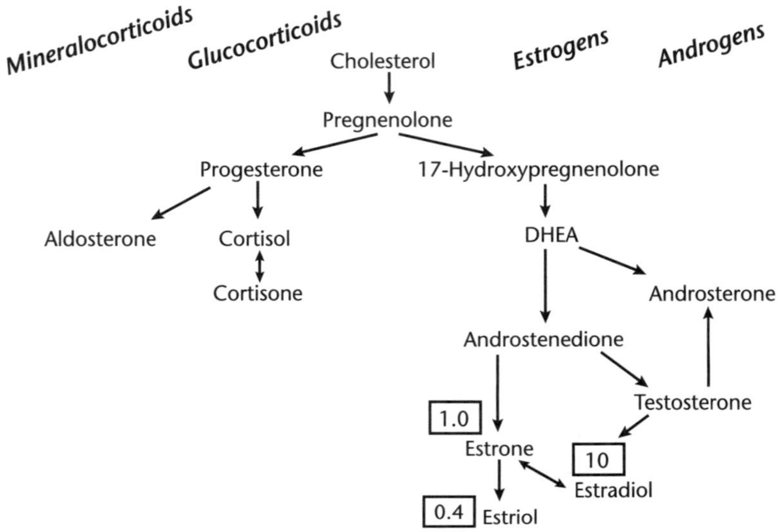

Measurement of Adrenal Hormones in Saliva

The adrenal hormones cortisol and DHEA are steroids produced in the adrenal cortex. They are actively involved in the body's development, growth, immune response, and cardiovascular function. Cortisol and DHEA affect carbohydrate, protein, and lipid metabolism, serve as anti-inflammatory agents, modulate thyroid function, and increase resistance to stress. Thus, fluctuating amounts of DHEA and cortisol may signal important alterations in adrenal function that can profoundly affect an individual's energy levels, emotional state, disease resistance, and general sense of well being.

Hypercortisolemia & Health Conditions

- Stress
- Aging
- Depression
- Schizophrenia
- Chronic Fatigue
- Insomnia
- Obesity
- Depressed immune function

About 1-10% of the steroids in the blood are in unbound or free form. The rest are bound to carrier proteins such as cortisol-binding globulin, sex hormone-binding globulin, and albumin. Subtle structural abnormalities due to genetic variation can lead to primary endocrine abnormalities because of impaired binding protein affinities. Adrenal hyper-function has been associated with impairment of cortisol binding [73]. Since only unbound steroids can freely diffuse into various target tissues in the body, they are the only hormones that are considered biologically active. Saliva testing measures the free-circulating, biologically active hormones [74]. The relationship of serum and salivary cortisol has been thoroughly examined [75]. Using single measurements of 11 p.m. salivary cortisol, elevated levels identified primary depressive patients with a sensitivity of 62.5% and a specificity of 75% [76]. Salivary cortisol concentrations at 11:00 p.m. have been shown to predict bone loss in elderly men [77] demonstrating the long-term effects of cortisol on bone mineral maintenance. Circadian variations are more marked in saliva than in plasma. The cortisol response to stress is developed at birth. Salivary cortisol determination in neonates has been proposed as a helpful measure to control for neonatal stress and birth asphyxia [78].

Circadian Rhythm of Cortisol

While variation in cortisol output is related to stress, the circadian rhythm of cortisol is linked to the sleep-wake cycle and controlled by the suprachiasmatic nucleus in the hypothalamus. Secretions are characterized by a steep increase in the morning, peaking at approximately 8 AM, followed by a gradual tapering off until about midnight, when circulating levels are at their lowest (Figure 10-5). The episodic secretion of cortisol is caused by the intermittent transformation of cortisol from its precursors in the adrenal cortex under the influence of ACTH [79]. This means that the brain is the primary site for control of the diurnal rhythm. The ratio of the maxima (8 AM) to the minima (midnight) is about 10 for saliva and only 3.3 for plasma. Patients with adrenal insufficiency show the greatest depression of cortisol output in the morning while the hypercortisolemic response of Cushing's syndrome is most prominent in the late evening [76].

FIGURE 10-5. Circadian Cortisol Rhythm

The area between the two curves is the normal range. When cortisol concentrations remain above normal during the afternoon and evening (arrows), deleterious effects of hypercortisolemia increase.

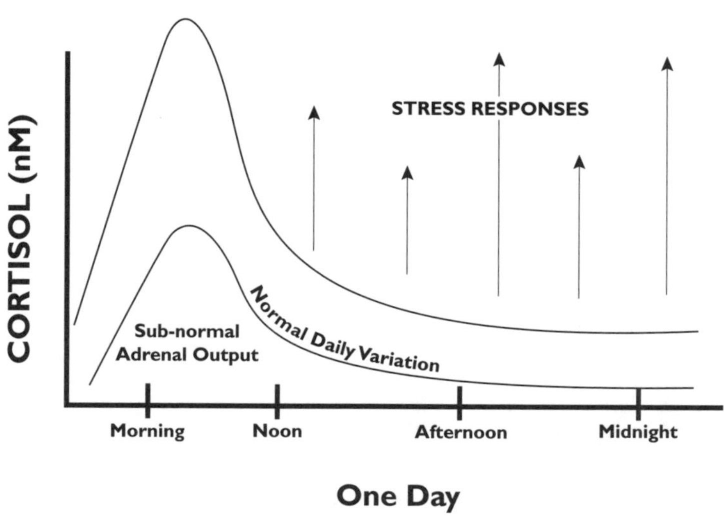

Patterns of Adaptive Responses

In response to ACTH, the adrenal gland normally produces much more DHEA than cortisol. The supply of precursors may become limiting, however, and under conditions of chronic real or perceived stress the adrenal glands enter a compensated phase in which the production of the stress hormones is altered. The quantitatively more demanding DHEA pathway is frequently the first to become limited. Because of such differences in response to ACTH, the production of DHEA usually declines with chronic stress, while, initially, cortisol remains elevated. The process is shown graphically in Figure 10-6 where the initial stress response is labeled "A1." As higher cortisol is required to shut down adrenal responses and bring ACTH into the normal range, the negative feedback of cortisol on the hypothalamus is lost. The resulting condition, known as "physiological hypercortisolemia," affects enzyme activities and leads to central adiposity [80]. Primate studies show effects of lowered serum phosphate and increased estrogenic responses with increased cortisol levels [81]. Monozygotic, female twin pairs showed a high degree of stability in AM and PM cortisol values collected daily over a 14-day period. Salivary cortisol was significantly higher for those siblings with a major history of depression than for their non-depressed twins [82]. Cushing's disease is characterized by constantly elevated levels in the upper right of quadrant "A1." Nighttime, salivary cortisol testing is an accurate way to screen for hypercortisolism in children

FIGURE 10-6. Stress Responses of Cortisol and DHEA
The values that are plotted are the average of noon and 5 PM levels for salivary cortisol and DHEA. The quadrants represent normal and abnormal patterns, with A1 to A4 representing progressive failure of the normal ACTH response.

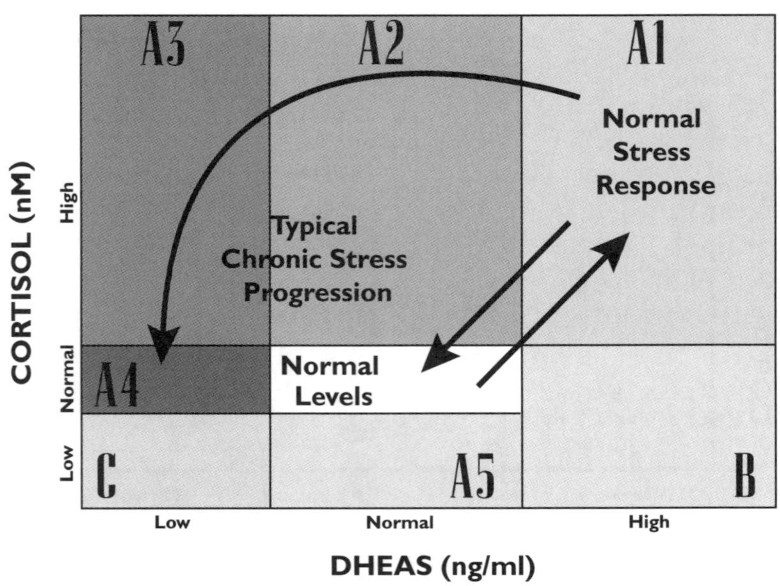

[83]. Stress-induced hormonal secretion is greater in older individuals than in their young counterparts, and the blood pressure increases with stress become greater with age [84].

Later phases of compensated response go through the progression from "A2" to "A5." The progression has been called "stress fixation." The arrow shows a progression from the upper right to upper left because DHEA is the first to decrease. Because of DHEA's much higher normal concentrations, the supply of precursors, especially pregnenolone, cannot be maintained at sufficient levels. Adrenal cortical output of DHEA falls from high to normal to low. Such a profile of low DHEA with high cortisol is typical of high achiever lifestyle such as the 38-year-old woman who commutes from Washington, DC to New York for her executive career and desires to become pregnant. As long as cortisol output is sustained at elevated levels, such intense demands for physiological function may be maintained, but the low DHEA has long-term negative effects. Stress-induced damage to the hippocampus, an area involved in memory processes and especially affected in Alzheimer's disease, is prevented by DHEA [85].

Some signs of stress fixation:

- Impaired energy production (insulin sensitivity drops)
- Reduction in glucose utilization and in amino acids due to gluconeo-genesis
- Increased bone resorption and osteoporosis
- Fat accumulation at the waist
- Increased protein breakdown
- Salt and water retention
- General immune suppression
- Increased rate of infection
- Yeast overgrowth
- Reduced vitality
- Insomnia
- Hunger
- Premenstrual syndrome

The same progression for cortisol may ensue if pregnenolone becomes exhausted. If the stress is prolonged, the production of both hormones falls into the sector labeled "C." Individuals affected with Addison's disease where the adrenals are unable to produce stress hormones have values that fall into the C sector. The more infrequent finding of elevated DHEA with normal or low cortisol (sector B) is found in hypothalamic dysfunction, and these individuals should avoid high stress occupations. Cognitive stress management intervention can reduce the rate of progression toward the stress fixation state. A 23% reduction in cortisol and a 100% increase in DHEA were found in individuals who demonstrated response to treatment by improved scales of guilt, hostility, burnout, anxiety, and stress [86].

The patient may be challenged with ACTH to determine if the reason for low cortisol and DHEA levels is a primary failure of the adrenal cortex response or low pituitary ACTH output. If response to the challenge is low, then inadequate precursor availability is suspected, and pregnenolone may be given to replace the missing precursor with less concern about inducing excessive corticosteroid production. Patients with chronic hypercortisolemia will show blunted ACTH responses because of the suppressive effect on the pituitary gland [87].

Salivary Secretory IgA

Secretory IgA (sIgA) forms an immune barrier protection against gastrointestinal tract infection [88]. Salivary levels of sIgA may be used to monitor this immune barrier function. Although sIGA production demands amino acid supply, dietary protein inadequacy has little effect on sIGA levels [89]. Chronic stress, however, decreases sIGA production [90]. The measurement of sIgA as a biomarker for functional impact of chronic stress greatly enhances the interpretation of stress hormone concentration data because single point elevations of cortisol may be only a normal, temporary stress response. Low sIgA is a sign of the impact of stress directly on the immune system, but elevated sIgA reveals another site of degeneration affected by stress.

The production of IgA is increased when dietary antigens are not blocked by the physical barrier in the small intestine. In human IgA nephropathy, there is a significant decrease in levels of IgA immune complexes after patients are put on a gluten-free diet. In patients with elevated IgA, gluten may act as a toxic lectin, increasing the permeability of the intestinal mucosa to various dietary antigens [91, 92]. Some individuals with only modest IgA deficits can have substantial mucosal permeability defects on milk challenge [93]. Also, IgA antibodies help to clear *C. albicans* from mucosal surfaces. IgA levels are lower in women who show positive cultures for *C. albicans.* [94]. This connection provides an explanation for the increased incidence of invasive candidiasis in highly stressed individuals.

Salivary Anti-Gliadin Antibodies (AGA)

Chronic stress is a factor in the onset of the gluten sensitivity characteristic of celiac disease. The corollary is also true; gluten sensitivity contributes a significant stress to the body and can add clinical value to adrenal hormone assessment. Detection of IgA-class anti-gliadin antibody provides an immunohistochemical marker of celiac disease latency and gluten sensitivity [95]. Early dietary treatment is important in conditions like celiac disease to avoid malnutrition and the development of malignant disorders. Any condition that causes deterioration of the epithelial layer of the small intestine can lead to increased gliadin antibodies. IgG anti-gliadin antibody concentrations were elevated in subjects with small intestinal bacterial overgrowth and positive luminal anti-gliadin antibodies [96].

Adrenal Support and Nutritional Factors in the Stress Response

Administration of DHEA can help overcome the loss of adrenal output, and the precursor of both DHEA and cortisol, pregnenolone, may be added to support adrenal function. Muscle protein breakdown caused by increased stress hormones can be reduced by administration of an amino acid solution [97]. Increased adrenal activity also leads to greater loss of B-complex vitamins that should be replaced by supplementation. Finally, vitamin C should be added to prevent tissue depletion, because it is rapidly diminished by the adrenal stress response. Vitamin C has an enhancing effect on adrenal glucocorticoid production [98, 99].

Omega-6 fatty acids are specifically depleted in individuals with high cortisol output [100]. Dietary supplements of linoleic or gamma-linolenic may be used to replenish tissue status of these fatty acids, but there is an important question of balance of omega-6 fatty acids with the omega-3 family in order to maintain the proper tissue responses to challenges. Patients with elevated cortisol may need to be evaluated for polyunsaturated fatty acid status (see Chapter 5 – Fatty Acids).

Vitamin C deficiency causes marked elevation in cortisol. The changes are of a magnitude that could lead to suppression of the inflammatory response, impairment of wound healing, and decrease in bone matrix formation. Such effects have been proposed as significant factors leading to oral health threats in vitamin C-deficient individuals [75]. Licorice consumption potentiates the effects of cortisol, and hypokalaemic hypertension can be induced by licorice addiction [101]. The effect is not a result of adrenocortical stimulation, however, but rather a reduction in the conversion of cortisol to cortisone [102].

As the degree of stress fixation becomes greater, these interventions become more important, and aggressiveness of nutrient support should increase. Other evaluations of overall nutrient status allow optimization of adrenal response and improvement of total body response to, and need for, the adrenal cortical hormones. Of course, cognitive stress reduction training is of value at any point in the process.

References

[1] Current unique and rare endocrine assays. 1998, Inglewood: Inter Science Institute.

[2] Horvath, K., *et al.*, Improved social and language skills after secretin administration in patients with autistic spectrum disorders. *J Assoc Acad Minor Phys*, 9(1). 9-15, 1998.

[3] Marrocco, W., *et al.*, Behavior of vitamin A, beta-carotene, retinol binding protein and prealbumin in the plasma of hypo- and hyperthyroid subjects. *Boll Soc Ital Biol Sper*, 60(4). 769-75, 1984.

[4] Pazos-Moura, C.C., *et al.*, Effect of iodine deficiency and cold exposure on thyroxine 5'-deiodinase activity in various rat tissues. *Am J Physiol*, 260 (2 Pt 1). E175-82., 1991.

[5] Voderholzer, U., *et al.*, Impact of experimentally induced serotonin deficiency by tryptophan depletion on sleep EEG in healthy subjects. *Neuropsychopharmacology*, 18(2). 112-24, 1998.

[6] Berdanier, C., Nutrients and gene exxpression. *Clinical aspects*. 1996, Boca Raton: CRC. 216.

[7] Eisenstein, A.B., *et al.*, Increased glucagon secretion in protein-fed rats: lack of relationship to plasma amino acids. *Am J Physiol*, 236(1). E20-7, 1979.

[8] Larsen, P.R. and M.J. Berry, Nutritional and hormonal regulation of thyroid hormone deiodinases. *Annu Rev Nutr*, 15. 323-52, 1995.

[9] Ruz, M., *et al.*, Single and multiple selenium-zinc-iodine deficiencies affect rat thyroid metabolism and ultrastructure. *J Nutr*, 129(1). 174-80, 1999.

[10] Utiger, R.D., Kashin-Beck disease--expanding the spectrum of iodine-deficiency disorders [editorial; comment]. *N Engl J Med*, 339(16). 1156-8, 1998.

[11] Santidrian, S., *et al.*, Effect of the thyroid status and protein-calorie malnutrition on the rate of myofibrillar protein degradation in mature male rats. *Rev Esp Fisiol*, 37(3). 309-16, 1981.

[12] Suzuki, Y., *et al.*, Plasma free fatty acids, inhibitor of extrathyroidal conversion of T4 to T3 and thyroid hormone binding inhibitor in patients with various nonthyroidal illnesses. *Endocrinol Jpn*, 39(5). 445-53., 1992.

[13] Nomura, T., *et al.*, The effect of serum free fatty acid on serum free thyroid hormone fractions in low T3 syndrome. *Nippon Naibunpi Gakkai Zasshi*, 63(6). 752-72., 1987.

[14] Pehowich, D.J., Thyroid hormone status and membrane n-3 fatty acid content influence mitochondrial proton leak. *Biochim Biophys Acta*, 1411(1). 192-200., 1999.

[15] Seven, A., *et al.*, Biochemical evaluation of oxidative stress in propylthiouracil treated hyperthyroid patients. Effects of vitamin C supplementation. *Clin Chem Lab Med*, 36(10). 767-70, 1998.

[16] Le Roith, D., Seminars in medicine of the Beth Israel Deaconess Medical Center. Insulin-like growth factors. *N Engl J Med*, 336(9). 633-40, 1997.

[17] Costigan, D.C., H.J. Guyda, and B.I. Posner, Free insulin-like growth factor I (IGF-I) and IGF-II in human saliva. *J Clin Endocrinol Metab*, 66(5). 1014-8, 1988.

[18] Nicklas, B.J., *et al.*, Testosterone, growth hormone and IGF-I responses to acute and chronic resistive exercise in men aged 55-70 years. *Int J Sports Med*, 16(7). 445-50, 1995.

[19] Sonntag, W.E., *et al.*, Pleiotropic effects of growth hormone and insulin-like growth factor (IGF)-1 on biological aging: inferences from moderate caloric-restricted animals. *J Gerontol A Biol Sci Med Sci*, 54(12). B521-38, 1999.

[20] Filho, J.C., *et al.*, Effect of protein intake on plasma and erythrocyte free amino acids and serum IGF-I and IGFBP-1 levels in rats. *Am J Physiol*, 277(4 Pt 1). E693-701, 1999.

[21] Chen, C., K.A. Noland, and D.N. Kalu, Modulation of intestinal vitamin D receptor by ovariectomy, estrogen and growth hormone. *Mech Ageing Dev*, 99(2). 109-22, 1997.

[22] Wu, S., M. Grieff, and A.J. Brown, Regulation of renal vitamin D-24-hydroxylase by phosphate: effects of hypophysectomy, growth hormone and insulin-like growth factor I. *Biochem Biophys Res Commun*, 233(3). 813-7, 1997.

[23] Raifen, R., Y. Altman, and Z. Zadik, Vitamin A levels and growth hormone axis. *Horm Res*, 46(6). 279-81, 1996.

[24] Li, Y., *et al.*, Dietary conjugated linoleic acids alter serum IGF-I and IGF binding protein concentrations and reduce bone formation in rats fed (n-6) or (n-3) fatty acids. *J Bone Miner Res*, 14(7). 1153-62, 1999.

[25] DeLany, J.P., *et al.*, Conjugated linoleic acid rapidly reduces body fat content in mice without affecting energy intake. *Am J Physiol*, 276(4 Pt 2). R1172-9, 1999.

[26] Parini, P., B. Angelin, and M. Rudling, Cholesterol and lipoprotein metabolism in aging: reversal of hypercholesterolemia by growth hormone treatment in old rats. *Arterioscler Thromb Vasc Biol*, 19(4). 832-9, 1999.

[27] O'Neal, D.N., *et al.*, The effect of 24 months recombinant human growth hormone (rh-GH) on LDL cholesterol, triglyceride-rich lipoproteins and apo [a] in hypopituitary adults previously treated with conventional replacement therapy. *Growth Horm IGF Res*, 9(3). 165-73, 1999.

[28] Davies, P.S., *et al.*, Effect of growth hormone on

height, weight, and body composition in Prader-Willi syndrome. *Arch Dis Child*, 78(5). 474-6, 1998.

[29] Ellegard, L., *et al.*, Low-dose recombinant human growth hormone increases body weight and lean body mass in patients with short bowel syndrome. *Ann Surg*, 225(1). 88-96, 1997.

[30] van der Sluis, I.M., *et al.*, Bone density and body composition in chronic renal failure: effects of growth hormone treatment. *Pediatr Nephrol*, 15(3-4). 221-8, 2000.

[31] Bondy, C.A., *et al.*, Clinical uses of insulin-like growth factor I [see comments]. *Ann Intern Med*, 120(7). 593-601, 1994.

[32] Holden, R.J., I.S. Pakula, and P.A. Mooney, Tumor necrosis factor-alpha: a continuum of liability between insulin- dependent diabetes mellitus, non-insulin-dependent diabetes mellitus and carcinoma (review). *Med Hypotheses*, 52(4). 319-23, 1999.

[33] Fain, J.A., *et al.*, Diabetes patient education research: an integrative literature review. *Diabetes Educ*, 25(6 Suppl). 7-15, 1999.

[34] Mensing, C., *et al.*, National standards for diabetes self-management education. Task Force to Review and Revise the National Standards for Diabetes Self-Management Education Programs. *Diabetes Care*, 23(5). 682-9, 2000.

[35] Schrezenmeir, J., Hyperinsulinemia, hyperpro-insulinemia and insulin resistance in the metabolic syndrome. *Experientia*, 52(5). 426-32, 1996.

[36] Montori, V.M., *et al.*, Fish oil supplementation in type 2 diabetes: a quantitative systematic review. *Diabetes Care*, 23(9). 1407-15, 2000.

[37] Morris, B.W., *et al.*, Chromium supplementation improves insulin resistance in patients with Type 2 diabetes mellitus [letter]. *Diabet Med*, 17(9). 684-5, 2000.

[38] Fox, G.N. and Z. Sabovic, Chromium picolinate supplementation for diabetes mellitus. *J Fam Pract*, 46(1). 83-6, 1998.

[39] Cam, M.C., B. Rodrigues, and J.H. McNeill, Distinct glucose lowering and beta cell protective effects of vanadium and food restriction in streptozotocin-diabetes. *Eur J Endocrinol*, 141(5). 546-54, 1999.

[40] Morishita, M., *et al.*, The dose-related hypoglycemic effects of insulin emulsions incorporating highly purified EPA and DHA. *Int J Pharm*, 201(2). 175-85., 2000.

[41] Shimizu, H., *et al.*, Long-term effect of eicosapentaenoic acid ethyl (EPA-E) on albuminuria of non-insulin dependent diabetic patients. *Diabetes Res Clin Pract*, 28(1). 35-40., 1995.

[42] Om, A.S. and K.W. Chung, Dietary zinc deficiency alters 5 alpha-reduction and aromatization of testosterone and androgen and estrogen receptors in rat liver. *J Nutr*, 126(4). 842-8, 1996.

[43] Orrell, R.W., *et al.*, Testosterone deficiency myopathy. *J R Soc Med*, 88(8). 454-6, 1995.

[44] Shippen, E. and W. Fryer, *The testosterone syndrome*. 1998, New York: M. Evans and Co., Inc. 230.

[45] Scopacasa, F., *et al.*, The relation between bone density, free androgen index, and estradiol in men 60 to 70 years old. *Bone*, 27(1). 145-9., 2000.

[46] Imani, B., *et al.*, Free androgen index and leptin are the most prominent endocrine predictors of ovarian response during clomiphene citrate induction of ovulation in normogonadotropic oligoamenorrheic infertility. *J Clin Endocrinol Metab*, 85(2). 676-82, 2000.

[47] Clemons, M. and P. Goss, Estrogen and the risk of breast cancer. *N Engl J Med*. 344(4). 276-85, 2001.

[48] Martucci, C.P. and J. Fishman, Impact of continuously administered catechol estrogens on uterine growth and luteinizing hormone secretion. *Endocrinology*, 105(6). 1288-92, 1979.

[49] Schneider, J., *et al.*, Antiestrogen action of 2-hydroxyestrone on MCF-7 human breast cancer cells. *J Biol Chem*, 259(8). 4840-5, 1984.

[50] Vandewalle, B. and J. Lefebvre, Opposite effects of estrogen and catechol estrogen on hormone-sensitive breast cancer cell growth and differentiation. *Mol Cell Endocrinol*, 61. 239-246, 1989.

[51] Bradlow, H.L., *et al.*, 2-hydroxyestrone: the 'good' estrogen. *J Endocrinol*, 150 Suppl. S259-65, 1996.

[52] Fishman, J. and C. Martucci, Biological properties of 16 alpha-hydroxyestrone: implications in estrogen physiology and pathophysiology. *J Clin Endocrinol Metab*, 51(3). 611-5, 1980.

[53] Fishman, J., *et al.*, Increased estrogen-16 alpha-hydroxylase activity in women with breast and endometrial cancer. *J Steroid Biochem*, 20(4B). 1077-81, 1984.

[54] Auborn, K., *et al.*, Estrogen metabolism and laryngeal papillomatosis: a pilot study on dietary prevention. Anticancer Res, 18(6B). 4569-73, 1998.

[55] Kirkman, L.M., *et al.*, Urinary lignan and isoflavonoid excretion in men and women consuming vegetable and soy diets. Nutr Cancer, 24(1). 1-12, 1995.

[56] Xu, X., *et al.*, Soy consumption alters endogenous estrogen metabolism in postmenopausal women. Cancer Epidemiol *Biomarkers Prev*, 9(8). 781-6, 2000.

[57] Nagata, C., *et al.*, Inverse association of soy product intake with serum androgen and estrogen concentrations in Japanese men. *Nutr Cancer*, 36(1). 14-8, 2000.

[58] Greaves, K.A., *et al.*, Consumption of soy protein reduces cholesterol absorption compared to casein protein alone or supplemented with an isoflavone extract or conjugated equine estrogen in ovariectomized cynomolgus monkeys. J Nutr, 130(4). 820-6, 2000.

[59] Lu, L.J., *et al.*, Increased urinary excretion of 2-hydroxyestrone but not 16alpha-hydroxyestrone in premenopausal women during a soya diet containing isoflavones. *Cancer Res*, 60(5). 1299-305, 2000.

[60] Haggans, C.J., *et al.*, The effect of flaxseed and wheat bran consumption on urinary estrogen metabolites in premenopausal women. *Cancer Epidemiol Biomarkers Prev*, 9(7). 719-25., 2000.

[61] Fowke, J.H., C. Longcope, and J.R. Hebert, Brassica vegetable consumption shifts estrogen metabolism in healthy postmenopausal women. *Cancer Epidemiol Biomarkers Prev*, 9(8). 773-9, 2000.

[62] Zeligs, M., Diet and estrogen status: The cruciferous connection. *J Med Food*, 1(2). 67-81, 1998.

[63] Michnovicz, J.J., H. Adlercreutz, and H.L. Bradlow, Changes in levels of urinary estrogen metabolites after oral indole-3-carbinol treatment in humans. *J Natl Cancer Inst*, 89(10). 718-23, 1997.

[64] Orgacka, H. and Z. Zbytniewski, Excretion of vanillic acid and homovanillic acid and tissue distribution of catecholamines and their metabolites in mice with various levels of pigmentation. *Endokrynol Pol*, 42(3). 471-9, 1991.

[65] Savio, T.J., *et al.*, Effect of horn flies on vanilmandelic acid excretion of dairy cattle. *J Dairy Sci*, 59(2). 318-20, 1976.

[66] Braverman, E. and C. Pfeiffer, Tyrosine: The Antidepressant, in *The Healing Nutrients Within*. 1987, Keats: New Canaan. p. 44-58.

[67] Mauron, J., Tyrosine and hypertension. *Bibl Nutr Dieta*, 38. 209-18, 1986.

[68] Dzurik, R., *et al.*, Blood pressure, 5-OH indoleacetic acid, and vanilmandelic acid excretion and blood platelet aggregation in hypertensive patients treated with ketanserin. *J Cardiovasc Pharmacol*, 7(Suppl 7). S29-31, 1985.

[69] Bhatt, H.R., Antagonistic effects of pentagastrin and cortisol on plasma level, urinary excretion and hepatic and renal uptake of vitamin B_{12} after intravenous injection in the cat [proceedings]. *J Physiol (Lond)*, 296. 88P, 1979.

[70] Feher, J.J. and R.H. Wasserman, Intestinal calcium-binding protein and calcium absorption in cortisol-treated chicks: effects of vitamin D3 and 1,25-dihydroxyvitamin D3. *Endocrinology*, 104(2). 547-51, 1979.

[71] Selye, H., *Stress without distress*. 1974, Philadelphia: J.B. Lippincott. 171.

[72] Frieden, E. and H. Lipner, *Biochemical Endocrinology of the Vertebrates*. 1971, Englewood Cliffs: Prentice-Hall. 164.

[73] Barragry, J.M., *et al.*, Defective cortisol binding globulin affinity in association with adrenal hyperfunction: a case report. *Acta Endocrinol (Copenh)*, 95(2). 194-7, 1980.

[74] Stahl, F. and G. Dorner, Responses of salivary cortisol levels to stress-situations. *Endokrinologie*, 80(2). 158-62, 1982.

[75] Aardal, E. and A.C. Holm, Cortisol in saliva--reference ranges and relation to cortisol in serum. *Eur J Clin Chem Clin Biochem*, 33(12). 927-32, 1995.

[76] Guechot, J., *et al.*, Physiological and pathological variations in saliva cortisol. *Horm Res*, 16(6). 357-64, 1982.

[77] Raff, H., *et al.*, Elevated salivary cortisol in the evening in healthy elderly men and women: correlation with bone mineral density. *J Gerontol A Biol Sci Med Sci*, 54(9). M479-83, 1999.

[78] Klug, I., *et al.*, Cortisol and 17-hydroxy-progesterone levels in saliva of healthy neonates. normative data and relation to body mass index, arterial cord blood ph and time of sampling after birth. *Biol Neonate*, 78(1). 22-6, 2000.

[79] Jusko, W.J., W.R. Slaunwhite, Jr., and T. Aceto, Jr., Partial pharmacodynamic model for the circadian-episodic secretion of cortisol in man. *J Clin Endocrinol Metab*, 40(2). 278-89, 1975.

[80] Samra, J.S., *et al.*, Effects of physiological hypercortisolemia on the regulation of lipolysis in subcutaneous adipose tissue. *J Clin Endocrinol Metab*, 83(2). 626-31, 1998.

[81] Kowalski, W.B., R.F. Valle, and R.T. Chatterton, Jr., Response of the primate secretory endometrium to subchronic hypercortisolemia. *J Soc Gynecol Investig*, 4(3). 152-9, 1997.

[82] Young, E.A., *et al.*, Similarity in saliva cortisol measures in monozygotic twins and the influence of past major depression. *Biol Psychiatry*, 48(1). 70-4, 2000.

[83] Gafni, R.I., D.A. Papanicolaou, and L.K. Nieman, Nighttime salivary cortisol measurement as a simple, noninvasive, outpatient screening test for Cushing's syndrome in children and adolescents. *J Pediatr*, 137(1). 30-5, 2000.

[84] Gotthardt, U., *et al.*, Cortisol, ACTH, and cardio-vascular response to a cognitive challenge paradigm in aging and depression. *Am J Physiol*, 268(4 Pt 2). R865-73., 1995.

[85] Bastianetto, S., *et al.*, Dehydroepiandrosterone (DHEA) protects hippocampal cells from oxidative stress-induced damage. *Brain Res Mol Brain Res*, 66(1-2). 35-41, 1999.

[86] Cruess, D.G., *et al.*, Cognitive-behavioral stress management buffers decreases in dehydro-

epiandrosterone sulfate (DHEA-S) and increases in the cortisol/DHEA-S ratio and reduces mood disturbance and perceived stress among HIV-seropositive men. *Psychoneuroendocrinology,* 24(5). 537-49, 1999.

[87] van der Pompe, G., M.H. Antoni, and C.J. Heijnen, Elevated basal cortisol levels and attenuated ACTH and cortisol responses to a behavioral challenge in women with metastatic breast cancer. *Psychoneuroendocrinology,* 21(4). 361-74., 1996.

[88] Takahashi, I. and H. Kiyono, Gut as the largest immunologic tissue. *JPEN J Parenter Enteral Nutr,* 23(5 Suppl). S7-12, 1999.

[89] Lim, G.M., G.F. Sheldon, and J. Alverdy, Biliary secretory IgA levels in rats with protein-calorie malnutrition. *Ann Surg,* 207(5). 635-40, 1988.

[90] Guhad, F.A. and J. Hau, Salivary IgA as a marker of social stress in rats. Neurosci Lett, 216(2). 137-40, 1996.

[91] Coppo, R., The pathogenetic potential of environmental antigens in IgA nephropathy. *Am J Kidney Dis,* 12(5). 420-4, 1988.

[92] Nagy, J., H. Scott, and P. Brandtzaeg, Antibodies to dietary antigens in IgA nephropathy. *Clin Nephrol,* 29(6). 275-9, 1988.

[93] Cunningham-Rundles, C., Analysis of the gastrointestinal secretory immune barrier in IgA deficiency. *Ann Allergy,* 57(1). 31-5, 1986.

[94] Schonheyder, H., *et al.*, IgA and IgG serum antibodies to Candida albicans in women of child-bearing age. *Sabouraudia,* 21(3). 223-31, 1983.

[95] Kaukinen, K., *et al.*, Small-bowel mucosal inflammation in reticulin or gliadin antibody-positive patients without villous atrophy. *Scand J Gastroenterol,* 33(9). 944-9, 1998.

[96] Riordan, S.M., *et al.*, Luminal antigliadin antibodies in small intestinal bacterial overgrowth. *Am J Gastroenterol,* 92(8). 1335-8, 1997.

[97] Hammarqvist, F., *et al.*, Stress hormone and amino acid infusion in healthy volunteers: short-term effects on protein synthesis and amino acid metabolism in skeletal muscle. *Metabolism,* 43(9). 1158-63, 1994.

[98] Kodama, M., F. Inoue, and T. Kodama, Intraperitoneal administration of ascorbic acid delays the turnover of 3H-labelled cortisol in the plasma of an ODS rat, but not in the Wistar rat. Evidence in support of the cardinal role of vitamin C in the progression of glucocorticoid synthesis. *In Vivo,* 10(1). 97-102, 1996.

[99] Kodama, M., T. Kodama, and M. Murakami, Vitamin C infusion treatment enhances cortisol production of the adrenal via the pituitary ACTH route. *In Vivo,* 8(6). 1079-85, 1994.

[100] Williams, L.L., *et al.*, Quantitative association between altered plasma esterified omega-6 fatty acid proportions and psychological stress. *Prostaglandins Leukot Essent Fatty Acids,* 47(2). 165-70, 1992.

[101] Epstein, M.T., *et al.*, Licorice raises urinary cortisol in man. *J Clin Endocrinol Metab,* 47(2). 397-400, 1978.

[102] Seelen, M.A., *et al.*, Hypertension caused by licorice consumption. *Nederlends Tijdschrift voor Geneeskunde,* 140(52). 2632-5, 1996.

Chapter 11

Urinary Lipid Peroxides, Sulfate, and Nitrate:

Combining Metabolic Marker Testing to Predict Nutrient Needs

Introduction

Urinary lipid peroxides, sulfate, and nitrate are easily (and inexpensively) measured. They provide useful data on specific functional nutrient needs of an individual, which can vary from person to person. They are not direct measures of nutrients themselves but they can be significantly affected by specific nutrient status and therefore provide positive information on individual nutrient insufficiencies. While not providing a comprehensive assessment of all nutrients required by the body, they do provide a good reflection of functional nutrient need in several critical areas of body function. When this information is combined with age and gender, a very useful nutrient supplement profile can be developed to address the unique nutrient needs of the individual.

The following is a brief review of these compounds with a rationale for their use as indicators of specific nutrient need. Although lipid peroxide testing was discussed in Chapter 9 – Oxidant Stress and urinary sulfate was discussed along with organic acids in Chapter 6, additional review of the use of these tests are presented here in the context of the special combination of urinary marker compounds.

Lipid Peroxides

Lipid peroxidation is the consequence of free radical reactions with the lipid or fatty components of the body. Lipid peroxides (LP) are generated in these reactions and have been associated with increased risk for atherogenicity and cancer [1]. Lipid peroxides are elevated in cardiovascular disease and diabetes [2-3] and have been used as a marker of oxidant stress and antioxidant defenses in the body [4-15]. Circulating LP in the serum are excreted in the urine [2-4, 6-8, 12, 16]. Both bodily fluids have been routinely measured in experimental animals and humans to assess lipid peroxidation and,therefore antioxidant status.

The traditional method for measurement of lipid peroxides in a sample is the thiobarbituric acid (TBA) reaction. Lipid peroxides in an acid environment degrade to malonaldehyde (MDA), which forms a colored adduct with TBA upon heating. These thiobarbituric acid reactive substances (TBARS) can then be measured spectrophotometrically or fluorometrically [4, 10, 16-18]. TBA can also react with other substances in the sample leading to a criticism of this method for its non-specificity of direct measurement of MDA only. Despite these limitations, the TBARS assay has been used repeatedly in hundreds of studies assessing oxidative stress and, in general, correlates well with other measures of free radical activity [4, 5, 7-11, 14, 15, 19-24].

Since urinary lipid peroxides reflect a level of oxidant stress for an individual at a particular point in time, this test is a useful functional indicator of need for compensatory antioxidant supplementation. A number of studies have shown a reduction in urinary lipid peroxides with antioxidant supplementation as well as increased lipid peroxidation with specific nutrient deficiencies [5, 11, 13, 20, 22, 25-42]. Using repeat measures, antioxidant supplementation could be adjusted to the individual's need at that point in time. Of course, this level of need could change over time depending on the individual's situation. Consequently, sequential testing of this marker of oxidant stress to adjust for changing needs for protective antioxidants would be advisable as part of an adult's annual physical check-up.

Sulfate

Inorganic sulfate is used in the body for detoxification of normal metabolic by-products, hormones, drugs, and other foreign compounds circulating in the blood. It is conjugated or chemically bound with these compounds as part of Phase II detoxification reactions that occur primarily in the liver. This conjugation step renders the molecules more water-soluble so they can circulate in the blood and be excreted from the body via the kidneys into the urine. Sulfate is primarily derived from the sulfur-containing amino acids in dietary protein – methionine and cysteine. Other Phase II conjugation compounds include glutathione, glycine, and glucuronic acid. Glutathione is a tripeptide containing cysteine and is, therefore, cysteine/methionine dependent. Sulfate is converted in the liver to the active co-substrate for conjugation adenosine 3'-phosphate 5'-phosphosulfate (PAPS).

Research indicates sulfate tissue levels can be influenced by certain dietary regimens. Protein calorie malnutrition, fasting, and low protein diets can deplete inorganic sulfate stores and glutathione levels due to the decreased availability of methionine and cysteine [43]. In addition, excessive exposure to foreign chemicals can deplete sulfate stores and lead to chronic illnesses [44, 45]. Sulfate levels are strictly determined by dietary protein intake and its subsequent digestion and assimilation of the amino acid components, cysteine and methionine. This process may become compromised in certain individuals, especially those on weight loss, low protein, or other low calorie diets. Malabsorption/maldigestion of protein and consequent amino acid imbalances are relatively common phenomena in the United States, particularly with the overuse and prevalence of antacids and acid blocking drugs.

Low urinary sulfate levels indicate depleted body stores and a functional need for supplementation of sulfur containing amino acids as well as support for other detoxification processes. Urinary sulfate directly reflects the circulating levels and availability of this important detoxification compound [44]. Evidence suggests that as the sulfation pathway is depleted, other Phase II conjugations can take place to compensate [46, 47]. These concepts have been well documented in the literature on acetaminophen detoxification [48-56]. Depletion of sulfate and glutathione significantly increases the toxicity of acetaminophen. This toxicity can be reversed with adequate supplementation of the precursor amino acids such as cysteine or its more easily absorbed form, n-acetylcysteine. Glycine or glutathione supplementation may also be useful as this would support alternative detoxification pathways that may also be under stress. As sulfate and glutathione levels are repleted in the body, urinary levels of sulfate are also normalized.

Nitrate

Nitric oxide (NO) is a recently discovered (1981), biologically active molecule, which has generated much enthusiasm in scientific circles. It was deemed "Molecule of the Year" in 1992 by Science magazine, and the 1998 Nobel Prize in Medicine and Physiology was awarded to its major researchers. NO regulates many different bodily processes in response to a variety of stimuli. NO is an important regulatory molecule and should not be considered either good or bad since both over- and under-production of NO can have deleterious effects on body function. Consequently, its regulation has been the subject of much study.

The essential amino acid arginine is converted to NO by the enzyme nitric oxide synthetase (NOS), which is present in two forms in the body, constitutive or inducible. The constitutive form is present in a fixed amount to provide a constant level of NO to a system. The inducible form is increased or generated via standard DNA/RNA protein synthetic cellular machinery in response to a specific stimulus. This allows the body to create additional amounts of NO in response to an increased demand.

Structurally NOS enzymes are very similar to the cytochrome P-450 enzymes used in the liver for detoxification. P-450 enzymes can be modulated by nutritional factors [57]. Similarly, various nutritional factors have been shown to influence NOS production [58-65] and the effects of NO [66-70].

Primary inducers of NOS are cytokines, cell signaling molecules produced in response to infection or inflammatory processes. NO is produced in large amounts in macrophages, a type of white blood cell involved in immune system defense. Macrophages are important factors in destroying cancers and infectious organisms. Nitric oxide can directly kill parasites and bacteria as well as fungal organisms [71]. All are kept in check by proper production of NO by immune system cells in many parts of the body [72]. This is why urinary tract infections are screened for using a dip stick that detects high levels of nitrate in the urine. This nitrate is being produced by macrophages in their fight against the bacterial invaders.

Adequate NO production is critical for proper intestinal mucosal immunity. Too much NO induction in an immune stimulated individual may damage the mucosal lining leading to malabsorption and intestinal dysfunction. GI luminal NO production has been shown to be 100 times higher in persons with ulcerative colitis [73]. Omega-3 fatty acids can suppress the activation of inflammatory prostaglandins associated with excess NO [66]. These, along with antioxidants, may be useful in controlling the consequences of excess NO output in tissues [68].

Nitric oxide is a potent vasodilator and is thought to be a fundamental regulator of vascular tone by its ability to stimulate arterial endothelial cells to relax the smooth muscle of the arteries. This effect is seen in the use of nitroglycerin for angina pain. Nitroglycerin is rapidly converted to NO, which relaxes the coronary arteries, increasing blood flow and reducing the chest pain. Suboptimal production of NO in the cardiovascular system has been linked to the development of hypertension. Supplemental arginine has been used to increase NO output and to reduce blood pressure and improve circulation [59, 74-77]. Garlic's beneficial effects on cardiovascular function are thought to be caused in part by its ability to stimulate NO production by activation of NOS [63, 64, 78, 79].

Arginine supplementation to raise NO levels has been shown in hypercholesterolemic experimental animals to lower cholesterol and reverse atherosclerotic plaquing of coronary arteries better than a cholesterol-lowering drug [80]. Endothelial function in the coronary arteries is impaired in hypercholesterolemic individuals, which leads to the formation of atherosclerotic lesions. Additional experimental data suggest this impaired function can be corrected by supplementation with arginine if the individual is low in arginine and/or NO synthesis [81-83]. Conversely, excessive NO production can be deleterious. Increased NO can lead to oxidant stress, which can damage coronary arteries leading to atherogenesis [84].

Nitric oxide and superoxide are produced concurrently in microbial induced immune activation and other inflammatory processes. These molecules react with one another to produce the very reactive pro-oxidant peroxynitrite. This potent free radical oxidant can damage cellular membranes as well as other cellular components [85]. Peroxynitrite also reacts with sulfhydryl groups, thereby depleting such compounds as glutathione [86], lipoic acid, cysteine, and other sulfhydryl containing enzymes and proteins. Peroxynitrite also depletes coenzyme Q10 and increases lipid peroxide formation. Anti-oxidant supplementation with vitamin C, E, and glutathione has been shown to reduce this oxidant stress imposed by peroxynitrite production [67]. NO will also bind to vitamin B12. This can lead to vitamin B12 deficiency, which can have serious ramifications for nervous system myelination, that requires vitamin B12 [87]. Supplementation of vitamin B12 could also help reduce NO levels in situations of overproduction of NO [88, 89].

Since NO is metabolized rapidly and is, therefore, difficult to measure directly, researchers have sought a way of determining NO production in the body. Nitric oxide produced by the body is rapidly converted to nitrite and nitrate, which are then excreted by the kidneys. Urinary nitrate is an expression of both dietary nitrate consumption and endogenous generation. If dietary consumption of nitrate is controlled or minimized and a person is free of urinary tract infections, which elevate nitrate, then total urinary nitrate output is a useful reflection of total NO production in the body. Many animal and human studies have used this evaluation as a measure of NO synthesis [90-96].

It should be noted that in individuals with overproduction of NO due to an induction of NOS, arginine supplementation may be contraindicated because of the potential for increased oxidant stress. A useful measure of this condition has been urinary nitrate.

Conclusion

Measurement of urinary lipid peroxides, sulfate, and nitrate provide useful information on functional need for nutrient supplementation. Lipid peroxides indicate the oxidant stress level of the individual and the need for adequate antioxidant protection. Sulfate levels reflect adequacy of sulfur amino acid and glutathione stores in the body, critical nutrients for detoxification, antioxidant protection, and structural maintenance. Low levels would provide justification for supplementation of cysteine, vitamins B6 and B5, and/or glycine. Nitrate in urine (under controlled conditions) directly reflects NO production. Too little or too much NO production can adversely affect healthy function of multiple systems in the body. Low nitrate or low NO can be raised with arginine supplementation. High nitrate or NO causes oxidant stress and indicates need for specific antioxidant such as flavonoids, vitamin B-complex, lipoic acid, vitamins C and E, and coenzyme Q10. Vitamin B12 can bind with NO and may be a useful adjunct in NOS up-regulated individuals where overproduction of NO is causing oxidant stress.

These urinary compounds constitute useful biomarkers of nutritional need that provide a logical scientific basis for determining individual nutritional requirements.

References

[1] Benzie, I.F., Lipid peroxidation: a review of causes, consequences, measurement and dietary influences. *Int J Food Sci Nutr*, 47(3). p 233-61, 1996.

[2] Nacitarhan, S., T. Ozben, and N. Tuncer, Serum and urine malondialdehyde levels in NIDDM patients with and without hyperlipidemia. *Free Radic Biol Med*, 19(6). 893-6, 1995.

[3] Ozben, T., S. Nacitarhan, and N. Tuncer, Plasma and urine malondialdehyde levels in non-insulin-dependent diabetic patients with and without microalbuminuria. *Int J Clin Lab Res*, 25(3). 162-4, 1995.

[4] de Zwart, L.L., *et al.*, Evaluation of urinary biomarkers for radical-induced liver damage in rats treated with carbon tetrachloride. *Toxicol Appl Pharmacol*, 148(1). 71-82, 1998.

[5] Bagchi, D., *et al.*, Cadmium-induced excretion of urinary lipid metabolites, DNA damage, glutathione depletion, and hepatic lipid peroxidation in Sprague-Dawley rats. *Biol Trace Elem Res*, 52(2). 143-54, 1996.

[6] Brown, E.D., *et al.*, Urinary malondialdehyde-equivalents during ingestion of meat cooked at high or low temperatures. *Lipids*, 30(11). 1053-6, 1995.

[7] Dhanakoti, S.N. and H.H. Draper, Response of urinary malondialdehyde to factors that stimulate lipid peroxidation in vivo. *Lipids*, 22(9). 643-6, 1987.

[8] Draper, H.H., *et al.*, Urinary malondialdehyde as an indicator of lipid peroxidation in the diet and in the tissues. *Lipids*, 19(11). 836-43, 1984.

[9] Hermanns, R.C., *et al.*, Urinary excretion of biomarkers of oxidative kidney damage induced by ferric nitrilotriacetate. *Toxicol Sci*, 43(2). 241-9, 1998.

[10] Janero, D.R., Malondialdehyde and thiobarbituric acid-reactivity as diagnostic indices of lipid peroxidation and peroxidative tissue injury. *Free Radic Biol Med*, 9(6). 515-40, 1990.

[11] Lee, H.S., D.W. Shoeman, and A.S. Csallany, Urinary response to in vivo lipid peroxidation induced by vitamin E deficiency. *Lipids*, 27(2). 124-8, 1992.

[12] Piche, L.A., H.H. Draper, and P.D. Cole, Malondialdehyde excretion by subjects consuming cod liver oil vs a concentrate of n-3 fatty acids. *Lipids*, 23(4). 370-1, 1988.

[13] Ravichandran, V. and R. Selvam, Increased lipid peroxidation in kidney of vitamin B-6 deficient rats. *Biochem Int*, 21(4). 599-605, 1990.

[14] Vaca, C.E., J. Wilhelm, and M. Harms-Ringdahl, Interaction of lipid peroxidation products with DNA. A review. *Mutat Res*, 195(2). 137-49, 1988.

[15] Wang, Y.H., *et al.*, Lipid peroxidation status as an index to evaluate the influence of dietary fats on vitamin E requirements of young pigs. *Br J Nutr*, 75(1). 81-95, 1996.

[16] Draper, H.H., L.G. McGirr, and M. Hadley, The metabolism of malondialdehyde. *Lipids*, 21(4). 305-7, 1986.

[17] Bagchi, D., J. Moser, and S.J. Stohs, Quantitative determination of urinary lipid metabolites by high pressure liquid chromatography as indicators of menadione-induced in vivo lipid peroxidation. *Arch Environ Contam Toxicol*, 26(3). 387-91, 1994.

[18] Guichardant, M., *et al.*, Malondialdehyde measurement in urine. *J Chromatogr B Biomed Appl*, 655(1). 112-6, 1994.

[19] Bagchi, M., *et al.*, Smokeless tobacco induced increases in hepatic lipid peroxidation, DNA damage and excretion of urinary lipid metabolites. *Int J Exp Pathol*, 75(3). 197-202, 1994.

[20] Drury, J.A., *et al.*, Does total antioxidant status relate to outcome in very preterm infants? *Clin Sci (Colch)*, 94(2). 197-201, 1998.

[21] Ekstrom, T., *et al.*, Lipid peroxidation in vivo monitored as ethane exhalation and malondialdehyde excretion in urine after oral administration of chloroform. *Acta Pharmacol Toxicol (Copenh)*, 58(4). 289-96, 1986.

[22] Lawrence, R.A. and S.G. Jenkinson, Effects of copper deficiency on carbon tetrachloride-induced lipid peroxidation. *J Lab Clin Med*, 109(2). 134-40, 1987.

[23] Nelson, G.J., *et al.*, The urinary excretion of thiobarbituric acid reactive substances and malondialdehyde by normal adult males after consuming a diet containing salmon. *Lipids*, 28(8). 757-61, 1993.

[24] Takahashi, M., *et al.*, Enhanced lipid peroxidation in rat gastric mucosa caused by NaCl. *Carcinogenesis*, 12(12). 2201-4, 1991.

[25] al-Turk, W.A., *et al.*, Dietary iron and 2,3,7,8-tetra-chlorodibenzo-p-dioxin induced alterations in hepatic lipid peroxidation, glutathione content and body weight. *Drug Chem Toxicol*, 11(1). 55-70, 1988.

[26] Awad, J.A., *et al.*, Detection and localization of lipid peroxidation in selenium- and vitamin E-deficient rats using F2-isoprostanes. *J Nutr*, 124(6). 810-6, 1994.

[27] Bagchi, D., *et al.,* Protective effects of antioxidants against endrin-induced hepatic lipid peroxidation, DNA damage, and excretion of urinary lipid metabolites. Free Radic Biol Med, 1993. 15(2): p. 217-22.

[28] Barja, G., *et al.*, Dietary vitamin C decreases endogenous protein oxidative damage,

malondialdehyde, and lipid peroxidation and maintains fatty acid unsaturation in the guinea pig liver. *Free Radic Biol Med,* 17(2). 105-15, 1994.

[29] Cadenas, S., *et al.*, Vitamin E decreases urine lipid peroxidation products in young healthy human volunteers under normal conditions. *Pharmacol Toxicol*, 79(5). 247-53, 1996.

[30] Harris, C.M. and W.L. Stone, The effects of in vitro lipid peroxidation on the activity of rat liver microsomal glutathione S-transferase from rats supplemented or deficient in antioxidants. *Life Sci,* 42(4). 415-20, 1988.

[31] Jain, S.K. and R. Wise, Relationship between elevated lipid peroxides, vitamin E deficiency and hypertension in preeclampsia. *Mol Cell Biochem,* 151(1). 33-8, 1995.

[32] Lee, H.S. and A.S. Csallany, The influence of vitamin E and selenium on lipid peroxidation and aldehyde dehydrogenase activity in rat liver and tissue. *Lipids*, 29(5). 345-50, 1994.

[33] Rayssiguier, Y., *et al.*, Magnesium and blood pressure. I. Animal studies. *Magnes Res,* 5(2). 139-46, 1992.

[34] Rayssiguier, Y., *et al.*, Magnesium and ageing. I. Experimental data: importance of oxidative damage. *Magnes Res*, 6(4). 369-78, 1993.

[35] Rayssiguier, Y., *et al.*, Dietary magnesium affects susceptibility of lipoproteins and tissues to peroxidation in rats. *J Am Coll Nutr*, 12(2). 133-7, 1993.

[36] Selvam, R. and V. Ravichandran, Effect of oral methionine and vitamin E on blood lipid peroxidation in vitamin B_6 deficient rats. *Biochem Int,* 23(6). 1007-17, 1991.

[37] Shaheen, A.A. and A.A. el-Fattah, Effect of dietary zinc on lipid peroxidation, glutathione, protein thiols levels and superoxide dismutase activity in rat tissues. *Int J Biochem Cell Biol*, 27(1). 89-95, 1995.

[38] Smith, J.T., *et al.*, A metabolic comparison of cysteine and methionine supplements in the diet of a rat. *J Nutr*, 113(2). 222-7, 1983.

[39] Sullivan, J.F., *et al.*, Enhanced lipid peroxidation in liver microsomes of zinc-deficient rats. *Am J Clin Nutr*, 33(1). 51-6, 1980.

[40] Venkatesan, N., *et al.*, Protection by taurine against adriamycin-induced proteinuria and hyperlipidemia in rats. *Proc Soc Exp Biol Med*, 215(2). 158-64, 1997.

[41] Wander, R.C., *et al.*, alpha-Tocopherol influences in vivo indices of lipid peroxidation in postmenopausal women given fish oil. *J Nutr,* 126(3). 643-52, 1996.

[42] Yoshikawa, T., H. Tanaka, and M. Kondo, Effect of vitamin E on adjuvant arthritis in rats. *Biochem Med,* 29(2). 227-34, 1983.

[43] Gregus, Z., T. Oguro, and C.D. Klaassen, Nutritionally and chemically induced impairment of sulfate activation and sulfation of xenobiotics in vivo. *Chem Biol Interact,* 92(1-3). 169-77, 1994.

[44] Kim, H.J., J.H. Cho, and C.D. Klaassen, Depletion of hepatic 3'-phosphoadenosine 5'-phosphosulfate (PAPS) and sulfate in rats by xenobiotics that are sulfated. *J Pharmacol Exp Ther*, 275(2). 654-8, 1995.

[45] Merikangas, K.R., *et al.*, Tyramine conjugation deficit in migraine, tension-type headache, and depression. *Biol Psychiatry*, 38(11). 730-6, 1995.

[46] Gregus, Z., *et al.*, Effect of glutathione depletion on sulfate activation and sulfate ester formation in rats. *Biochem Pharmacol*, 37(22). 4307-12, 1988.

[47] Gregus, Z., *et al.*, Availability of glycine and coenzyme A limits glycine conjugation in vivo. *Drug Metab Dispos*, 20(2). 234-40, 1992.

[48] Corcoran, G.B., *et al.*, Effects of N-acetylcysteine on the disposition and metabolism of acetaminophen in mice. *J Pharmacol Exp Ther*, 232(3). 857-63, 1985.

[49] Corcoran, G.B., *et al.*, Acetaminophen sulfation deficit in obese rats overfed an energy-dense cafeteria diet. *Endocr Res*, 13(2). 101-21, 1987.

[50] Dills, R.L. and C.D. Klaassen, Effect of reduced hepatic energy state on acetaminophen conjugation in rats. *J Pharmacol Exp Ther*, 238(2). 463-72, 1986.

[51] Glazenburg, E.J., *et al.*, Effects of variation in the dietary supply of cysteine and methionine on liver concentration of glutathione and "active sulfate" (PAPS) and serum levels of sulfate, cystine, methionine and taurine: relation to the metabolism of acetaminophen. *J Nutr*, 113(7). 1363-73, 1983.

[52] Gregus, Z., *et al.*, Sulfation of acetaminophen and acetaminophen-induced alterations in sulfate and 3'-phosphoadenosine 5'-phosphosulfate homeostasis in rats with deficient dietary intake of sulfur. *Drug Metab Dispos*, 22(5). 725-30, 1994.

[53] Kim, H.J., *et al.*, Homeostasis of sulfate and 3'-phosphoadenosine 5'-phosphosulfate in rats after acetaminophen administration. *J Pharmacol Exp Ther,* 261(3). 1015-21, 1992.

[54] Patel, M., B.K. Tang, and W. Kalow, Variability of acetaminophen metabolism in Caucasians and Orientals [see comments]. *Pharmacogenetics,* 2(1). 38-45, 1992.

[55] Price, V.F. and D.J. Jollow, Effects of sulfur-amino acid-deficient diets on acetaminophen metabolism and hepatotoxicity in rats. *Toxicol Appl Pharmacol,* 101(2). 356-69, 1989.

[56] Reicks, M., R.J. Calvert, and J.N. Hathcock, Effects of prolonged acetaminophen ingestion and dietary methionine on mouse liver glutathione. *Drug Nutr Interact,* 5(4). 351-63, 1988.

[57] Bland, J.S., *et al.*, a Medical Food-Supplemented Detoxification Program in the Management of Chronic Health Problems. *Altern Ther Health Med,* 1(5). 62-71, 1995.

[58] Masters, B.S., Nitric oxide synthases: why so

complex? *Annu Rev Nutr,* 14. 131-45, 1994.

[59] Boger, R.H., S.M. Bode-Boger, and J.C. Frolich, The L-arginine-nitric oxide pathway: role in atherosclerosis and therapeutic implications. *Atherosclerosis,* 127(1). 1-11, 1996.

[60] Bode-Boger, S.M., *et al.*, Exercise increases systemic nitric oxide production in men. *J Cardiovasc Risk,* 1(2). 173-8, 1994.

[61] Chen, X. and T.J. Lee, Ginsenosides-induced nitric oxide-mediated relaxation of the rabbit corpus cavernosum. *Br J Pharmacol,* 115(1). 15-8, 1995.

[62] Chen, X., S. Salwinski, and T.J. Lee, Extracts of Ginkgo biloba and ginsenosides exert cerebral vasorelaxation via a nitric oxide pathway. *Clin Exp Pharmacol Physiol,* 24(12). 958-9, 1997.

[63] Das, I., N.S. Khan, and S.R. Sooranna, Nitric oxide synthase activation is a unique mechanism of garlic action. *Biochem Soc Trans,* 23(1). 136S, 1995.

[64] Das, I., N.S. Khan, and S.R. Sooranna, Potent activation of nitric oxide synthase by garlic: a basis for its therapeutic applications. *Curr Med Res Opin,* 13(5). 257-63, 1995.

[65] De Nicola, L., *et al.*, Enhancement of nitric oxide synthesis by L-arginine supplementation in renal disease: is it good or bad? *Miner Electrolyte Metab,* 23(3-6). 144-50, 1997.

[66] Carbonell, T., *et al.*, Fish oil and oxidative stress by inflammatory leukocytes. *Free Radic Res,* 27(6). 591-7, 1997.

[67] Halliwell, B., *et al.*, Interaction of nitrogen dioxide with human plasma. Antioxidant depletion and oxidative damage. *FEBS Lett,* 313(1). 62-6, 1992.

[68] Van der Vliet, A., *et al.*, Interactions of peroxynitrite with human plasma and its constituents: oxidative damage and antioxidant depletion. *Biochem J,* 303(Pt 1). 295-301, 1994.

[69] Rubbo, H., *et al.*, Nitric oxide regulation of superoxide and peroxynitrite-dependent lipid peroxidation. Formation of novel nitrogen-containing oxidized lipid derivatives. *J Biol Chem,* 269(42). 26066-75,1994.

[70] Larrick, J.W., Metabolism of arginine to nitric oxide: An area for nutritional manipulation of human disease? *J Optimal Nutr,* 3(1). 22-31, 1994.

[71] Jones-Carson, J., *et al.*, Gamma delta T cell-induced nitric oxide production enhances resistance to mucosal candidiasis [see comments]. *Nat Med,* 1(6). 552-7, 1995.

[72] Lundberg, J.O., *et al.*, High nitric oxide production in human paranasal sinuses. *Nat Med,* 1(4). 370-3, 1995.

[73] Lundberg, J.O., *et al.*, Greatly increased luminal nitric oxide in ulcerative colitis [see comments]. *Lancet,* 344(8938). 1673-4, 1994.

[74] Bode-Boger, S.M., *et al.*, L-arginine stimulates NO-dependent vasodilation in healthy humans-- effect of somatostatin pretreatment. *J Investig Med,* 47(1). 43-50, 1999.

[75] Bode-Boger, S.M., *et al.*, L-arginine-induced vasodilation in healthy humans: pharmacokinetic-pharmacodynamic relationship. *Br J Clin Pharmacol,* 46(5). 489-97, 1998.

[76] Bode-Boger, S.M., *et al.*, L-arginine induces nitric oxide-dependent vasodilation in patients with critical limb ischemia. A randomized, controlled study. *Circulation,* 93(1). 85-90, 1996.

[77] Boger, R.H., *et al.*, Biochemical evidence for impaired nitric oxide synthesis in patients with peripheral arterial occlusive disease [see comments]. *Circulation,* 95(8). 2068-74, 1997.

[78] Das, I., J. Hirani, and S. Sooranna, Arginine is not responsible for the activation of nitric oxide synthase by garlic. *J Ethnopharmacol,* 53(1). 5-9, 1996.

[79] Pedraza-Chaverri, J., *et al.*, Garlic prevents hypertension induced by chronic inhibition of nitric oxide synthesis. *Life Sci,* 62(6). 71-7, 1998.

[80] Boger, R.H., *et al.*, Dietary L-arginine reduces the progression of atherosclerosis in cholesterol-fed rabbits: comparison with lovastatin. *Circulation,* 96(4). 1282-90, 1997.

[81] Boger, R.H., *et al.*, Asymmetric dimethylarginine (ADMA): a novel risk factor for endothelial dysfunction: its role in hypercholesterolemia. *Circulation,* 98(18). 1842-7, 1998.

[82] Boger, R.H., *et al.*, Dietary L-arginine decreases myointimal cell proliferation and vascular monocyte accumulation in cholesterol-fed rabbits. Atherosclerosis, 136(1). 67-77, 1998.

[83] Jeserich, M., *et al.*, Reduced plasma L-arginine in hypercholesterolaemia [letter] [see comments]. *Lancet,* 339(8792). 561, 1992.

[84] Loscalzo, J. and G. Welch, Nitric oxide and its role in the cardiovascular system. *Prog Cardiovasc Dis,* 38(2). 87-104, 1995.

[85] Kaur, H. and B. Halliwell, Evidence for nitric oxide-mediated oxidative damage in chronic inflammation. Nitrotyrosine in serum and synovial fluid from rheumatoid patients. *FEBS Lett,* 350(1). 9-12, 1994.

[86] Clancy, R.M., *et al.*, Nitric oxide reacts with intracellular glutathione and activates the hexose monophosphate shunt in human neutrophils: evidence for S-nitrosoglutathione as a bioactive intermediary. *Proc Natl Acad Sci U S A,* 91(9). 3680-4, 1994.

[87] Flippo, T.S. and W.D. Holder, Jr., Neurologic degeneration associated with nitrous oxide anesthesia in patients with vitamin B_{12} deficiency. *Arch Surg,* 128(12). 1391-5, 1993.

[88] Greenberg, S.S., *et al.*, Hydroxocobalamin (vitamin B12a) prevents and reverses endotoxin-induced hypotension and mortality in rodents: role of nitric oxide. *J Pharmacol Exp Ther,* 273(1). 257-65, 1995.

[89] Akaike, A., *et al.*, Protective effects of a vitamin B_{12} analog, methylcobalamin, against glutamate

cytotoxicity in cultured cortical neurons. *Eur J Pharmacol*, 241(1). 1-6, 1993.

[90] Stichtenoth, D.O., *et al.*, Urinary nitrate excretion is increased in patients with rheumatoid arthritis and reduced by prednisolone. *Ann Rheum Dis*, 54(10). 820-4, 1995.

[91] Tsukahara, H., *et al.*, Assessment of endogenous nitric oxide formation in newborns: measurement of urinary nitrite and nitrate concentrations. *Biol Neonate*, 72(5). 322-6, 1997.

[92] Tsukahara, H., *et al.*, Urinary nitrite/nitrate excretion in infancy: comparison between term and preterm infants. *Early Hum Dev*, 47(1). 51-6, 1997.

[93] Tsukahara, H., *et al.*, Age-related changes of urinary nitrite/nitrate excretion in normal children. *Nephron*, 76(3). 307-9, 1997.

[94] Winlaw, D.S., *et al.*, Urinary nitrate excretion is a noninvasive indicator of acute cardiac allograft rejection and nitric oxide production in the rat. *Transplantation*, 58(9). 1031-6, 1994.

[95] Sever, R., T. Cook, and V. Cattell, Urinary excretion of nitrite and nitrate in experimental glomerulonephritis reflects systemic immune activation and not glomerular synthesis. *Clin Exp Immunol*, 90(2). 326-9, 1992.

[96] Campillo, B., *et al.*, Serum and urinary nitrate levels in liver cirrhosis: endotoxemia, renal function and hyperdynamic circulation. *J Hepatol*, 25(5). 707-14, 1996.

Chapter 12

Integration of Multiple Profiles in Patient Care

Comprehensive Nutritional Evaluation

Since essential nutrients are indispensable for the function of all cells and tissues, the symptoms and signs resulting from deficiencies of various nutrients exhibit a high degree of overlap. For example, conditions such as malaise, depression, or diarrhea can be affected by deficiencies of thiamin, methionine, or zinc. Thus, it is generally difficult to know which nutrient groups to evaluate based on initial clinical observations for most patients.

The ideal evaluation would provide information on the status of all 40+ essential and conditionally essential nutrients, plus factors that regulate nutrient needs. Interventions could then be designed to support all areas of insufficiency necessary for improving metabolic responses. Such a test profile presents the initial obstacles of high cost and numerous specimens. The large set of data from the results must be reviewed, evaluated, and documented. The evaluation of dozens of interacting parameters can be daunting and time-consuming. Most clinicians surmount these obstacles by stepwise integration of single test results or profiles in their clinical practice and use of the expertise of colleagues and laboratory experts to guide their decisions in each case.

Stepwise integration should begin with the single profile that yields the most comprehensive and useful insight into critical nutrient needs. For example, the enzyme stimulation assays discussed in Chapter 2 – Vitamins provide information about a single vitamin, while profiles of amino acids or organic acids yield data for multiple nutrients. Based on patient history and clinical assessment, one of the multiple analyte profiles is typically selected first. Other tests may be added to expand the range of nutrients initially evaluated or as follow-up testing to confirm or elaborate abnormalities found in an initial routine evaluation. A common approach is to order the profile of organic acids in urine as a sensitive, broad-range initial test that helps guide specific clinical interventions. The evaluation may then be expanded with specific antioxidant marker analysis or fatty acid profiling. Mineral or amino acid profiling can enhance confidence and provide reference points for monitoring the success of interventions.

Another rational approach is to deal with issues that may have overriding impact on nutrient supplementation, such as problems with digestion, assimilation, and immune competence, before focusing on other specific nutrient deficiencies. This approach uses digestive and microbial stool testing, food-specific serum IgG levels, and other measures of digestive function to alter food intake and enhance digestive capacity. Specific nutrient deficiencies that may be present can be made worse by food restrictions. For example, food sensitivity testing frequently leads to elimination of several classes of high protein foods. The reduced intake of high protein foods can exacerbate amino acid deficiencies, and without adequate amino acids, restoration of gut immune barrier integrity can be difficult.

This type of circular dependency is best dealt with by simultaneous evaluations of food sensitivities, amino acid status, and degree of intestinal microbial overgrowth. Thus, the integration of data from multiple profiles provides a much more complete picture of areas of most critical need.

As the complexity of laboratory data increases, interpreting results to arrive at a clinical solution becomes more difficult. In response, laboratory experts familiar with the implications of the results have developed algorithms to assist in the interpretation. The most common example of such expert reliance is the computation of relative risk from cholesterol data. Of course, the people at the laboratory don't see the patient, so their recommendations are always subject to the attending physician's final revision based on the patient's medical history and physical exam. Nutrient recommendations based on integration of multiple profiles are derived from knowledge of how prioritized or weighted multiple markers may be used to increase confidence in determining a patient's nutrient needs and guide decisions about appropriate supplementation levels.

Multiple Markers

The clinical relevance of laboratory data ranges from pathognomonic factors that define the presence of a disease to risk factors that suggest the likelihood for future disease. For example, the finding of elevated fasting glucose defines the presence of diabetes, while the finding of elevated total cholesterol indicates a possible increased risk of heart disease. Both of these tests are performed routinely because they have demonstrated value in improving patient outcomes. Their usefulness, however, is greatly enhanced when other related parameters are measured. Examples of tests that help to define the nature of some primary evaluations are shown in Table 12-1.

Some measures of nutritional status likewise fall into categories of reinforcing tests. There are web-like interactions between nutrients, toxicants, and metabolic controls, but the most useful common denominator in the context of this work is nutrient deficiency.

A Case Study - Multiple Markers Within a Single Profile

Ms. B was described as a very bright workaholic who presented with a sense of "brain fog" that had dramatically increased in recent weeks. She also reported greater difficulty with mood swings around her menstruation with severe depression, anxiety, and suicidal thoughts. She had chronic fatigue for two or three years. An overnight urine sample was sent for quantitative organic acid profiling, and the results revealed corroborating evidence of underlying

TABLE 12-1. Test Combinations to Guide Patient Management

Clinical Challenge	Primary Evaluation	Reinforcing Tests
Diabetes	Blood glucose	Glucose tolerance Insulin C-peptide Glycoproteins
Heart disease risk	Cholesterol	Triglycerides LDL & HDL cholesterol Small, dense LDL Testosterone, Insulin Lipid peroxides, Vitamin E
Peripheral neuropathy	Low serum vitamin B_{12}	Methylmalonate Homocysteine Complete blood count
Slow wound healing	Low serum zinc	RBC trace element profile Fatty acid profile (LA/DGLA)
Rapid aging	Elevated lipid peroxide	Serum vitamin E Serum coenzyme Q10 RBC mineral profile

metabolic difficulty (Tables 12-2 to 12-5)

In the first set, two features relate to nutrient requirements (Table 12-2). Elevated citrate is a marker for arginine deficiency and elevated malate and pyruvate are found in deficiencies of coenzyme Q_{10} (Chapter 6 – Organic Acids). This data suggests that arginine and coenzyme Q_{10} are needed. Deficiency of the B-complex vitamins can also lead to elevated citrate, but there are usually multiple citric acid cycle intermediates elevated in such cases. In addition, there are

TABLE 12-2. Ms. B's Central Energy Pathway Intermediate Organic Acids in Urine

Set 1	μg/mg Creatinine		Limits
Citrate	2303	H	500 - 2300
cis-Aconitate	226		5 - 250
Isocitrate	527		50 - 800
α-Ketoglutarate	22.8		3 - 25
Succinate	24.5		5 - 35
Fumarate	0.82		0.2 - 1.2
Malate	7.8	H	≤ 6.0
Lactate	22		4 - 30
Pyruvate	0.97	H	≤ 0.7

other markers for the B-vitamins discussed below.

The compounds of Set 2 (Table 12-3) are elevated in specific nutrient deficiencies as discussed in Chapter 6 – Organic Acids. Thiamin (vitamin B_1) is the key nutrient required to clear α-ketoisovalerate. Methylmalonate is a specific marker for vitamin B_{12} inadequacy and hydroxymethylglutarate is the compound from which coenzyme Q_{10} is synthesized in the liver. The deficiency of hydroxymethylglutarate corroborates our findings from Set 1, and our suspicion that coenzyme Q_{10} deficits are contributing to the brain fog reported by the patient is increased.

TABLE 12-3 Ms. B's B-Vitamin-Insufficiency-Marker Organic Acids in Urine

Set 2	μg/mg Creatinine		Limits
α-Ketoisovalerate	0.31		≤ 1.5
α-Keroisovalerate	2.38	H	≥ 2.0
α-Keto-β-methylvalerate	0.85		≥ 1.2
Methylmalonate	4.1	H	≥ 3.0
β-Hydroxyisovalerate	9.2		≥ 20
Hydroxymethylglutarate	<0.1	L	0.2 - 1.0

Abnormally high values for three other compounds indicate that Ms. B is having difficulty with her elimination of toxic compounds (Table 12-4). Elevated orotate is a sensitive marker for arginine deficiency. Orotate is the product of an alternate pathway for clearing ammonia when the urea cycle is inefficient, and chronic sub-clinical ammonia accumulation is a possibility. The decision to supplement with arginine because of high citrate in Set 1 (Table 12-2) is greatly strengthened. In addition, the clearance of orotate is a magnesium-dependent process, so there is evidence of a need for this commonly deficient mineral.

TABLE 12-4. Ms. B's Detoxification Marker Organic Acids in Urine

Set 3	μg/mg Creatinine		Limits
Orotate	224	H	≤ 180
Pyroglutamate	98.5	H	≥ 80
Sulfate	50.9	L	≥ 180

A separate issue is raised by the rest of the pattern found in Table 12-4. Glutathione and inorganic sulfate are critical for hepatic detoxification. Sulfate is the excretory form of sulfur from most sulfur-containing compounds in the body. Since glutathione makes up a major fraction of total body sulfur due to the presence of cysteine in its structure, it should not be surprising that depletion of glutathione correlates with low urinary sulfate. Pyroglutamate elevation signals loss of glutathione in the amino acid recovery pathway in the kidney (see Chapter 6 – Organic Acids), which contributes to depletion of the total body glutathione pool. In this case, the sulfate is very low and the pyroglutamate is elevated. These two markers together strongly indicate glutathione depletion and gives extra cause for concern over the issue of detoxification capacity that may lead to endotoxemia. Since we now have evidence of impaired detoxification capacity, the question of toxin load becomes more important. Indications of one significant source of toxins, intestinal microbial overgrowth, are found in compound Set 4 in this case, the bacterial overgrowth marker organic acids (Table 12-5).

TABLE 12-5. Some of Ms. B's Bacterial Overgrowth Marker Organic Acids in Urine

Set 4	μg/mg Creatinine		Limits
p-Hydroxyphenylacetate	20		≤ 50
Tricarballylate	5.1	H	≥ 1.8
Dihydroxyphenylpropionate	0.05		≥ 0.9

The three compounds in Set 4 are some of the metabolic markers of intestinal dysbiosis, showing significant elevation of tricarballylate. Tricarballylate is not produced in human metabolism but is found in high concentrations in some ruminants treated with powerful antibiotics. The compound does not seem to be highly toxic, but it avidly binds to magnesium, preventing absorption of the mineral. Animals with tricarballylic aciduria develop magnesium deficiency. Although the condition has been little investigated in humans, the finding of such greatly abnormal concentrations in this case adds to the suspicion of magnesium deficiency. Thus, we find additive signs pointing to magnesium inadequacy with high tricarballylate and the orotate accumulation noted above. If microbial metabolites, such as the highly toxic p-cresol, had been elevated in Ms. B's urine, concern about toxic loading from intestinal dysbiosis would have been greatly increased.

Patterns of nutrient insufficiencies contributing to specific degenerative diseases vary among individuals. Some common characteristics are summarized in Tables 12-6 through 12-8. Most patients with complex degenerative disorders show various degrees of severity in multiple health threat categories.

TABLE 12-6. Abnormal Laboratory Results Frequently Found in Organic Acid Profiles
Arrows indicate whether the analyte is elevated (↑) or depressed (↓) as a sign of the health threat.

	Health Threat				Other Related Tests	
Organic Acid	Mitochondrial Dysfunction	B-Complex Deficiency	Detoxification Challenge	Neuroendocrine Imbalance	Test*	Result
Adipate, Suberate, Ethylmalonate	↑				FAP	Palmitic↑, Stearic↑
Citric Acid Cycle Intermediates	↑				AAP eMin	EAA↓, Gln↓ Multiple elements↓
Pyruvate		↑			eMin	Mg↓
α-Ketoglutarate		↑				
BCKA		↑			AAP	BCAA↓
Methylmalonate		↑			FAP	Odd-chain↑
α-Hydroxybutyrate		↑			eMin	Cr, V↓
HMG	↑				FatSol Vits	CoQ_{10}↓
Pyroglutamate			↑		AAP	Met, Cys, Tau↓
Sulfate			↓		AAP	Met, Tau↓
Dysbiosis Markers			↑		AAP, IgG	BCAA↓, Many IgG↑
Benzoate			↑		AAP	Gly, Ser↓
HVA, VMA				↑↓	AAP	Tyr, Phe↓
5-HIA				↑↓ (SSRI)	AAP	Trp↓

* AAP = amino acids (plasma); FAP = fatty acids (plasma); eMin = erythrocyte minerals; FatSol; = vitamins A, E, β-Carotene, and Q_{10} (serum)

TABLE 12-7. Abnormal Laboratory Results Frequently Found in Amino Acid Profiles

	Health Threat			Other Related Tests	
Amino Acid	Dietary Deficiency/ Maldigestion	Chronic Inflammatory Disorder	Behavior Disorder	Test*	Result
Arginine		↓		OAU	Citrate, orotate↑
BCAA	↓			AAP	Glutamine↑
Phenylalanine	↓		↓		
Tryptophan	↓		↓		
Methionine	↓			Detox	Hepatic capacities ↓
Threonine	↓				
Glycine, Serine	↓			OAU	Benzoate high↑

* OUA = organic acids (urine); AAP= amino acids (plasma); Detox = detoxification capacity tests

TABLE 12-8. Abnormal Laboratory Results Frequently Found in Fatty Acid Profiles

Fatty Acid	Slow Growth, Behavioral Disorder, Brain Development, Reproduction, Skin (EFA Deficiency)	Heart Disease, Depression, Chronic Inflammatory Disorder (n6 Imbalance)	GI Disorders, Poor Wound Healing, Hormonal Abnormalities	Cancer
ALA	↓		↑*	
EPA	↓	↓	↓	
DHA	↓			
GLA	↓	↓	↑*	
DGLA	↓	↓	↓	
AA		↑		
Oleic				↑
Palmitic		↑		
Stearic				↓
Elaidic		↑		

* For GI disorders, the elevations result as a manifestation of zinc deficiency.

Advanced Study of the Case – The Power of Simultaneous Multiple Profiles

We can proceed to investigate the metabolic basis of Ms. B's brain fog, fatigue, and suicidal tendencies because the profile of organic acids in urine was not the only one that was performed. The following profiles will be discussed as an illustration of how the results can shed light on the problem and allow us to determine the most critical areas of nutrient demand in this phase of Ms. B's life. The profiles did not include hormone, food allergy, or other immune system evaluations. Additional digestive system and detoxification assessments could provide valuable insight. For those cases where stepwise investigations into metabolic causes can be done quickly, satisfying individual nutrient demands is a worthy first effort that frequently yields restoration of health. After all, the formation of all classes of hormones and other metabolic controls, including those that modulate the immune system are dependent on the chain of nutrients that serve as co-catalysts, antioxidants, and detoxification substrates. While the primary purpose of the profiles included in this case is evaluation of nutrient status, the nature of the analysis, especially in the case of organic acids in urine, allows insight into nutrient-related metabolic systems as well.

The array of markers for functional vitamin adequacy in the organic acid profile covers most vitamins except the fat-soluble group. A separate profile of fat-soluble antioxidant vitamins extracted from serum is performed (Table 12-9).

TABLE 12-9. The Fat-Soluble Antioxidant Vitamin Profile

Antioxidants	μg/L		Limits
Coenzyme Q_{10}	0.73	L	0.8 - 1.5
Vitamin E	13.1		12 - 40
Vitamin A	0.59		0.5 - 1.2
β-Carotene	< 0.2	L	0.4 - 3.5

Here is evidence from an independent assay that supports the assessment of functional coenzyme Q_{10} inadequacy by showing that Ms. B's circulating levels are below normal (Table 12-9). In addition to its electron transport role, coenzyme Q_{10} performs a general membrane antioxidant function in concert with the other antioxidants in this profile. We see that β-carotene is undetectable and, although vitamins E and A are not "L" (below their respective limits), note that the levels are just over the stated lower limits. The pattern of inadequate antioxidant protection found here leads us to suspect that the rate of oxidant damage to tissue components is too high. The suspicion is confirmed by finding elevated levels of lipid peroxides in serum (Table 12-10). With these results, we have a point of reference for the current rate of oxidative damage against which future testing may be compared. Any interventions that improve antioxidant capacity should cause a lowering of the serum lipid peroxide levels. Lipid peroxide levels are found elevated in some cases even though the fat-soluble antioxidant vitamins are all normal. In such cases, the next most likely area of insufficiency is that of trace elements.

TABLE 12-10. Serum Lipid Peroxide Test Results

Serum Lipid Peroxides	μmol/ml		Limits
Lipid peroxides	1.6	H	≤ 1.0

Mineral assessment was performed by measuring concentrations in erythrocytes (Table 12-11). Remember that we already have indications of functional deficiency of magnesium and there is an additional functional deficiency of zinc revealed with the fatty acid profile (Table 12-12). An independent assay of circulating levels of these elements in erythrotcytes (Table 12-11) gives significant weight to conclusions about deficiencies indicated by metabolic markers. The important findings of low circulating levels of essential nutrients help to identify nutrient deficiency. Erythrocyte magnesium of 25 ppm is enough to explain the molecular basis of symptoms of mood swings and depression in this patient. Look closely here to see how many other trace elements are near the low limit. We find that, in addition to low levels of selenium and zinc, chromium, molybdenum, and vanadium are within 10% of their low limits, indicating a potential problem with trace element absorption and assimilation. Low stomach acid output is the most frequently occurring factor that leads to general trace element malabsorption.

TABLE 12-11. The Erythrocyte Mineral Profile

Element Analysis – Erythrocyte	PPM Packed Cells		Limits
Chromium	0.72		0.7 - 2.7
Copper	0.88		0.5 - 2.0
Magnesium	25	L	40 - 70
Manganese	0.09L		0.1 - 0.4
Molybdenum	0.03		0.03 - 0.16
Potassium	479		200 - 1000
Selenium	0.31	L	0.4 - 2.0
Vanadium	2.06		2.0 - 7.0
Zinc	7.9	L	10 - 40

Pertinent features of the fatty acid profile are shown in Table 12-12. The pattern displays the frequently found deficiency of alpha linolenic acid for which flaxseed oil is recommended. Note the relationship of the two omega 6 fatty acids, linoleic (C18:2) and gamma linolenic acid (C18:3). The former is converted to the latter by the action of the strictly zinc-dependent enzyme, delta-6-desaturase (see Chapter 5 – Fatty Acids). Failure of adequate conversion is due to zinc deficiency and results in such a pattern of normal linoleic with low gamma linolenic. The pattern is frequently strengthened by simultaneously finding low dihomogammalinoleic

TABLE 12-12. Erythrocyte Fatty Acid Profile Data

Fatty Acids in Erythrocytes	μM		Limits
Alpha linolenic	0.9	L	1 - 3
Eicosapentaenoic	6.4		3.5 - 8
Docosapentaenoic	14		10 - 30
Docosahexaenoic	28		15 - 35
Linoleic	42		40 - 70
Gamma linolenic	0.8	L	1 - 3
Eicosadiencoic	10.4		7.5 - 25
Dihomogammalinolenic	5		5 - 9

acid (DGLA), which in this case is right on the lower limit. This is the pattern mentioned above showing functional deficiency of zinc. The dietary polyunsaturated fatty acids (PUFAs) need to be used as fish and black currant oils while correcting the zinc deficiency in such cases. The adequate EPA and DHA levels found in the current case, however, mean that a balance of flaxseed and fish oils is the best choice to restore fatty acid balance.

The amino acid profile for Ms. B shows all amino acids within normal limits except for slightly low values for leucine and lysine (Table 12-13). Note especially the arginine level. Here we find a value within normal limits, while multiple metabolic markers indicated a functional deficit of the nutrient. In addition, arginine is not considered an essential nutrient for adults. This example illustrates how functional markers assist identification of critical nutrient status. The simultaneous elevations of citrate and orotate described above override the fact that the circulating arginine concentration is within the limits that the laboratory has approximated as "normal." The body status of arginine needs to be improved to meet metabolic demands, not just to raise plasma arginine levels. Sometimes a clinical trial of free form amino acids will produce a favorable response when patterns of low-normal plasma amino acids are found as in this case. This is especially worth considering with Ms. B who presented with multiple symptoms (fatigue, loss of mental clarity, and depression) that are known to respond to amino acid replacement.

This case presents an example of the integration of data from multiple profiles to establish a plan of metabolic management. When multiple indicators of functional amino acid inadequacies are present, a combination amino acid supplementation formula could be useful for restoration of optimal nutritional status even when most amino acids in plasma are in the "normal" range. Such interventions

TABLE 12-13. A Portion of The Profile of Free Amino Acids in Plasma

Plasma Amino Acids	μM		Limits
Arginine	37		35 - 160
Histidine	84		65 - 150
Isoleucine	56		50 - 160
Leucine	100	L	105 - 250
Lysine	125	L	140 - 250
Methionine	22		20 - 60
Phenylalanine	46		45 - 140
Threonine	120		85 - 250
Tryptophan	50		45 - 120
Valine	178	L	180 - 480

to restore optimal function rather than simply assure normal concentrations are a common theme in nutritional medicine. The clinician is challenged to use laboratory limits in new ways because nutrient concentration values that are normal may still be inadequate.

The patterns found in this case cover all essential nutrient categories. They are frequently observed and illustrate the general concept of strengthening conclusions based on clinical laboratory data by using multiple markers from different metabolic profiles. The varieties of patterns for abnormal results are endless, but the experienced practitioner develops a keen sense of where to look for reinforcing results. Some laboratories offer reports that perform pattern recognition with computer software, greatly shortening the learning curve for the novice multi-profile user. It is wise to verify the expertise of those who write such algorithms so that valid interpretations are assured.

Algorithms for Supplementation from Metabolic Profiles

The design of a customized nutritional supplementation plan that will help restore a patient's normal metabolic function is a goal of nutrient use in molecular medicine. Laboratory results can be used to guide variations in nutrient supplementation ranges. An algorithmic framework that may serve to translate test results into nutrient supplementation regimes is presented in Table 12-14. Each nutrient is shown with relevant laboratory evaluation. The supplementation levels range from basic "insurance formula" to therapeutic intervention amounts for adults.

For example, the finding of elevated lipid peroxides with low serum vitamin A justifies higher supplementation of vitamin A than low vitamin A levels with normal lipid peroxides. Testing can reveal circumstances where certain nutrients should be explicitly excluded. These cases show a "0" intervention level in the tables. For example, because of the potential for toxic effects from vitamin A, an elevated serum level mandates removal of this nutrient from any supplementation given unless there are compelling reasons to maintain elevated serum concetrations. If all test results for a given nutrient are negative, default amounts (listed for "No abnormalities" in the tables) will prevent deficiencies during periods of dietary inadequacy. The permutations that are shown illustrate the use of coincident abnormalities to signal varying supplementation. These algorithms include the most directly relevant tests, but they exclude the impact of other factors such as an increased body burden of toxic metals.

TABLE 12-14. Customizing Dietary Supplement Interventions Based on Laboratory Evaluations
The specimen tested is indicated by the letter in parentheses following the test, where s = Serum; p = Plasma; u = Urine; and e = Erythrocyte.

Vitamin A

	Test		I.U./d
1	Vitamin A (s)	H	0
2	No abnormalities		0-3000
3	Lipid Peroxides (s, u)	H	3000-5000
4	Vitamin A (s)	L	5000-7000
5	3 and 4 True		7000-10000

β-Carotene

	Test		mg/d
1	Vitamin A (s)	H	0
2	β-Carotene	L	--
3	No abnormalities		0-5000
5	1 and 2 True		5000-10000
6	2 True and 1 NOT True		10000-25000

Vitamin E

	Test		mg/d
1	Vitamin E (s)	H	0
2	No abnormalities		50-100
3	Arachidonic acid (p)	H	100-300
4	Vitamin E (s)	L	300-600
5	Lipid Peroxides	H	300-600
6	3 and 4 True		500-1000

Vitamin B1 (Thiamin)

	Test		mg/d
1	No abnormalities		0-5
2	αKβMeVal (u)	H	5-10
3	αKiVal (u)	H	5-10
4	αKiCap (u)	H	5-10
5	Pyruvate (u)	H	5-10
6	αKG (u)	H	5-10
7	Any three or more True		30-50

Vitamin B2 (Riboflavin)

	Test		mg/d
1	No abnormalities		0-5
2	1, 2, or 3, for Vitamin B1		5-10
3	Succinate (u)	H	5-10
4	Pyruvate (u)	H	5-10
5	αKG (u)	H	5-10
6	Oleic acid (p, e)	H	10-20
7	3, 4, and 5 True		30-50

Vitamin B3 (Niacin)

	Test		mg/d
1	No abnormalities		0-5
2	Keto Acids (u)	H	20-70
3	Lactate (u)	H	20-70
4	2 AND 3		70-120
5	Total Cholesterol (s)	H	200-400
6	LDL Cholesterol (s)	H	200-400
7	Total/HDL Cholesterol	H	200-400
8	6 AND 7 True		400-600
9	3, 5, AND 7 True		600-800

Vitamin B6 (Pyridoxin)

	Test		mg/d
1	Homocysteine (u)	H	30-50
2	Leucine (p)	H	10-30
3	Isoleucine (p)	H	10-30
4	Glutamine (p)	H	10-30
5	No abnormalities		0-10
6	2, 3, and 4 True		30-50

Biotin

	Test		mcg/d
1	No abnormalities		0-800
2	Vaccenic (p, r)	H	1000-2000
3	β-Hydroxyisovalerate (u)	H	2000-5000

TABLE 12-14. Customizing Dietary Supplement Interventions Based on Laboratory Evaluations Cont.
The specimen tested is indicated by the letter in parentheses following the test, where s = Serum; p = Plasma; u = Urine; and e = Erythrocyte.

Iron

	Test		mg/d
1	Iron (s)	H	0
2	% Saturation (s)	H	0
3	Iron (s)	L	0-12
4	No abnormalties		0-12
5	3 and Iron-binding Capacity (s)	H	5-10
6	3 True and NOT 2 True		15-30

Zinc

	Test		mg/d
1	Zinc (e)	H	0
2	No abnormalties		0
3	Zinc (e)	L	0-10
4	LA/DGLA	H	0-10
5	4 True and Lipid Peroxides (s)	H	5-10
6	5 True and Zinc (e)	L	15-30

N-Acetylcysteine

	Test		mg/d
1	No abnormalities		0
2	Sulfate (u)	L	200-300
3	Pyroglutamate (u)	H	200-300
4	2 and 3 True	H	300-400
5	4 True and Taurine (p)	H	400-600
6	4 True and Methionine (p)	H	400-600

Coenzyme Q_{10}

	Test		mg/d
1	No abnormalities		0
2	Serum Q_{10} (s)	H	30
3	Hydroxymethylglutarate (u)	H	30
4	Succinate (u)	H	60
5	2 and Lipid peroxides (s)	H	120
6	2, 3, and 5 True (p)	H	300

Flax Oil (Alpha Linolenic Acid)

	Test		gm/d
1	ALA (p, e)	H	0
2	ALA (p, e)	L	3
3	2 True and EPA (p,e)	H	3
4	2 True and EPA (p, e)	L	6

Tryptophan

	Test		gm/d
1	Trp (p, u)	H	0
2	Trp (p, u)	L	20-50
3	5-HIA (u)	H	50-100
4	2 and 3 True		100-200

Selenium

	Test		mcg/d
1	No abnormalities		0-50
2	Lipid Peroxides (s)	H	50-100
3	Selenium (r)	L	100-200
4	Mercury (r)	H	200-400
5	2 and 3 True		200-400

Carnitine

	Test		mg/d
1	No abnormalities		0
2	Adipate (u)	H	300-600
3	Suberate (u)	H	300-600
4	Ethylmalonate (u)	H	300-600
5	2 and 3 True		600-900

Pattern Recognition

All clinical assessment is a matter of pattern recognition. Patterns of signs and symptoms make syndromes and lead to pathognomonic disease definition. In molecular medicine, patterns can lead to greater specificity of nutrient deficiency and aggressiveness of intervention for a given case. Sometimes multiple profiles of apparent disparate functions are found to form links that direct the clinician's attention to the more pertinent system failure. For example, concurrent findings of multiple food allergies, along with metabolic markers of intestinal bacterial overgrowth and elevated markers of free radical pathology may be combined to reach a diagnostic conclusion. Any one of the three testing areas might lead to correct interventions, but together they point to the nature of the problem as an inflammatory bowel with degeneration of both immune and physical barriers and resulting increased antioxidant demand.

The following clusters of data illustrate scenarios that amplify clinical insight about nutritional interventions (Table 12-15). The next section presents an illustration of how such clusters are used with patient history and physical examination in specific patients.

TABLE 12-15. Five illustrations of how multiple profile data provides improved clinical insight. Each cluster of data presents a different scenario with differing starting point observations. The tables should be read from top to bottom as a temporal sequence of tests that were performed. EAA = Essential Amino Acid

Cluster 1

Test	Finding	Indication
IgG Food Profile	Multiple high antibody levels	Specific food avoidance or rotation
Organic Acids (u)	Multiple bacterial metabolic markers	Probiotic and bacteriostatic agents
Lipid Peroxides	Elevated	Strict hypoallergic diet with probiotics, glutamine, zinc, pantothenic acid, and full spectrum antioxidant support

Cluster 2

Test	Finding	Indication	Clinical Impression
Mineral Profile (e)	Multiple low trace elements	Betaine-HCL and mineral asparates	Hypochlorhydria
Organic Acids (u)	Multiple bacterial markers	Restore orthobiosis	Hypochlorhydria reinforced – protein maldigestion encouraging bacterial growth
Amino Acids (p)	Multiple low EAA	Free-form, EAA supplement	Chronic hypochlorhydria leading to amino acid depletion, but amino acid supplementation should be started slowly to assure assimilation and prevent further stimulation of intestinal bacterial growth

TABLE 12-15. Cont.

Cluster 3

Test	Finding	Indication	Clinical Impression
RBC Mineral Profile	Low-normal zinc with low selenium	Zinc not profoundly depleted	Possible need for selenium
Fatty Acid Profile	High LA/DGLA	n-6 fatty acids needed	Functional zinc deficiency
Antioxidant Profile	Normal	Fat soluble vitamins not profoundly depleted	
Lipid Peroxide	Elevated	Antioxidant protection	Aggresive repletion of both selenium and zinc is needed to overcome functional insufficiency of both elements and interim primrose oil and vitamin E are needed

Cluster 4

Test	Finding	Indication	Clinical Impression
Organic Acids (p)	High benzoate, p-hydroxyphenylacetate, and tricarballylate	Bacterial overgrowth	Challenged hepatic detoxification
Amino Acids (p)	Low glycine and Ser	Support with glycine or EAA mixture	Limited glycine conjugation capacity with increased demand for excretion of microbial toxins
Organic Acids (u)	High β-hydroxybutyrate	Impaired insulin function	Poor gluconeogenic support of plasma glucose
Organic Acids (u)	Elevated adipate and suberate	Supplement carnitine	Simultaneous mitochondrial, dysbiotic, and detoxification challenges

Cluster 5

Test	Finding	Indication	Clinical Impression
DMSA Challenge 24 hr. urine	Elevated Cd and Hg	Reduce toxic metal esposure	Possible oral or IV chelation and amalgam removal
Amino Acids (p)	Low Met and Taur	Need for sulfur amino acids Impaired insulin function	Restore cellular metal complexing ability
Organic Acids (u)	Elevated pyroglutamate	Glutathione depletion	Aggressive sulfur amino acid repletion needed with repeat testing of sulfur compound status before implementing aggressive chelation or amalgam removal

Integrating Clinical Impressions with Multiple Laboratory Profiles

Mr. L's Amino Acids

History and Observations

Mr. L was a forty-seven year old man who presented with a history of sinus congestion and depression. He noted a family history of depression (father and brother) and related that his first bout of severe depression, which required a six-month course of standard medications, occurred after he had put himself on a low calorie weight loss diet. As a child, he had a history of "hay fever" and recurrent ear infections, which were treated with multiple antibiotics. Asthma, which seemed to be associated with environmental and food changes, was being controlled with occasional use of a steroid inhaler.

Additional family history included attention deficit disorder and chronic sinusitis in both of Mr. L's sons, ages 6 and 12; his parents' deaths were due to coronary artery disease and cancer. Mr. L experienced lactose intolerance with symptoms of bloating and diarrhea when he consumed milk or ice cream. He had hemorrhoids about every six months with rectal bleeding. He also had a recurrence of numbness and weakness in the extremities and leg cramps at night, which were successfully treated with calcium supplements. He complained of increasing symptoms of joint and muscle pain and headaches. The most annoying symptoms were associated with a more recent degeneration of nasal membrane integrity. Symptoms of excessive mucus formation, stuffy nose, hay fever, and sinus problems were all reported as "severe." He had been on antibiotics five or six times in the past year.

Impressions

- 4.5 decades of potential nutritional deficit
- Dysbiosis endotoxins?
- Familial metabolic lesions?
- Essential fatty acids?
- Magnesium?
- Deficiencies of B_6?, B_{12}?
- Delayed food reactions?

Impressions and Laboratory Data

As the running list of impressions listed in the margin shows, the initial impression added up to a suspicion of multiple nutrient insufficiencies with potential toxin loads that may exceed detoxification capacities. Initial laboratory evaluations included standard serum chemistries, food-specific IgG antibody levels, plasma amino acids, urinary organic acids, and erythrocyte minerals. Foods showing significantly elevated IgG levels were egg, kidney bean, navy bean, and peanut. Moderate levels were found for milk and string bean, and almond, pineapple, and lima bean showed low levels. The amino acids tyrosine, taurine, glycine, and serine were low in fasting plasma and the sulfate to creatinine ratio was low, along with multiple elevations of bacterial products in urine. Mineral insufficiencies were indicated by low erythrocyte magnesium and zinc. His serum cholesterol was slightly elevated.

Interventions

Mr. L was instructed about the use of an elimination/rotation diet in which all foods with severe and moderate IgG reactions were eliminated. He was placed on supplementation of a high-potency multivitamin-mineral, additional B-complex vitamins, and a combination mineral product containing calcium, magnesium, and zinc. A rice-based hypoallergenic medical food was used to support detoxification function, and a broad-spectrum lactobacillus formula, berberine, and grapefruit seed extract with calcium undecylenate were used to reestablish orthobiosis.

Four week follow up

At a four-week follow up visit, Mr. L reported much improvement in GI symptoms with some improvement of sinus symptoms. The rice-based formula was causing some reactions, so it was replaced with fructo-oligosaccharide and glutamine to assist intestinal healing. Scoring of a health assessment questionnaire had dropped from 34 to 11, indicating significant improvement. At this time he was also advised to start a custom amino acid powder that was produced according to the description in Chapter 4 – Amino Acids. At a twelve-week follow up, Mr. L reported that he is "feeling great." The sinus congestion was relieved within a few days of starting the amino acids. His mood was steady and his energy level was greatly improved.

Long-term follow up

Long-term follow up visits were done to check for responses to additional, normal life stressors. At six months, all previous areas of complaint were still much improved, and his serum cholesterol was within normal limits. The health questionnaire score increased slightly due to symptoms of seasonal allergies. At nine months, Mr. L reported feeling "much, much better." He stated, "Many problems are gone for good." Follow-up testing revealed all IgG reactions were markedly decreased and urinary sulfate and plasma amino acids had all increased to normal or near-normal limits.

Discussion

This patient is typical of individuals who have been described as "the walking wounded" who attempt to live with annoying and mildly debilitating symptoms. None of the symptoms would lead to a conclusion of overt nutrient deficiency. Laboratory evaluation revealed the multiple molecular nature of the problem. This "walking wounded" patient can now be described as walking with intestinal hyperpermeability and small intestinal bacterial overgrowth accompanied by an immune system burden of IgG-antigen complexes. Nutrient insufficiencies are concurrent in this situation impeding the return of wellness by restricting hepatic detoxification capacity. The elimination of IgG-reactive foods and the apparent restoration of the intestinal barrier integrity helped reduce the chronic challenge to the immune system. The main presenting symptom involving the mucosal membranes was not completely resolved until the introduction of an extra ten grams of free form amino acids and glutamine daily. This step may be considered to have two critical effects. First, the repair of the tall columnar epithelial cells of the gut is enhanced by the improved availability of non-essential amino acids, especially glutamine, which is used for energy and regeneration by those cells. Glutamine sparing is accomplished by supplying essential amino acids with cofactors, α-KG, and pyridoxal-5-phosphate. The second effect is that of increased growth hormone activity with increasing plasma amino acid levels. Mucosal membrane repair is expected to accelerate under the action of growth hormone.

The stages of healing in this case may be diagrammed as Gut -> Nose -> Brain. The insomnia and depression were the final symptoms to clear. The origin of mental improvement likely involved the reduction of toxins flowing from the gut and the improvement of neurotransmitter precursor availability because of improved amino acid intake and assimilation with lowered detoxification demands.

Mr. F's Stages of Health

STAGE I

At the age of 34, Mr. F had experienced soft tissue pain for 12 years and joint pain for the past two years. The pain was being controlled with Flexeril® and NSAIDs. For many years, he had experienced difficulty with sleep and multiple symptoms of inflammatory bowel syndrome. The onset coincided with his starting a stressful business after graduating from college.

Impressions

- Chronic inflammatory response?
- B-complex and Mg depletion?

A lactose/mannitol test showed normal carbohydrate assimilation with hyperpermeability. A hydrogen breath test and stool cultures revealed bacterial overgrowth. Very high levels of IgG class antibodies to milk, eggs, and chicken, all of which were frequently consumed, were found.

- Food allergic responses?
- Dysbiosis endotoxins?
- Delayed food reactions identified

The bowel situation was improved with regular consumption of steamed vegetables, supplementation of antioxidant vitamins, and avoidance of challenging foods. The favorable organisms, *Lactobacilli* and *Bifidobacteria*, were also supplemented. Within ten weeks the joint and soft tissue pain was virtually absent without medication, and bowel and sleep habits became normal. Over the next year he also noticed a substantial lengthening of time between upper respiratory infections.

- Reduced endotoxins
- Normal immune response restored

STAGE II

After several months of improved pain status, Mr. F started experiencing increasing bouts of profound fatigue and a persistent lethargy. Testing of organic acids in urine at this point revealed very low levels of citrate and fumarate with simultaneous low VMA, the neurotransmitter catabolite, and elevated levels of the *Clostridium* metabolic marker, dihydroxyphenylpropionate. Profiling of trace elements in hair showed elevation of mercury, even though mercury was undetected in a 24-hour urine collection indicating long-term

- Nutrient-related metabolic effects now seen.
- B-vit's and AA's low
- New opportunist growing
- Insidious poison detected

low level exposure. The organic acid profile also revealed a low sulfate, indicating depletion of hepatic conjugation pathways. A fasting plasma amino acid analysis revealed multiple low levels of the essential amino acids plus low tyrosine and taurine.

Impressions

- Nutrient levels raised to meet individual requirements
- Orthobiosis finally restored

The supplement program was shifted to include a custom formulated free form amino acid powder fortified with tyrosine and taurine. N-acetyl cysteine (NAC) was added to boost glutathione production and assist in the mobilization of mercury. This routine, with careful control of meal size, reduced his fatigue. A final effort to identify endotoxin sources with fungal testing led to the finding of greatly elevated IgG levels to *Pullularia pullalans*. Treatment with the antifungal agent, Lamisil®, while maintaining probiotic organisms, resulted in remission of persistent musculoskeletal pain.

Mrs. O's Fibromyalgia

Chief Complaints – Medical History:

A 50-year-old white female, Mrs. O, presented at the clinic with postmenopausal symptoms of severe hot flashes and depression. She had been diagnosed 10 years ago with fibromyalgia characterized by severe muscle aches and pain, fatigue, and depression. The only treatment that had seemed to help her was Prozac, which controlled her pain although, with the advent of menopause, depression was still a factor. She had been taking hormone replacement, Premarin and progesterone, since November 1998. This helped with the hot flashes but also made her very irritable. At the time she first visited the clinic, the fibromyalgia pain had returned to a significant degree, and she was tired all the time, depressed, retaining water, and wanted to get off her medications as she felt they were affecting her adversely. Her health assessment questionaire score was 83, reflecting significant adverse symptom presentation. She commented to one of the doctors that she was giving herself 8 months -- if this treatment didn't work, then she would end it all as life was not worth living feeling this way.

Test Results:

Blood chemistries and complete blood count were normal except for elevated cholesterol. The organic acid analysis indicated multiple functional B vitamin deficiencies that can significantly affect the body's ability to generate energy. The amino acid panel revealed significant essential amino acid deficiencies (8 out of 10). The IgG food allergy panel indicated adverse reactions to milk and wheat.

Treatment Protocol:

At the clinic that treated Mrs. O, part of the routine treatment protocol for all diagnosed fibromyalgia patients is a series of vitamin and mineral intravenous (IV) infusions, two per week for one month. This has been found to significantly accelerate the recovery of these patients. In addition, the fibromyalgia patient is put on a regimen of oral nutrients designed to correct the biochemical imbalances detected by the testing. For this patient this included a high potency multiple vitamin and mineral, extra B-complex, chromium, omega 3 fatty acids, extra magnesium in liquid form, betaine HCl and pepsin as a digestive/absorptive aid, black cohosh and soy isoflavones to help with the hot flash symptoms. A custom blended amino acid formula was made up based upon the amino acid deficiencies seen in the test. A medical food product designed to treat systemic inflammatory processes and a probiotic supplement to help normalize bowel microflora. 5-Hydroxytryptophan (5-HTP) was recommended before bed. Since the patient was also taking Prozac, this was slowly tapered off as the 5-HTP was increased. Methylsulfonyl methane (MSM) was recommended for its anti-inflammatory properties. Milk and wheat were removed from her diet for four weeks to determine potential adverse immunologic consequences of elevated IgG levels.

Follow-up: 2 weeks

The IV infusions were discontinued because they were painful for her and they took longer than normal. She developed an itchy rash that went away upon stopping the MSM. She had stopped all hormone replacement and feels worse than ever. However, she understood she may have felt worse before she felt better due to the changes her body had to go through to get healthy. All other nutrient supplements were continued.

Follow-up: 6 weeks

The IV infusions are no longer painful and she experienced much more energy and was pain free. She had not taken Prozac for four weeks and felt very up-beat emotionally. Her hot flashes had stabilized and were infrequent and mild. She feels the supplements have helped a lot. She has lost several pounds and is no longer bloated. She was started on an oral magnesium solution, which has helped many fibromyalgia patients. She said she loved the magnesium supplement since it relaxes her and made her feel "happy."

Follow-up: 12 weeks

Patient continued feeling well. IVs were reduced to once a month for maintenance. The clinicians discussed reducing her supplement schedule to maintenance levels. Her health assessment score was down from 83 to 12, relecting greatly reduced pain and discomfort.

Discussion:

While this patient had a particularly successful outcome, most fibromyalgia patients notice significant benefit in their symptoms with this therapeutic approach. When the aggressive IV protocol was added, the percentage of successful outcome at this clinic increased substantially.

Conclusion

The clinical situations and cases presented in this chapter serve to illustrate the following:

1. Modern degenerative diseases are of complex origin and the progression of symptoms is dependent on each patient's history. Complex cases that defy classical diagnoses and therapy may be amenable to molecular medicine approaches that depend on identification of nutrients, toxins, or metabolic controls that are at the root of the problem.
2. Complex cases must be approached in a way that restores normal function to cells, tissues, and organs. Simple, safe means such as nutrient supplements, food derivatives, and probiotic organisms, when used in the appropriate levels for a given patient, can restore normality to biochemical processes that control cell function.
3. Laboratory evaluations can identify the interventions that are required for restoration of normal function.

Appendix A: Clinical Guide to Test Selection

Initial diagnosis can guide test selection, although deficiency of any essential nutrient can exacerbate almost any condition. Leg cramps can be caused by calcium and magnesium deficiencies, and the condition also is associated with vitamin E deficiency. Leg cramps, however, have no particular connection with intestinal dysbiosis. Other conditions like aging can result from many diverse nutritional factors, so a single nutrient category starting point is more difficult to define.

Diseases/Conditions	Serum A., E., β-Carotene, CoQ10 Lipid Peroxides	Organic Acids in Urine	Amino Acids	Fatty Acids	Minerals	Toxic Heavy Metals	Detoxification Capacity	Intestinal Microbial Markers	Hormone Profile	IgG Food Sensitivities
Acne	x	x		x						
Aging	x	x	x	x	x	x	x		x	
Alcoholism		x								
Allergies		x								x
Anemia		x			x					
Autism	x	x						x		
Cardiomyopathy	x	x		x						
Carpel Tunnel syndrome		x								
Cataracts	x	x								
Constipation		x						x		
Depression		x	x				x	x		
Diminished propioception	x									
Down's syndrome	x	x	x					x		
Dry membranes		x		x						
Dry, scaly skin	x			x	x					
Dry, breaking hair	x									
Elevated serum cholesterol		x								
Eye disorders	x									
Failure to thrive			x	x		x				
Fatigue	x									
Fatty liver		x								
Fibrocystic breasts	x									
Frequent colds, flu, etc.	x									
Gait disturbance	x									
Glossitis		x								
Hair loss		x								
Headache		x								

Clinical Guide to Test Selection Cont.

Diseases/Conditions	Serum A., E., β-Carotene, Q10 Lipid Peroxides	Organic Acids in Urine	Amino Acids	Fatty Acids	Minerals	Toxic Heavy Metals	Detoxification Capacity	Intestinal Microbial Markers	Hormone Profile	IgG Food Sensitivities
Heart disease	x			x	x				x	
Hepatic insufficiency		x								
Hot flashes	x									
Hypertension			x		x				x	
Hypochlorhydria		x								
IBD										x
Impaired healing	x									
Impotence	x									
Infertility	x	x								
Insomnia	x									
Kidney stones		x								
Learning disorders		x								
Leg cramps	x				x					
Loss of alveolar bone	x									
Lung disorders	x									
Memory failure		x								
Night blindness	x									
Ophthalmoplegia	x									
Parkinson's disease		x								
Peripheral neuropathy		x								
Phlebitis	x									
Premature aging	x									
Premenstrual syndrome	x									
Schizophrenia		x								
Sleep disorders			x						x	
Sterility	x									
Stunted growth	x									
Weakened immune system	x									
Weakness		x								

Appendix B: Nutrient Evaluations Related to Standard Serum Chemistries

Test	Potential Clinical Indications	Related Nutrient Evaluations
Electrolyte & Urea Panel Na, K, Cl, CO2, Anion gap BUN, Creatinine	Metabolic acidosis Diuresis	Organic acid profile RBC mineral profile
Glucose	Diabetes Hypoglycemia	Trace element profile, Fatty acid profile Organic acid profile, Amino acid profile
Protein Panel Total protein, Albumin, Globulin, A/G ratio	Protein loss edema Myeloma	Amino acid profile Fatty acid profile, Antioxidants
Calcium	Tertiary hyperparathyroidism Diuretic therapy	Bone loss marker, Hair mineral profile RBC mineral profile
Enzyme Panel AST (SGOT), ALT (SGPT), Alk, Phosphatase, Total Bilirubin	Liver disease Heart disease	Organic acid profile, Amino acid profile Cardiac health profile
Lipid Panel Total Chol, Triglycerides, HDL Chol, LDL (CALC), Chol/HDL and LDL/HDL ratio	Hyperlipidemia Hypocholesterolemia	Fatty acid profile Coenzyme Q_{10}
Iron Panel Iron, TIBC, Iron saturation	Iron deficiency Iron excess	Trace element profile Antioxidants
Thyroid Panel Total T3, T3 Uptake, T4, T7 TSH	Primary hypothyroidism Primary hyperthyroidism	Organic acid profile Amino acid profile

Appendix C: Cardiovascular Health Profiling

Cardiovascular health profiling is an enhancement of cardiovascular risk assessment by laboratory methods. The evaluation of total nutritional status as presented in this book applies to cardiac health, but the most critical tests for heart tissue are those that evaluate magnesium, vitamin E, coenzyme Q_{10}, and oxidative damage markers such as lipid peroxides. In order to evaluate the earliest factors that initiate degeneration of vascular and cardiac tissue, hormones must be tested. Insulin insensitivity is one of the reasons that serum triglycerides and cholesterol levels rise. In men, falling testosterone levels is associated with almost every major aspect of heart disease.

Test Category	Analyte	Abnormality Associated with Increased Risk or Decreased Health	Significance
Blood Lipid Components	Total cholesterol	↑	Atherosclerotic risk factor
	HDL Cholesterol (direct)	↓	Atherosclerotic risk factor
	LDL Cholesterol	↑	Atherosclerotic risk factor
	Triglycerides	↑	Atherosclerotic risk factor
	Lipoprotein	↑	Atherosclerotic risk factor
Lipoprotein Ratios	LDL/HDL	↑	Atherosclerotic risk factor
	Total Cholesterol/HDL	↑	Atherosclerotic risk factor
Oxidant Stress Indicators	Coenzyme Q10	↓	Myocardial protection
	Vitamin E (serum)	↓	Myocardial protection
	Lipid Peroxides (serum)	↑	Myocardial protection
	Homocysteine (serum)	↑	Atherosclerotic risk factor
Other Important Cardiovascular Indicators	C-Reactive Protein	↑	Thrombogenic risk factor
	Fibrinogen	↑	Thrombogenic risk factor
	RBC Magnesium	↓	Myocardial protection
	Insulin	↑	Blood Lipids
	Testosterone	↓	Myocardial protection

Cardio-protection or risk reduction parameter	Effect of Testosterone
Total Cholesterol	Decrease
LDL Cholesterol	Decrease
HDL Cholesterol	Increase
Thromboxanes A2 Recptor	Increase
Fibrinolytic Activity	Increase
Atithrombin III Activity	Increase
Angina	Decrease
Systolic Blood Pressure	Decrease

Appendix D:
Summary of Drug-Induced Nutrient Depletions

This table is presented in two sections: Prescription drugs and OTC drugs. Within each section drugs are listed by category of action.

Prescription Drugs

Drug Category	Nutrient Depletions	Explanation
Female Hormones		
1. Estrogen Replacement: Premarin, Estrace, Menest, Estratest, etc.	Vitamin B_6, Folic Acid	Estrogen metabolism interferes with absorption of both folic acid and vitamin B_6. Since B_6 is involved with synthesis of serotonin, depletion can cause anxiety, depression, sleep disturbances, and irritability. Anemia from folate depletion causes weakness and low energy. Low levels of folic acid are associated with increased incidence of birth defects, cervical dysplasia, and elevated homocysteine, which is a major risk factor for cardiovascular disease.
2. Oral Contraceptives	Vitamin B_2, Vitamin B_6, Vitamin B_{12}, Vitamin C, Folic Acid, Zinc, Magnesium	Studies show reduced serum levels of nutrients listed. Depletions of vitamins B_6, B_{12}, and C are not as frequently seen with the use of newer low dose estrogen OCs). * See above for problems associated with folic acid depletion.
Diuretics		
3. Hydralazine Diuretics: Apresoline, Apresazide Hydralazine, Ser-Ap-Es	Vitamin B_6, Magnesium	These diuretics block an enzyme, which can cause vitamin depletion. Pyridoxine deficiency can cause depression and/or nerve damage causing numbness or tingling of hands or feet. Magnesium is also lost in the urine.
4. Loop Diuretics: Lasix, Bumex, Edecrin	Magnesium, Vitamin B_1, Vitamin B_6, Potassium, Zinc	Diuretic-induced magnesium and potassium deficiencies can cause increased irregularities in heartbeat blood pressure. Increased urination may also cause depletion of vitamins B_1 and B_6.
5. Thiazide Diuretics: Diuril, Enduron, Hctz: Lozol, Zaroxolyn, etc.	Magnesium, Potassium, Sodium, Zinc	Urinary depletion of magnesium and potassium can exacerbate irregular blood pressure and cardiac function. Zinc depletion can suppress wound healing and immune function. Hyponatremia could also develop.
6. Potassium-Sparing Diuretics: Dyazide, Maxzide	Magnesium, Potassium, Sodium, Zinc	Inhibits enzyme necessary for folic acid synthesis. Chronic use can lead to folic acid depletion. Diuresis can also cause calcium depletion. CAUTION: taking potassium with potassium-sparing diuretics could cause hyperkalemia (symptoms = weakness, impairment of speech cognition).
Cholesterol Lowering Drugs		
7. HMG-CoA Reductase Inhibitors: Lescol, Mevacor, Pravacol, Zocor	Coenzyme Q_{10}	These drugs block the liver enzyme necessary for synthesis of cholesterol but also block the body's ability to manufacture CoQ_{10}. CoQ_{10} has many important functions including cellular energy production and regulation of blood pressure and cardiac function.
8. Bile Acid Sequestrant: Questran, Colestid	Vitamins A, D, E, K, and B_{12}, β-Carotene, Folic Acid, Iron, Fat	Nutrient depletions caused by interfering with absorption. Doses of nutritional supplements should be taken as far away from dosing of the drug as possible. Fat absorption is also inhibited.

Drug	Nutrients Depleted	Comments
Anticonvulsants		
9. Phenobarbital and Barbiturates	Vitamin D, Calcium, Folic Acid	Long-term use interferes with vitamin D metabolism and may reduce the absorption of calcium. Folic acid levels are lowered in both plasma and erythrocytes.
10. Phenytoin, Dilantin	Vitamin D, Calcium, Folic Acid, Vitamin B_{12}	Decreases vitamin D availability, reducing absorption of calcium. Vitamin B_{12} and folic acid absorption are also decreased. Symptoms of bone disease, anemia, and neurological problems can develop as well as gum and periodontal disease.
Anti-Inflammatory Drugs		
11. Carbamazepine, Tegretol	Folic Acid	Inhibits folate absorption; long-term use could create anemia.
12. Corticosteroids: Prednisone, Aristocort, Cortef, Decadron, Medrol	Calcium, Vitamin D, Potassium, Selenium, Zinc	These drugs reduce levels of vitamin D and decrease the absorption of calcium, resulting in bone loss and skeletal problems. Long-term use may also deplete potassium, selenium, and zinc.
13. Gout Medications: Colchicine, Colbenemid	β-Carotene, Vitamin B_{12}, Sodium, Potassium	Colchicine inhibits the absorption of all these nutrients. Changes in pH cause GI symptoms and B_{12} malabsorption. Symptoms from nutrient depletions can include weakness and peripheral neuritis.
14. Sulfasalazine, Azulfidine	Folic Acid	Intestinal absorption of folic acid is inhibited, which can lead to anemia-related weakness and low energy.
15. Indomethacin, Indocin	Vitamin C, Folic Acid, Amino Acids, Iron	Decreases absorption of both vitamin C and folic acid. Increases rate of gastric emptying which decreases absorption of amino acids. Indocin can cause iron deficiency due to blood loss.
16. Other NSAIDs: Ibuprofen, Naproxen, Sulindac	Folic Acid	These anti-inflammatory drugs competitively inhibit the enzymatic synthesis of folic acid. Long-term use could lead to anemia. Low levels of folic acid are associated with increased incidence of birth defects, cervical dysplasia, and elevated homocysteine, a major risk factor for cardiovascular disease.
Antibiotics		
17. Antibiotics	Acidophilus, Bifidus, All B-Vitamins, Vitamin C, Vitamin K	Antibiotics also kill the beneficial bacteria. The "friendly bacteria" produce vitamins B_2, B_3, B_6, B_{12}, K, biotin, folic acid, pantothenic acid, and a number of natural antibiotics in our intestines. Destroying acidophilus & bifidus bacteria causes nutrient depletions and decreases the immune response.
18. Tetracycline Antibiotics	Calcium, Magnesium, Zinc, Vitamin B_6, Vitamin B_{12}	Tetracyclines chelate with calcium, magnesium, and zinc. Long term use can cause mineral depletions. Tetracyclines also interfere with the absorption of vitamins B_6 and B_{12}.
19. Tuberculosis Drugs: Isoniazid, INH	Niacin, Vitamin B_6, Calcium, Vitamin D, Vitamin E	Isoniazid binds to pyridoxine, depleting vit. B_6. The conversion of tryptophan to niacin is also inhibited, the synthesis of vitamin D is reduced, and the absorption of calcium is inhibited. Vitamin E absorption is also decreased.
20. Neomycin	Calcium, Magnesium, Nitrogen, Sodium, Iron, Potassium, Vitamin A, Vitamin B_{12}, Vitamin K	Neomycin induces maldigestion and malabsorption by three mechanisms: inhibition of pancreatic enzymes, precipitation of bile salts, and mucosal damage.

Drug	Nutrient Depleted	Comments
21. Trimethoprim, Bactrim, Septra, Trimpex	Folate	Trimethoprim is a mild folate antagonist with only minimal risk. However, long-term use and/or high dose usage may create a deficiency, especially in patients who are in a marginal nutritional state to begin with.
Anti-Diabetic Drugs		
22. Sulfonylureas: Diabeta, Glynase, Tolinase Micronase, Dymelor	Coenzyme Q_{10}	Inhibition of the NADH-oxidase enzyme can lead to a coenzyme Q_{10} deficiency.
23. Biguanides: Glucophage	Vitamin B_{12}	Competitive inhibition of vitamin B_{12} absorption could cause depletion in some individuals.
Ulcer Medications		
24. H-2-Receptor Antagonists Axid, Pepcid AC, Tagamet, Zantac	Vitamin B_{12}, Folic Acid, Vitamin D, Iron, Zinc, Protein	Malabsorption of dietary cobalamin, iron, and folic acid by H-2 antagonists may contribute to nutrient depletions. Altered pH may also reduce absorption of calcium, vitamin D, and zinc. Altering gastric pH also interferes with digestion of protein. Folic acid depletion is associated with increased incidence of birth defects, cervical dysplasia, and elevated homocysteine, a major risk factor forcardiovascular disease.
25. Proton Pump Inhibitors: Prilosec, Prevacid	Vitamin B_{12}, Protein	By altering the gastric pH, these drugs may cause malabsorption of vitamin B_{12}. Probable interference with protein digestion.
Heart Medications		
26. Digoxin Lanoxin	Calcium, Magnesium	Causes increased urinary excretion of both calcium and magnesium. Magnesium deficiencies increase likelihood of cardiac dysrhythmias and atrial fibrillation.
27. Beta Blockers: Inderal, Lopressor, Tenormin, Viskin, etc.	Coenzyme Q_{10}	These drugs antagonize the activity of the enzymes involved in the synthesis of coenzyme Q_{10}. Deficiency can cause heart, blood pressure, and immune system-related problems.
28. Potassium Chloride Supp. Kaon, Micro K, Slow-K, etc.	Vitamin B_{12}	Many chemotherapy drugs cause nausea, vomiting, and significant damage to gastric and intestinal mucosa. These factors cause decreased appetite and malabsorption leading to a wide variety of nutrient depletions.
Psychiatric Medications		
29. Tricylic Antidepressants: Elavil, Pamelor, Tofranil Asendin, Norpramin, Etrafon, Sinequan, etc.	Coenzyme Q_{10}, Vitamin B_{12}	Tricyclics inhibit enzymes necessary for production of coenzyme Q_{10} which can cause cardiovascular symptoms. Both Elavil & Tofranil cause depletion of vitamin B_2 by interfering with absorption. Deficiency can cause skin, neurological and energy problems.
30. Major Tranquilizers: Thorazine, Navane, Mellaril, Prolixin, etc.	Vitamin B_2, Coenzyme Q_{10}	These drugs inhibit the absorption of vitamin B_2 & coenzyme Q_{10}. Depletion of these vitamins can cause skin, neurological, and energy-relate problems.
Other Medications		
31. Chemotherapy Drugs	Most Nutrients	Slow release of potassium chloride salts alters intestinal pH, which decreases absorption of vitamin B_{12}. Depletion can cause weakness and tiredness associated with anemia.
32. Theophylline	Vitamin B_6	Inhibits enzyme pyridoxal kinase causing vitamin B_6 depletion.
33. Coumadin	Vitamin K	Interferes with the enzyme responsible for the synthesis of vitamin K.
34. Zidovadine, Perovir AZT	Copper, Zinc	Drug causes specific depletion of both copper and zinc.

OTC Drugs

Drug	Nutrient Depleted	Comments
1. Aspirin	Vitamin C, Folic Acid, Potassium, Zinc	Drug most likely to produce vitamin C deficiency in normal individuals. Chronic use can cause iron depletion due to blood loss in GI tract. Aspirin depletes folic acid by displacing bound serum folate. Causes urinary loss of potassium. Depletion-related symptoms: weakness & low energy from anemia. See NSAIDs below for folic acid depletion risk factors.

Drug	Nutrients	Comments
2. Other NSAIDs: Ibuprofen, Naproxen	Folic Acid	These anti-inflammatory drugs competitively inhibit the enzymatic synthesis of folic acid. Now available as OTCs, easy access and long-term use could lead to folic acid depletion and anemia. Low levels of folic acid are associated with increased incidence of birth defects, cervical dysplasia, and elevated homocysteine which is a major risk factor to cardiovascular disease.
Laxatives		
3. Mineral Oil, Agoral, Haley's M-O	Vitamins A, D, E, and K β-Carotene	Inhibits absorption - fat soluble nutrients dissolve in the mineral oil and are lost when the oil is excreted.
4. Phenolphthalein: Correctol, Ex-Lax, Feen-a-Mint	Calcium, Potassium, Vitamin D	Causes decreased nutrient absorption due to increased intestinal motility and mucosal permeability.
5. Bisacodyl, Dulcolax, Antacids	Potassium	Intense peristalsis and rapid bowel emptying can cause hypokalemia.
Antacids		
6. Magnesium and Aluminum	Calcium, Phosphate, Copper, Iron, Magnesium, Potassium, Zinc, Protein	An alkaline pH inhibits the absorption of these nutrients. Chronic use can lead to skeletal problems due to calcium & phosphate depletion. The digestion of protein is also affected.
7. Sodium Bicarbonate: Alka-Seltzer, Baking Soda		An alkaline pH inhibits the absorption of these nutrients. Chronic use can lead to skeletal problems due to calcium & phosphate depletion. Sodium bicarbonate-altered intestinal pH specifically inhibits the absorption of folic acid. The digestion of protein also is inhibited.
Ulcer Medications		
7. Axid, Pepcid, Tagamet, Zantac	Vitamin B12, Folic Acid, Vitamin D, Zinc	Malabsorption of dietary cobalamin, iron, and folic acid by H-2 antagonists may contribute to nutrient depletions. Altered pH may also reduce absorption of calcium, vitamin D, and zinc. Folic acid depletion is associated with increased incidence of birth defects, cervical dysplasia, and elevated homocysteine, a major risk factor for cardiovascular disease.

Index

A

D

G

H

Q

R

S

T

U

V

W

X

Y

Z